Lecture Notes in Statistics 162

Springer
New York
Berlin
Heidelberg
Barcelona
Hong Kong
London
Milan
Paris
Singapore
Tokyo

Constantine Gatsonis
Robert E. Kass
Bradley Carlin
Alicia Carriquiry
Andrew Gelman
Isabella Verdinelli
Mike West (Editors)

Case Studies in Bayesian Statistics

Volume V

Constantine Gatsonis
Center for Statistical Sciences
Brown University
Box G-A416
Providence, RI 02912
USA

Robert E. Kass
Department of Statistics
Carnegie Mellon University
Baker Hall 229
Pittsburgh, PA 15213
USA

Bradley Carlin
Division of Biostatistics
University of Minnesota
Box 303 Mayo Building
Minneapolis, MN 55455
USA

Alicia Carriquiry
Department of Statistics
Iowa State University
222 Snedcor Hall
Ames, IA 50011-1210
USA

Andrew Gelman
Department of Statistics
Columbia University
618 Mathematics Building
New York, NY 10027
USA

Isabella Verdinelli
Department of Statistics
Carnegie Mellon University
Baker Hall 232
Pittsburgh, PA 15213
USA

Mike West
Institute of Statistics and Decision Sciences
Duke University
214A Old Chemistry Building
Durham, NC 27708-0251
USA

Printed on acid-free paper.

Camera-ready copy provided by the editors.
Printed and bound by Sheridan Books, Inc., Ann Arbor, MI.
Printed in the United States of America.

9 8 7 6 5 4 3 2 1

ISBN 0-387-95169-5 SPIN 10786836

Springer-Verlag New York Berlin Heidelberg
A member of BertelsmannSpringer Science+Business Media GmbH

Preface

The 5th Workshop on Case Studies in Bayesian Statistics was held at the Carnegie Mellon University campus on September 24-25, 1999. As in the past, the workshop featured both invited and contributed case studies. The former were presented and discussed in detail while the latter were presented in poster format. This volume contains the three invited case studies with the accompanying discussion as well as ten contributed papers selected by a refereeing process. The majority of case studies in the volume come from biomedical research. However, the reader will also find studies in education and public policy, environmental pollution, agriculture, and robotics.

INVITED PAPERS

The three invited cases studies at the workshop discuss problems in educational policy, clinical trials design, and environmental epidemiology, respectively.

1. In *School Choice in NY City: A Bayesian Analysis of an Imperfect Randomized Experiment* **J. Barnard, C. Frangakis, J. Hill**, and **D. Rubin** report on the analysis of the data from a randomized study conducted to evaluate the New York School Choice Scholarship Program. The focus of the paper is on Bayesian methods for addressing the analytic challenges posed by extensive non-compliance among study participants and substantial levels of missing data.

2. In *Adaptive Bayesian Designs for Dose-Ranging Drug Trials* **D. Berry, P. Mueller, A. Grieve, M. Smith, T. Parke, R. Blazek, N. Mitchard, C. Brearley**, and **Michael Krams** use Bayesian updating and decision analysis to develop a new class of designs for clinical trials, combining a dose-finding and a confirmatory stage. The authors describe a system for implementing their design in trials of the efficacy of neuroprotective drugs and report the results of extensive simulations of the performance of their design.

3. In *Modeling the Impact of Traffic-Related Air Pollution on Childhood Respiratory Illness* **N. Best, K. Ickstadt, R. Wolpert, S. Cockings, P. Elliott, J. Bennett, A. Bottle**, and **S. Reed** use Bayesian spatial, point-process regression models to study the relation between traffic-related air pollution and hospital admission rates for

respiratory illnesses in children. The analysis incorporates complex estimates of exposure to traffic polution, addresses the possibility of confounding by socioeconomic and other factors, and accounts for spatial dependence possibly arising from spatially varying unmeasured factors.

CONTRIBUTED PAPERS

Ten of the contributed papers were selected for inclusion in this volume after a refereeing process. They are arranged in the alphabetical order of the first named author.

1. **P. Abbit** and **F. Breidt** use data from soil surveys in Iowa to estimate quantiles of soil texture profiles in *A Hierarchical Model for Estimating Distribution Profiles of Soil Texture.*

2. **J. Carlin** uses a small factorial experiment resulting in a 3x2x2 table to highlight the problems of carrying out Bayesian inference with small sample sizes in *Assessing the Homogeneity of Three Odds Ratios: A Case Study in Small-sample Inference.*

3. **R. Evans, S. Hui** and **J. Sedransk** present a Bayesian analysis of data on Bone Mineral Density of the spine for a cohort of women in **Modeling Rates of Bone Growth and Loss Using Order-Restricted Inference and Stochastic Change Points.**

4. **S. Fernandez, R. Fernando, A. Carriquiry**, and **B. Gulbrandtsen** analyze data from a dog pedigree and use MCMC methods to estimate genotype probabilities connected with progressive retinal atrophy in *Estimating genotype probabilities in complex pedigrees.*

5. **P. Hoff, R. Halberg, A. Sverdlovsky, W. Dove**, and **M. Newton** discuss another application of Bayesian methods to genetics in *Identifying Carriers of a Genetic Modifier using Nonparametric Bayesian methods.* The specific context of their analysis is the study of the *Min* mutation in the adenomatous polyposis coli (APC) gene.

6. **S. MacEachern** and **M. Peruggia** develop numerical and graphical explotatory tools for a Bayesian analysis and use them to build a regression model relating body to brain weight in *Bayesian Tools for EDA and Model Building: A Brainy Study.*

7. **P. Sebastiani, M. Ramoni** and **P. Cohen** model sensory inputs of a mobile robot at Markov Chains and show how the use of a Bayesian

clustering method permits the robot to learn different types of activities in *Bayesian Analysis of Sensory Inputs of a Mobile Robot.*

8. **H. Seltman** uses a two-state hidden Markov Chain to study the secretion of cortisol in **Hidden Markov Models for Analysis of Biological Rhythm Data.**

9. **S. Schmidler, J. Liu** and **D. Brutlag** discuss Bayesian models for the prediction of protein structure, which incorporate information based on non-local sequence interactions in *Bayesian Protein Structure Prediction.*

10. **C. Tsai** and **K. Chaloner** discuss the derivation of the sample size needed to ensure a high probability for reaching consensus among clinicians in *Using Prior Opinions to Examine Sample Size in Two Clinical Trials.* The prior opinions analyzed in the paper were elicited from clinicians about two regimens for Pneumocystis Carinii Pneumonia prophylaxis in patients with advanced HIV disease.

ACKNOWLEDGEMENTS

A number of individuals provided tremendous help for the Workshop and this volume. We would like to thank Michelle Agie, Rose Krakovsky, Mari Alice McShane, Margie Smykla and Heidi Sestrich for their work on the arrangements for the Workshop, Heidi Sestrich for editorial support, Yangang Zhang, Jung-Ying Tzeng, Fang Chen and Xiaohua Zhang for compiling the author and subject index of the volume, and several referees for their reviews of contributed case studies. Our thanks also go to Carnegie Mellon University, Duke University, the National Science Foundation and the National Institutes of Health for financial support for the Workshop. Above all we would like to thank the workshop participants for their contribution to the continued success of this series.

Constantine Gatsonis
Brad Carlin
Alicia Carriquiry
Andrew Gelman
Rob Kass
Isabella Verdinelli
Mike West

List of Contributors

Abbitt, Pamela J., Department of Statistics, North Carolina State University

Ansolabehere, Stephen, Department of Political Science, Massachusetts Institute of Technology

Barnard, John, Department of Statistics, Harvard University

Bennett, James, Department of Epidemiology and Public Health, Imperial College School of Medicine

Berry, Donald A., Department of Biostatistics, The University of Texas M.D. Anderson Cancer Center

Best, Nicola G., Department of Epidemiology and Public Health, Imperial College School of Medicine

Blazek, Richard, Pfizer Central Research

Bottle, Alex, Department of Epidemiology and Public Health, Imperial College School of Medicine

Breidt, F. Jay, Department of Statistics, Colorado State University

Brutlag, Douglas L., Department of Biochemistry, Stanford University

Carlin, John B., Murdoch Children's Research Institute, University of Melbourne, and the University of South Florida

Carriquiry, Alicia L., Dept. of Animal Science and Lawrence H. Baker Center for Bioinformatics and Biological Statistics, Iowa State University

Chaloner, Kathryn, School of Statistics, University of Minnesota

Cockings, Samantha, Department of Epidemiology and Public Health, Imperial College School of Medicine

Cohen, Paul, Department of Computer Science, University of Massachusetts

Dominici, Francesca, School of Hygiene & Public Health, Johns Hopkins University

Dove, William F., McArdle Laboratory for Cancer Research, University of Wisconsin-Madison

Elliott, Paul, Department of Epidemiology and Public Health, Imperial College School of Medicine

Evans, Richard B., Department of Veterinary Diagnostic and Production Animal Medicine, Iowa State University

Fernandez, Soledad A., Depts. of Statistics and Animal Science, Iowa State University

Fernando, Rohan L., Dept. of Animal Science and Lawrence H. Baker Center for Bioinformatics and Biological Statistics, Iowa State University

Frangakis, Constantine, School of Hygiene & Public Health, Johns Hopkins University
Gitelman, Alix I., Department of Statistics, Oregon State University
Grieve, Andy P., Pfizer Central Research
Gulbrandtsen, Bernt, Dept. of Breeding and Genetics, Danish Institute of Animal Science
Halberg, Richard B., McArdle Laboratory for Cancer Research, University of Wisconsin-Madison
Hill, Jennifer, School of Social Work, Columbia University
Hoff, Peter D., Department of Statistics, University of Washington
Hui, Siu L., Regenstrief Institute, Indiana University Center for Aging Research
Ickstadt, Katja, Department of Mathematics, Darmstadt University of Technology
Junker, Brian W., Department of Statistics, Carnegie Mellon University
Krams, Michael, Pfizer Central Research
Land, Stephanie R., National Surgical Adjuvant Breast and Bowel Project, Pittsburgh Cancer Institute, and the Department of Biostatistics, University of Pittsburgh
Liu, Jun S., Department of Statistics, Stanford University
MacEachern, Steven N., Department of Statistics, Ohio State University
Mitchard, Neil, Pfizer Central Research
Müller, Peter, Institute of Statistics and Decision Sciences, Duke University
Newton, Michael A., Department of Statistics and Biostatistics, University of Wisconsin-Madison
Parke, Tom, Tessella Support Services plc.
Peruggia, Mario, Department of Statistics, Ohio State University
Ramoni, Marco, Children's Hospital Informatics Program, Harvard Medical School
Reed, Sean, Department of Epidemiology and Public Health, Imperial College School of Medicine
Rubin, Donald B., Department of Statistics, Harvard University
Samet, Jonathan, School of Hygiene & Public Health, Johns Hopkins University
Schmidler, Scott C., Department of Statistics, Stanford University
Sebastiani, Paola, Department of Mathematics and Statistics, University of Massachusetts
Sedransk, J., Department of Statistics, Case Western Reserve University

Seltman, Howard, Department of Statistics, Carnegie Mellon University
Shedlovsky, Alexandra, McArdle Laboratory for Cancer Research, University of Wisconsin-Madison
Smith, Michael, Pfizer Central Research
Tsai, Chin-Pei, Department of Epidemiology and Public Health, Yale University School of Medicine
Wieand, H. Samuel, National Surgical Adjuvant Breast and Bowel Project and the Department of Biostatistics, University of Pittsburgh
Wolpert, Robert L., Institute of Statistics and Decision Sciences, Duke University

Contents

Invited Papers

Contributed Papers

INVITED PAPERS

School Choice in NY City: A Bayesian Analysis of an Imperfect Randomized Experiment

John Barnard
Constantine Frangakis
Jennifer Hill
Donald B. Rubin

ABSTRACT The precarious state of the educational system existing in the inner-cities of the U.S., including its potential causes and solutions, has been a popular topic of debate in recent years. Part of the difficulty in resolving this debate is the lack of solid empirical evidence regarding the true impact of educational initiatives. For example, educational researchers rarely are able to engage in controlled, randomized experiments. The efficacy of so-called "school choice" programs has been a particularly contentious issue. A current multi-million dollar evaluation of the New York School Choice Scholarship Program (NYSCSP) endeavors to shed some light on this issue. This study can be favorably contrasted with other school choice evaluations in terms of the consideration that went into the randomized experimental design (a completely new design, the Propensity Matched Pairs Design, is being implemented) and the rigorous data collection and compliance-encouraging efforts. In fact, this study benefits from the authors' previous experiences with the analysis of data from the Milwaukee Parental Choice Program, which, although randomized, was relatively poorly implemented as an experiment.

At first glance, it would appear that the evaluation of the NYSCSP could proceed without undue statistical complexity. However, this program evaluation, as is common in studies with human subjects, suffers from unintended, although not unanticipated, complications. The first complication is non-compliance. Approximately 25% of children who were awarded scholarships decided not to use them. The second complication is missing data: some parents failed to complete fully survey information; some children did not take pre-tests; some children failed to show up for post-tests. Levels of missing data range approximately from 3 to 50% across variables. Work by Frangakis and Rubin (1999) has revealed the severe threats to valid estimates of experimental effects that can exist in the presence of noncompliance and missing data, even for estimation of simple intention-to-treat effects.

The technology we use to proceed with analyses of longitudinal data from a randomized experiment suffering from missing data and non-compliance involves the creation of multiple imputations for both missing outcomes and missing true compliance statuses using Bayesian models. The fitting of Bayesian models to such data requires MCMC methods for missing data. Our Bayesian approach allows for analyses that rely on fewer assumptions than standard approaches.
These analyses provide evidence of small positive effects of private school attendance on math test scores for certain subgroups of the children studied.

1 Prologue

Every day policy decisions are made that may have a great impact on our lives based on quantitative analyses of social science data. Rigorous mathematical statisticians are sometimes wary of participating in social science analyses because social science data sets are nearly always messy relative to those in the physical or biological sciences even when statisticians are involved in the design of the study. Human subjects can be capricious, randomized experiments can rarely be performed, and the measures are often only loosely tied to the phenomena of interest, as well as being intrinsically noisy. However, this should not lessen the statistician's responsibility to model, as rigorously as possible, the science of the problem.

The Bayesian paradigm, because of its flexibility, is a powerful way to conceptualize how to approach such messy problems from the design through the analysis stage. Using this paradigm as a guide does not necessarily imply performing formal Bayes calculations at each step because these might be impossibly demanding in the time frame or with available resources. However it does mean that we design a study with the eventual Bayesian analyses in mind, where "design" here is defined broadly to include not only the plan for assigning treatments to individuals, but also evaluation issues such as the plan for what types of data will be collected and in what manner. We want to design to minimize problems at the end without being blind to the eventual complications that will nearly certainly arise. Rather, we optimally would like to frame these complications as aspects of the broadly defined phenomenon of interest and then build them into our Bayesianly-inspired template for the study and its data.

Of particular importance, knowing which issues create the most problems for our ultimate Bayesian analysis and which variables would be most useful for modeling these helps guide our design. In fact, many of the benefits of practical importance in a study such as this arise through the design:

deciding how to minimize the complications and whether these complications can be incorporated into the analyses. If there are no complications, the payoff to being Bayesian is typically relatively small. In our setting, the evaluation of a program that may have major impact on lives of our children, there will be an emphasis on these design aspects.

In this way, this application may stand in contrast to many Bayesian applications that often focus on analyses of existing datasets, thereby showcasing clever modeling and computation, but neglecting issues of how the data were obtained, or how the data collection was influenced by the Bayesian analyses to be conducted and used to draw practical conclusions. We present as equally important aspects of the study: (1) our assessment of the most important complications involved (non-compliance with treatment assignment, missing outcomes, and missing covariates), and (2) our attempts to minimize these complications and to accommodate eventual incorporation of them into the analysis (for instance inclusion of survey questions intended to help the modeling of these complications).

Our analysis does not represent a completely satisfactory job of simultaneously handling all the complications. As an example, we do not model the multivariate nature of the outcomes; we fit separate models for each outcome examined (reading and math test scores). Further work will gradually expand this initial model to incorporate the complicated structure of this experiment and the "response" to school choice that it measures. The more the model becomes inclusive of the complications, the more we will be able to take advantage of the elements we incorporated in our initial design that anticipated this structure.

2 Introduction

Over the past few years, interest in school choice has escalated. Congress and many state legislatures have considered school voucher proposals that enable families, particularly low-income families, to choose among a wide range of schools, public and private, religious and secular. In 1990 the Wisconsin legislature enacted a pilot program that gave public students access to secular private schools in the City of Milwaukee; then in 1996 the legislature expanded this program to include religious schools. After surviving a constitutional challenge, the program went into effect in the fall of 1998. A similar program in Cleveland, enacted by the Ohio legislature, began its third year of operation in the fall of 1998. At the federal level, a pilot program for the District of Columbia received congressional approval in the summer of 1998, but was vetoed by President Clinton.

Special interest groups, political leaders and policy analysts on all sides of the ideological spectrum have offered arguments both for and against the continuation and/or expansion of these school choice programs. Supporters of school choice assert that low-income, inner-city children learn more in private schools; critics retort that any perceived learning gains in private schools are due to the selected nature of private-school families. Proponents suggest that families develop closer communications with schools they themselves choose; critics reply that when choices are available, mismatches often occur and private schools expel problem students, adding to the educational instability of children from low-income, inner-city families. Champions of choice suggest that a more orderly educational climate in private schools enhances learning opportunities, whereas opponents declare that private schools select out the "best and the brightest," leaving behind the most disadvantaged. Voucher advocates argue that choice fosters racial and ethnic integration; critics, meanwhile, insist that private schools balkanize the population into racially and ethnically homogeneous educational environments.[1]

Few of these disputes have been resolved, in part because very few voucher experiments have been attempted. Although many published studies compare public and private schools, they have been consistently criticized for comparing dissimilar populations. Even when statistical adjustments are made for background characteristics, it remains unclear whether findings reflect actual differences between public and private schools or simply differences in the kinds of students and families attending them[2].

Though this problem has plagued educational research for years, it is not insurmountable. The best solution is to implement numerous large-scale controlled randomized experiments. Randomized experiments, though standard in other fields, have only recently found their way into educational studies, such as the Tennessee Star experiment, which found that smaller classes have positive effects on test scores among students in kindergarten

[1]Recent works making a case for school choice include Brandl (1998); Coulson (forthcoming); Cobb (1992); and Bonsteel and Bonilla (1997). A collection of essays that report mainly positive school-choice effects are to be found in Peterson and Hassel (1998). Works which critique school choice include Ascher, Fruchter, and Berne (1996); Carnegie Foundation for the Advancement of Teaching (1992); Gutmann (1987); Levin (1998); Fuller and Elmore (1996); Rasell and Rothstein (1993); Cookson (1994).

[2]Major studies finding positive educational benefits from attending private schools include *cole:hoff:kilg:1982, chub:moe:1990, neal:1997. Critiques of these studies have been prepared by Goldberger and Cain (1982); Wilms (1985).

and first grade (Mosteller, 1995). Until now, however, randomized designs have not been carefully used to study the validity of competing claims about school choice.

In this article, we describe a case study of a randomized experiment conducted in New York City made possible by the School Choice Scholarships Foundation (SCSF), a privately-funded school choice program. The SCSF program provides the first opportunity to estimate the impacts of a school choice pilot program that has the following characteristics: a lottery that allocates scholarships randomly to applicants, which has been administered by an independent evaluation team that can guarantee its integrity; baseline data on student test performance and family background characteristics collected from students and their families prior to the lottery; data on a broad range of characteristics collected from as much as 83 percent of the test group and control group one year later. Because it has these qualities, the SCSF program is an ideal laboratory for studying the effects of school choice on outcomes such as parental satisfaction, parental involvement, school mobility, racial integration and, perhaps most noteworthy, student achievement.

The school choice initiative in New York is described in Section 3 followed by study objectives and implementation in Sections 4 and 5. The innovative randomized design developed for this study is presented in detail in Section 6. Section 7 introduces the template of the imperfect randomized experiment and the corresponding notation is given in Section 9. The model is described in Section 10; technical details of the computations are reserved for Appendix A. Results of the analysis are discussed in Section 11.

3 School Choice Scholarships Foundation (SCSF) Program

In February 1997 SCSF announced that it would provide 1,300 scholarships to low-income families currently attending public schools. These scholarships were worth up to $1,400 annually, and could be used for up to three years to help pay the costs of attending a private school, either religious or secular. SCSF received initial applications from over 20,000 students between February and late April 1997.

In order to become eligible for a scholarship, children had to be entering grades one through five, live in New York City, attend a public school at the time of application, and come from families with incomes low enough to qualify for the federal government's free school lunch program. To qualify, students and an adult member of each family had to attend verification sessions where SCSF program administrators documented family income and children's public-school attendance.

Because of the large number of initial applications, it was not feasible to invite everyone to these verification sessions. To give all families an equal chance of participating, therefore, a preliminary lottery was used to determine who would be invited to a verification session. Only these families were then included in the final lottery that determined the allocation of scholarships among applicants.

The final lottery, held in mid-May 1997, was administered by Mathematica Policy Research (MPR); SCSF announced the winners. Within the guidelines established by SCSF, all applicants had an equal chance of winning the lottery. SCSF decided in advance to allocate 85 percent of the scholarships to applicants from public schools whose average test scores were below the city-wide median (henceforth labeled "low-score" schools). Consequently, applicants from these schools, who represented about 70 percent of all applicants, were assigned a higher probability of winning a scholarship.

Subsequent to the lottery, SCSF helped families find placements in private schools. By mid-September 1997, SCSF reported that 1,168 scholarship recipients, or 75 percent of all those offered a scholarship, had successfully gained admission to some 225 private schools.

4 Objectives of the Study

The evaluation of the School Choice Scholarship Foundation (SCSF) was conducted by MPR; the co-principal investigators were David Myers, MPR, and Paul Peterson, Harvard University (henceforth the evaluation team will be referred to solely as MPR for simplicity). The evaluation provides answers to three questions. First, what is the impact of being offered a scholarship on student and parent outcomes? Second, what is the impact of using a scholarship (participating in the scholarship program)? That is, what is the value-added of using a scholarship over and above what families and children would do in the absence of the scholarship program (which could include either public or private school attendance)? Third, what is the impact of attending a private school on student and parent outcomes? That is, would students who attend public schools do better academically if they attended private schools? Each of these questions may be answered by using information collected for the SCSF evaluation. Until this evaluation, no one study has addressed these three questions. Furthermore, this study may produce highly credible evidence concerning these questions because we randomly assigned families to a treatment group (offer of a scholarship) and a control group.

5 Implementation

In order to evaluate the voucher program, SCSF collected data on family demographics, parents' opinions on matters relating to their children's education, and student test scores, both prior to the lottery and one year later; one of the conditions for participating in the program was agreement to provide confidential baseline and follow-up information. MPR also made extensive efforts to encourage cooperation with the study guidelines as will be discussed in greater detail in the following sections.

5.1 Issues in the Implementation of the SCSF Evaluation

A critical issue in the design, implementation, and analysis of a random assignment experiment, such as the evaluation of the SCSF program, concerns deviations from the perfectly controlled experiment effected by families and children. We have identified four such behaviors:

1. Some families offered a scholarship did not subsequently accept the scholarship and attend a private school.

2. Some families not offered a scholarship sent their children to a private school anyway[3].

3. Some families invited to attend data collection and testing sessions one year after the baseline survey did not show up.

4. Some parents and students did not complete all items in their questionnaire, and some students did not complete enough items in the standardized reading and math assessments to be given a score.

---

[3]Classifying this behavior as a deviation assumes that the treatment is defined as private school attendance and that the range of private schools attended by the treatment group is similar to the private schools attended by the control group. This issue will be discussed in greater detail in Section 11.

The first two of these behaviors will henceforth be referred to under the general rubric of "noncompliance," the last two as "missing data." For the SCSF evaluation, ensuring compliance with the assigned treatment was largely out of the control of the evaluation team. If we define treatment as private school attendance, clearly the team could neither force winners to use their scholarships, nor could they keep those who did not win from attending private school. The SCSF did, however, provide services to help scholarship winners find appropriate private schools, which may have helped compliance rates. If we define treatment as participation in the scholarship program, then the only form of non-compliance is scholarship winners deciding not to participate in the program (clearly those who did not win could not obtain a scholarship or receive help from program administrators in finding a school). Again, provision of help in finding private schools for scholarship winners probably may have lessened noncompliance.

In social science studies it is generally difficult for evaluators to have much control over noncompliance of the control group with respect to participating in program services. They cannot prevent members of the control group from going out and finding similar services if they are available in the community; sometimes the services may be more or less intensive than those offered by the program being studied. It is also unclear that evaluators should want to prevent such actions. If we want the study to answer a public policy question (e.g. "Should we make available Program A? Will it make a difference in this community?"), the correct control should probably represent the other services the target population has available to them. However, in this case the issue is often "Do students learn more in private schools?".

Evaluators generally have more control, potentially, over the amount or kinds of missing data that occur. Below, we describe the procedures used to minimize missing data.

5.2 *Collection of Baseline Data*

During the verification sessions at which eligibility was determined, MPR asked students to take the Iowa Test of Basic Skills (ITBS) in reading and mathematics. Students in kindergarten applying for a scholarship for first grade did not take the test (see Section 5.5). Each student's performance was given a national percentile ranking. While their children were taking tests, MPR asked parents to complete questionnaires that would provide information on their satisfaction with the school their child was currently attending, their involvement in their child's education, and their background characteristics. Discussions between the evaluation team and some of the authors regarding what questions to include on the baseline survey focused not only on what types of covariates were expected to be predictive of the primary outcomes of interest, but also what might be predictive of compliance behavior and propensity towards non-response. This was done in anticipation of structuring non-compliance and missing data into our eventual Bayesian analysis.

Although grandmothers and other relatives and guardians also accompanied children to verification sessions, in over 90 percent of the cases it was a parent who completed the questionnaire. MPR held the sessions at private schools, where students took the tests in classroom settings. In nearly all cases, private school teachers and staff proctored the tests and were under the supervision of MPR staff. The verification sessions took place during March, April, and early May 1997 on weekends and vacation days.

5.3 *Collection of Follow-Up Data in 1998*

The first follow-up data collection was completed in summer 1998. MPR invited each of the 1,960 families in the treatment group and the control group to attend testing sessions. Most of the testing sessions were held on weekends during spring 1998. MPR held the testing sessions at private schools and parents were asked to complete a questionnaire that included many of the same items that were part of the baseline questionnaire. Students in grades 3-5 were given a questionnaire. The response rates for the first follow-up data collection are shown in Table 1.1. The overall response rate for the parent survey was 84 percent for the scholarship families and 80 percent for families in the control group. To achieve these high response rates, MPR used two forms of incentives. First, they offered all families in the control group a chance to win a scholarship for $1,400 for three years, but to be eligible, families and their children were required to attend a testing session. To preserve the integrity of the control group, we[4] randomly selected about 100 winners for the second year of scholarships. Second, a variable incentive scheme allowed many control group families that attended a testing session to receive an incentive of $75 on average (some were offered $50 and others were offered $100).

[4]This was actually performed by colleague Neal Thomas, see Hill, Rubin, and Thomas (1999).

Scholarship Users	89%
Scholarship Decliners	66%
Treatment Group Total	84%
Control Group Total	80%

TABLE 1.1. Response Rates on the First Follow-Up Parent Survey

5.4 *Item Nonresponse*

To minimize item nonresponse in the survey questionnaires, staff at each data collection session reviewed the questionnaires for completeness as parents and students returned them at the end of the testing session. In cases where many items appeared to have been left incomplete, staff asked the parents and students to try to complete the items. If a parent or child did not understand the item, staff would work with them so that they might be able to provide a response. Sometimes, one parent would refuse to answer about the other parent if they were no longer living in the home. In Table 1.2, we illustrate the variability in item nonresponse rates that occurred in the baseline survey. It becomes quite clear upon reviewing these results that often there was little information concerning a child's father. For example, among the parent questionnaires, more than 35 percent of them were missing information about fathers' educational attainment and almost 60 percent were missing information about fathers' employment. In contrast, missing values were present for about seven percent of the responses concerning mothers' education and mothers' employment.

5.5 *Additional Complications with the Data*

Two additional complications with the data are noteworthy. The first is that no pre-test scores were obtained for applicants in kindergarten because: (1) these children would most likely never have been exposed to a standardized test hence considerable time would have been spent instructing the children on how to take a test, and (2) there was concern that separating such young children from their guardians in this new environment with unfamiliar teachers might lead to discipline or behavioral issues. This creates a structural missingness in the data that is distinct from the standard types of missing data encountered, and thus needs to be handled differently. Second, we do not yet have complete compliance data for the multi-child families. For this reason, the analyses in this paper are limited to results for the 1250 "single-child" families (i.e. families that only had one child partici-

Item Description	% Response
Female guardian's highest level of education	95
Female guardian's ethnicity	94
Female guardian's country of birth	88
Number of years female guardian has lived at current residence	97
Female guardian's employment status	95
Female guardian's religion	94
How often female guardian attends religious services	96
Male guardian's highest level of education	83
Male guardian's ethnicity	81
Male guardian's country of birth	72
Number of years male guardian has lived at current residence	60
Male guardian's employment status	76
Male guardian's religion	71
How often male guardian attends religious services	63
Number of children under 18 living at home	94
Number of children at home attending a public elementary or high school	93
Number of children at home attending a religious private elementary or high school	58
Number of children at home attending a non-religious private elementary or high school	55
Whether there's a daily newspaper in the child's home	90
Whether there's an encyclopedia in the child's home	86
Whether there's a dictionary in the child's home	95
Whether there are more than 50 books in the child's home	92
The main language spoken in the home	92
Whether anyone in the home receives assistance through food stamps	93
Whether anyone in the home receives assistance through welfare (AFDC or public assistance)	89
Whether anyone in the home receives assistance through social security	77
Whether anyone in the home receives assistance through Medicaid	87
Whether anyone in the home receives assistance through Supplemental Security Income (SSI)	79
Yearly income of household before taxes	92

TABLE 1.2. Response Rates by Item for Baseline Parent Questionnaire

pating in the lottery) who were in grades 1-4 at the time of the spring of 1997 application process.

6 Design

Although the lottery used to award scholarships naturally created a randomized design, it also precluded blocking on variables selected purely for their assumed predictive power. Randomization within certain subgroup classifications (e.g. ethnicity) might have appeared inequitable to the public[5]. Another complication was that evaluation funding only allowed for 1000 treatment families and 1000 control families to be followed. How to choose the control families from the reservoir of over 4000 families who participated in the lottery but did not win a scholarship became the focus of the design issues and led to the development of a new experimental design, the Propensity Matched Pairs Design (PMPD). The PMPD is a design which creates matched pairs using the popular propensity score matching technique developed by Rosenbaum and Rubin (1983).

6.1 The Lottery and its Design Implications

The original plan for the lottery included three stages.

1. Interested families would submit applications to the program.

Over 20,000 families participated in the initial application stage. For administrative purposes, applications were batched by the date received into five time periods.

2. All potentially eligible families would be invited to a half-day of screening, which would include confirmation of eligibility, pretesting of children, and completion of a survey regarding the family's relevant background characteristics.

This plan was followed for the first batch of applicants. However, due to a variety of logistical constraints, coupled with the overwhelming response

[5]The randomization was slightly constrained as will be described in more detail in this section. However, in one case this was done to ensure higher representation from a more disadvantaged population, and this policy was clearly stated in advertisements for the program. The other "blocks" – application wave and family size – were present for logistical reasons concerning data collection and allocation of a fixed number of scholarships. In general, in this type of program, administrators would like to keep these types of deviations from a pure lottery to a minimum.

to the program, not all potentially eligible families were screened in the next four waves. Sampling of applicants had to be performed in order to reduce the number invited to participate in the screening stage. To keep the aggregate probability of receiving a scholarship equal across the time periods, the probability of receiving a scholarship amongst those screened had to be increased to offset the reduced probabilities of being invited to a screening session.

3. Families who completed the screening and whose eligibility was confirmed would be allowed into the final lottery.

Over 5000 families participated in the final lottery. In accordance with the goals of the SCSF program, applicants from "low-score" schools (i.e., schools whose average test scores were below the city-wide median) were given a higher chance of winning a scholarship than those from "high-score" schools (schools whose average test scores were above the city-wide median). Families from "low-score" schools were to represent 85% of those winning scholarships. This oversampling took place during the lottery for those who applied in the first wave (since there was no sampling performed at the screening stage). In the second through fifth waves, however, the differential selection of those from high versus low-score schools was largely accomplished in the sampling at the *screening* stage. The implication of this difference is that the treatment and control groups in the last four waves are balanced on the low/high variable whereas the treatment and initial control groups (i.e., those who did not win a scholarship) from the first wave are unbalanced on the low/high variable as well as variables correlated with this variable.

6.2 *Multi-child Families*

The SCSF program was set up so that all eligible siblings of scholarship winners were also offered scholarships. Because of this, families are the unit of randomization, and all matching and subsampling took place at the family level. Since covariate data were collected not only at the family level, but also at the student level, the set of these variables is somewhat different for the families in which more than one child applied to the program ("multi-child" families). That is, since our units of observation are families, yet some data are collected at the student level, multi-child families have more information than single-child families, so the variable "reading test score", for instance, cannot mean the same thing for all families.

For families with more than one child applying, new family variables were created. These variables were computed across all family members applying. For each family, the average and standard deviation of continuous variables were calculated for initial test scores, age, education expectations and grade level. The mean and standard deviation are based on available values; if only one value is available for a multi-child family, then the standard deviation is missing. For the majority of multi-child families, which are two child families, the original values can be derived from the mean and standard deviation. Binary variables (e.g., low/high and sex) were recoded as 1 if all responding children in the family responded negatively, 3 if all responding children responded positively, and 2 if responses were mixed. Indicators for the presence of any missing data among all family members for each variable were also created.

6.3 *PMPD Versus Randomized Block*

The study design provides an opportunity to test empirically the performance of the PMPD. In the first application lottery, in which all apparently eligible applicants were invited to be screened, the ratio of eligible nonwinners (control families) to winners (treatment group families) is approximately five to one, an ideal situation for the PMPD. In the second through fifth waves, however, which had smaller control groups due to the limits placed on how many families were invited to be screened, the groups are more nearly equal in size. This latter scenario is more appropriate (given the study design) for a randomized block experiment, with time periods (waves) serving as blocks. Implementing both designs concurrently allows for an empirical comparison of efficiency. However, the PMPD has a more difficult setting in which to achieve balance because of the initial imbalance on the low/high variable and other baseline covariates correlated with it.

6.4 *Design Implementation*

The implementation of the two designs proceeded as follows. The data can be conceptualized as being divided into four subgroups based on family size (single vs. multiple children) and design (PMPD vs. randomized block). The initial sample sizes,[6] further broken down by time period, are

[6]These are the sample sizes after removal of 100 families randomly chosen from the control group to receive scholarships for the following academic year, and 100 for the year after that. The additional

Family Size	Treatment	PMPD	Randomized Block					Total
			2	3	4	5	Subtotal	
Single	Scholarship	404	115	67	82	192	456	860
	Control	2626	72	65	87	135	359	2985
Multi	Scholarship	147	44	27	31	75	177	324
	Control	969	27	23	33	54	137	1106

TABLE 1.3. Initial Sample Sizes (unit is a family)

Family Size	PMPD	Rand.Block	Total
Single	353	323	646
Multi	147	177*	354
Overall	500	500	1000

* Only 137 available in control group.

TABLE 1.4. Target Sizes for Both Scholarship and Control Samples

displayed in Table 1.3.

The goal was to equalize sample sizes across treatment groups and then, if possible, across blocks, including across single versus multi-child families. It was apparent that we would only be able to approximate this goal in the stratified study. The limiting factor is the number of multi-child control families (137).

Because of financial constraints, we could only follow-up 2000 study participants (a "participant" is a family), and thus some random sub-sampling of lottery winners was performed. Because we had very similar numbers of lottery winners in each design, we targeted a similar number of control families in each design, as seen in Table 1.4.

Propensity Matched Pairs Design

The strategy for the PMPD was to match 500 sub-sampled scholarship winners from the first time period to 500 controls from the same time period, with separate matching for single and multiple-child families. As a consequence of the dataset being split into two parts (single versus multi-child families), all matching takes place within family size categories. This exact matching on family size produces perfect balance for this variable, which implicitly treats family size as the most important matching variable.

---

scholarship offers were used as incentives to increase participation in the follow-up data collection process. New winners were announced following the second and third follow-up testing visits.

Determinations had been made by the evaluators as to the relative "importance" of the remaining covariates. As described further in Section 6.5, importance is judged by a combination of the initial imbalance of a covariate across treatment groups and the perceived strength of the predictive relationship of it to post-randomization outcome measures, which include: the primary outcomes themselves, noncompliance behavior (referring to whether or not a family uses an offered scholarship), attrition from the study, and other types of missing data.

After family size, the most important variable by this definition was judged to be the binary variable for low versus high-test-score school, because it was thought to be highly correlated with the outcomes, and because of the imbalance that occurred in the first time period due to its use in determining lottery winners. It is closely followed in importance by grade level and initial test scores. The remaining covariates are ranked as: ethnicity, mother's education, participation in special education, participation in a gifted and talented program, language spoken at home, welfare receipt, food stamp receipt, mother's employment status, educational expectations, number of siblings (includes children not eligible because of age), and an indicator for whether the mother was foreign born. The final propensity score models, presented in Sections 6.12 and 6.13, were chosen based on the balance created in these variables' distributions across treatment groups. Identification of special variables and the overall ranking of the covariates informed decisions regarding which variables might be appropriate for exact matching, which should receive special treatment in the propensity score method, and what tradeoffs to make in terms of the resulting balance.

The ranking of the variables can be helpful in implementing the propensity score methodology; however, correlations among the variables diminish the importance of the ordering chosen. Therefore the specific ordering chosen may not have a major impact on the creation of matched pairs and should not be viewed as an assumption required for successful implementation.

Sub-Sampling for the Randomized Block Design

We randomly sub-sampled from the cells of the randomized block design to arrive at the final sample sizes, which met the limitation of 1000 families per design. The number sub-sampled were selected to equalize the number of scholarship and control families within blocks, and the number of families across blocks.

1. 133 original single-child lottery winners were randomly withheld

Family Size	Treatment	PMPD	Randomized Block					Total
			2	3	4	5	Subotal	
Single	Scholarship	353	72	65	82	104	323	676
	Control	353	72	65	82	104	323	676
Multi	Scholarship	147	44	27	31	75	177	324
	Control	147	27	23	33	54	137	284
	Total	1000					960	1960

TABLE 1.5. Final Sample Sizes

for the randomized block design: 43 in time period two, 2 in time period three, 88 in time period five

2. 36 single-child eligible controls were randomly withheld from randomized block design: 5 in time period four, 31 in time period five

The final sample sizes are displayed in Table 1.5.

6.5 *General Propensity Score Methodology*

Propensity score matching was introduced by Rosenbaum and Rubin (1983) as a means of creating better balance in observational studies, thereby allowing for valid causal inference under the assumption of strongly ignorable treatment assignment, i.e., treatment assignment on the basis of the covariates being used to estimate the propensity score. Matching is used as a way of alleviating the biases that can be created by self-selection. As documented in a variety of places, e.g., (Rubin, 1973, 1979; Roseman, 1998), the combination of matching and regression adjustment is typically far superior to either technique alone for controlling bias in observational studies. Not only does matching reduce bias created by the self-selection into treatment groups that occurs in observational studies, it increases efficiency in randomized experiments, such as the one in this study. The extra payoff from matching mostly arises when the linear model underlying regression adjustment is not entirely correct.

Methods for estimating propensity scores are well-documented and, in the case of no missing data, quite straightforward (Rosenbaum and Rubin, 1984). When missing data exist, as they do in this study, extensions of the general methodology (D'Agostino and Rubin, 1999) can be implemented. The goal is to balance closely all covariates and patterns of missing data across the treated and matched control groups.

6.6 *Complete Data*

In the case of complete data, the general strategy is to calculate a "propensity score" for each study participant. This score represents a participant's chance or "propensity" of receiving the treatment (e.g., a scholarship offer),

$$P(Z = 1 \mid X) , \tag{1.1}$$

where Z denotes treatment assignment and X denotes all of the measured covariates (recall, here, fully observed). This probability is straightforward to estimate using logistic regression or linear discriminant techniques.

Matching on the Propensity Score

The propensity scores can be regarded as defining a new covariate value for each individual, which is a function of all of the covariates potentially correlated with the outcomes. In practice the logits of these estimated probabilities are often used because they are linear in the covariates. Balancing this new covariate generally has the effect of improving the balance of all the other covariates that went into its estimation. A good way to balance propensity scores when the treatment group is much smaller than the control reservoir is to match on propensity scores. Procedurally, this can be accomplished by sorting the treatment group members by their propensity scores and then, one by one, finding for each treated subject, the control group member who has the closest score. Once a match has been made, the chosen control group member is removed from the control reservoir so it cannot be chosen again (Cochran and Rubin, 1973). This is called nearest remaining neighbor, or nearest available, matching.

Nearest Available Mahalanobis Matching Within Propensity Score Calipers

The Mahalanobis metric (or distance) between a treatment group member with vector covariate values X_t and a control group member with covariate values X_c (the same set of variables for both), is

$$(X_t - X_c)^T \Sigma^{-1} (X_t - X_c) , \tag{1.2}$$

where Σ is the variance-covariance matrix for these variables, for which, in practice, we substitute the pooled sample variance-covariance matrix. A combination of propensity score matching and matching based on the Mahalanobis metric using a subset of variables has many of the advantages of each method (Rubin and Thomas, 1996). The combination has been shown to be often superior to either technique used on its own (Rosenbaum and Rubin, 1985). With this refinement, as before, propensity scores are calculated for all study participants and then treatment participants are ordered by their propensity scores. Each treatment group member in turn will be initially "matched" to a subset of the control reservoir members whose scores are no more than c propensity score units (e.g., $c = 0.10$ propensity score standard deviations) away from the treatment member's propensity score. Thus the initial matches must fall within a $2c$ length propensity score caliper, symmetric about that treatment group member's score[7]. Mahalanobis matching is used to choose a "nearest neighbor" within this subset of study participants with respect to several special covariates. The control group member whose values, X_c, of the special covariates minimize the distance from the values, X_t, of the special covariates for the treatment member, is chosen from the subset of controls who fall within the caliper. We include only the continuous covariates most predictive of the outcome variables in the Mahalanobis metric, as discussed in Section 6.5.

Special Variables

The more predictive a covariate is of the outcomes of interest, the more crucial is the balance of this covariate across treatment groups. For example, controlling for a covariate (e.g., by balancing) that is uncorrelated with the outcomes plays no useful role, whereas controlling for one that is highly correlated with the outcome will play a crucial role for precise estimation.

Covariates that evaluators are most concerned about balancing receive special treatment in one of two ways. When feasible, exact matches can be required for the most critical of these variables. For instance, if sex were deemed to be the most important variable to balance, when looking at matches for a female treatment group member, no males would be considered. It it is only possible to exact match on discrete variables and only desirable to match on one or two of these. For an example of exact matching in a propensity score context see Rosenbaum and Rubin (1984). Recall that in this study we exact match on family size.

As an alternative to, or in addition to, this exact matching, the Mahalanobis matching within propensity score calipers can be constrained to only a chosen few variables considered more important to balance than the others. Mahalanobis matching is most effective when applied to a small number of essentially continuous covariates (Rosenbaum and Rubin, 1985; Gu and Rosenbaum, 1993). Matching within propensity score calipers attempts to improve balance for all of the covariates, whereas Mahalanobis

[7]This technique is described and illustrated in the context of a real life example in Rosenbaum and Rubin (1985)

matching within calipers attempts to achieve close pair matches on the few special covariates.

6.7 *Advantages over ANCOVA (Analysis of Covariance) adjustments*

We have already mentioned the benefits of using matching in addition to ANCOVA (regression adjustments) for both bias reduction and precision of estimation (in Section 6.5). There is another benefit of matching relative to regression adjustment. Adjusting for covariate differences after the experiment has the disadvantage that researchers could settle on the "best" model solely by choosing the one that best supports their a priori biases regarding the issue in question. Matching, on the other hand, uses only covariate balance as a diagnostic; outcomes are not even included in the model, nor are they often even available at the time of matching, as in our application. Therefore, no such researcher bias can occur in the selection of the propensity score model.

6.8 *Diagnostics*

There are a variety of combinations of the above techniques that will each yield "matched" treatment and control groups. The estimation of the propensity score alone could be accomplished by numerous models, depending on what variables are included and what interactions or non-linear terms are added. Diagnostics, which compare the treatment and control groups with respect to the distributions of the covariates, help the researcher determine which matched control group is superior. Since the goal of the matching is balanced groups, the adequacy of a model or procedure can be judged by treatment versus control group comparisons of sample moments of the joint distribution of the covariates, primarily means and variances, but also correlations. It is often helpful at this stage to have a ranking of covariates in order of perceived importance, beyond just the few selected to be "special" variables. Such a ranking, as described for this study in Section 6.4, can help the researcher choose between models with good overall balance that have slight tradeoffs in terms of more or less exceptional balance on specific variables.

6.9 *True Versus Estimated Propensity Scores*

A surprising fact about the use of propensity scores is that, in general practice, the use of the estimated propensity score typically results in more precise estimates than the use of the "true" population propensity score. This is especially true when the treatment and control groups are relatively similar initially; the logic is as follows. There are two types of errors that can result from estimates of treatment effect. The first involves systematic biases, which occur when, in expectation, the two groups differ on important characteristics. The second involves conditional biases, which refer to the random differences between groups that average to zero over repeated samples but are nonetheless present in any given sample. Both population and estimated propensity scores effectively reduce the systematic bias in samples; but estimated propensity scores more effectively reduce sample-specific randomly generated bias (Rubin and Thomas, 1992). Because a randomized lottery was held to determine scholarship receipt, there is no systematic bias, so estimated propensity scores, in contrast to population propensity scores, work to reduce conditional bias.

6.10 *Incomplete Data*

Techniques to estimate propensity scores in the presence of missing data have been proposed by D'Agostino and Rubin (1999). The type of strategy that is optimal depends upon how the missing data were generated and the relationship of this missingness to the outcomes of interest.

The SCSF program study starts from the advantageous position of a randomized design, within which incomplete baseline data is less problematic than in the case of an observational study. The goal is simply to get the best possible balance on all covariates that we expect to be predictive of outcomes. To the extent that the "missingness" of our covariates is predictive of outcomes, we want propensity score models that include information about the missing data mechanisms (e.g., indicators for the missingness of a particular variable) in order to balance the missingness across treatment groups better than it would be balanced by chance alone. If we believe that this missingness is predictive of the outcomes, then this balance has efficiency implications for our inferences about treatment effects, just as better balance on any other covariate improves efficiency of estimation. In addition, missingness will be used to model compliance status.

As an example, in the SCSF program there were single mothers in the study who refused to fill out the part of the application survey pertaining to the father of the child. The missingness of these variables could be viewed as a proxy measure for the strength of the relationships in the family and so was hypothesized a priori to be predictive of the outcomes. Therefore this missingness variable was included in our propensity model so that we could try to improve its balance across treatment groups.

The other missingness indicator chosen by evaluators as important in this study was that corresponding to mother's education. Investigators think that a missing response to this question reflects a mother's attitude towards education, which could be predictive of educational outcomes, compliance behavior, or subsequent missing data.

The techniques appropriate for including missing data mechanisms in a model are more complicated than those we discussed in Section 6.6. We used a computer program written by Neal Thomas to implement the technique developed by D'Agostino and Rubin (1999), which relies on the ECM algorithm (Meng and Rubin, 1993) to calculate propensity scores for each subject, including those with missing covariate values. The ECM algorithm is a variant of the standard EM algorithm which is used in situations where the maximization step is computationally awkward. It replaces the M-step with two or more conditional maximization (CM) steps, each of which has a straight-forward solution.[8]

The Mahalanobis matching within propensity score calipers in the SCSF program was modified for missing covariate values as follows. If possible, for the matched control, the same missing pattern was required. If no such matched control was found, we exact matched on the design variable low/high school, which was fully observed. If a matched control still was not found, we would have matched on the propensity score alone; however, this situation never occurred.

6.11 Relative Strengths of Designs – Diagnostics

We can judge the relative strengths of our designs through diagnostics that measure balance in various ways. Results from the PMPD are contrasted with results from both the randomized block design (2nd through 5th time

[8]For the general location model (often used with missing data e.g., Little and Rubin (1987) and Schafer (1997)), one CM-step gets maximum likelihood estimates for the parameters in the normal distributions conditional on the parameters for the log-linear model (cell probabilities for the contingency table) and a second CM-step obtains estimates for the log-linear model conditional on the parameters of all of the multivariate normal distributions. More CM-steps are often used within the log-linear model portion to avoid running the Iterative Proportional Fitting (IPF; Bishop, Fienberg, and Holland (1975)) to convergence at each iteration of the ECM algorithm.

periods), a simple random sample chosen from the control reservoir in the first time period, and a stratified random sample also chosen from the control reservoir in the first time period. The stratified random sample was randomized within low/high school categories; 85% of the children were chosen to be from low-score schools and 15% from high-score schools. This comparison was chosen because it represents the most likely alternative to the PMPD design that MPR would have implemented.

6.12 *Single Child Families*

Following the criteria discussed in Section 6.4, a model for the propensity score was chosen. The contingency table for the categorical variables ethnicity (Hispanic/Black/other), religion (Catholic/other), participation in gifted program, participation in special education, and winning a scholarship, is constrained by a log-linear model that allows for two-way interactions. The continuous portion of the general location model places an additive model across contingency table cells on the means of the following variables: language (spanish/english), whether or not father's work status is missing, participation in food stamp program, participation in Aid to Families with Dependent Children (AFDC), low/high school, mother's birth location (U.S./Puerto Rico/other), sex, number of eligible children in household, income, mother's education, math scores and grade level. Mahalanobis matching was done in 0.10 calipers of (linear) propensity score standard deviations on the two test score variables and the grade level variable; the low/high variable also played a special role in the Mahalanobis matching as described in Section 6.10. For algorithmic efficiency, indicator variables for discrete variables that are fully observed (such as low/high), and any of their interactions, can be treated as continuous with no loss of generality. This is preferable as it reduces the effective dimensionality of the model.

The resulting balance for variables designated by the evaluation team to be most predictive of outcomes[9] is given in Table 1.6. In the table, "z-stat" stands for the z-statistic corresponding to the difference in means between

[9]The list of all variables included in the final analysis is displayed in Table B.1 in Appendix B.

Variable	Application Wave 1 Simple Random Sample	Stratified Random Sample	PMPD	Waves 2-5 Randomized Block
low/high	-0.98	0.00	0.11	0.21
grade level	-1.63	0.03	-0.03	-0.39
reading score	-0.38	0.65	0.48	-1.05
math score	-0.51	1.17	0.20	-1.37
ethnicity	1.80	1.68	1.59	1.74
mom's education	0.16	0.14	0.09	1.67
special education	0.31	1.66	-0.17	0.22
gifted program	0.42	-1.16	-0.13	0.75
language	-1.06	-0.02	-1.03	-0.44
afdc	-0.28	0.49	0.83	-1.57
food stamps	-1.08	-0.27	0.94	-1.31
mother works	-1.26	-0.30	-1.18	0.40
educ. expectations	0.50	1.79	0.57	0.19
children in household	-1.01	-1.75	0.41	-1.02
birth location	0.49	0.73	-1.40	-0.69
length of residence	0.42	0.71	0.66	-0.78
dad's work missing	1.09	0.70	0.00	0.16
religion	-1.84	-0.19	-0.74	-0.80
sex	0.88	1.22	0.76	0.53
income	-0.38	-0.62	0.74	-1.21
age as of 4/97	-1.57	0.18	-0.47	-0.87

TABLE 1.6. Balance: Single-Child Families

the two groups for a covariate[10]. The results for the PMPD are compared to the results for the randomized block design and to the results for the stratified random sample (stratified on low/high school) of the same size from the pool of all potential matching subjects.

Overall, the resulting balance from the PMPD is quite good. Compared to the randomized block design, the PMPD has lower absolute z-scores for 16 variables, higher z-scores for only 5. It is beaten by the simple random sample for 6 variables and by the stratified random sample for 9 variables (there is one tie). In addition, the gains when PMPD beats its competitors

[10]This is calculated for each covariate, x, as

$$\frac{\bar{x}_t - \bar{x}_c}{\sqrt{\hat{\sigma}_t^2/n_t + \hat{\sigma}_c^2/n_c}}$$

where t and c subscripts denote sample quantities from the treatment and control groups, respectively.

are generally larger than the gains of the competitors over PMPD. The superior performance of the stratified random sample also reflects the gains which can be made when a control reservoir of this size is available for choosing the control group to be followed.

Propensity score theory predicts a gain in efficiency for differences in covariate means over simple random sampling by a factor of approximately two (Rubin and Thomas, 1992, 1996). We have constructed half-normal plots of the Z-statistics displayed in Table 1.6 which were standardized by the usual two-sample variance estimate, which assumes random allocation to treatment groups. Therefore, we expect these Z-statistics to follow the standard normal distribution when the assumptions of random allocation are true (thus the Z-statistics are expected to fall on the solid line with slope 1 in each diagram). If the observations fall above the line with slope 1, they originate from a distribution with *larger* variance than we are using to standardize the differences, because they are systematically more dispersed than the corresponding quantiles of the standard normal. If they fall below that line, they originate from a distribution with *smaller* variance than we are using to standardize the differences because they are systematically less dispersed than the the standard normal.

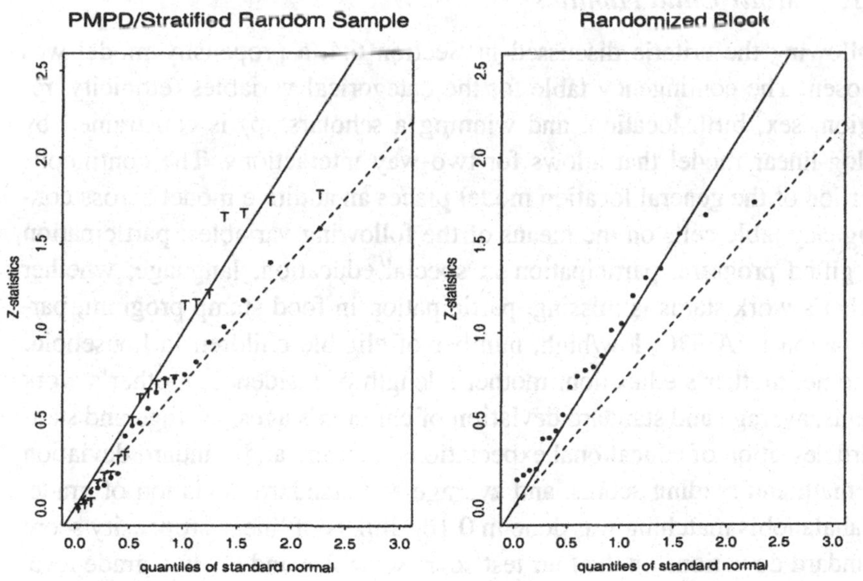

FIGURE 1.1. Half-Normal Plots of Z-Statistics for Single-Child Families

For Figure 1.1, the solid line in each panel corresponds to the normal distribution with variance 1 and the the dotted line in each panel corresponds to the normal distribution with variance $1/2$. The dots in the left and right panels represent the Z-statistics from the PMPD and randomized block designs respectively. This figure thus reveals that the gains predicted by Rubin and Thomas (1992) for the propensity score matching are fairly closely achieved for the study of single-child families. These results can be contrasted with those from the randomized block experiment, which are consistent with the standard normal distribution. The stratified random sample (displayed as "T" points) is the best of the alternatives but still fails to achieve the efficiency gains of the PMPD. We excluded the simple random sample as it is an unlikely alternative given the initial low/high imbalance in the first application wave.

Since the variance in the difference in means is reduced by a factor of two, this is equivalent, for some analyses, to increasing the sample size by a factor of two for these variables. This principle holds, for instance, for any linear combination of the measured covariates, however, in practice outcome variables are not perfectly predicted by these variables, resulting in a less dramatic improvement in efficiency (Rubin and Thomas, 1996).

6.13 *Multi-Child Families*

Following the criteria discussed in Section 6.4, a propensity model was chosen. The contingency table for the categorical variables (ethnicity, religion, sex, birth location, and winning a scholarship) is constrained by a log-linear model that allows for two-way interactions. The continuous portion of the general location model places an additive model across contingency table cells on the means of the following variables: participation in gifted program, participation in special education, language, whether father's work status is missing, participation in food stamp program, participation in AFDC, low/high, number of eligible children in household, income, mother's education, mother's length of residence, mother's work status, average and standard deviation of children's ages, average and standard deviation of educational expectations, average and standard deviation of math and reading scores, and average and standard deviation of grade. Mahalanobis matching was done in 0.10 calipers of linear propensity score standard deviations on the four test score variables and the two grade level variables; the low/high variable also played a special role in the Mahalanobis matching as described in Section 6.10.

The resulting balance of the design as compared with both the corresponding randomized block design and a stratified random sample of

Variable	Application Wave 1 Simple Random Sample	Stratified Random Sample	PMPD	Waves 2-5 Randomized Block
low/high	-3.81	0.00	-0.98	0.15
avg. grade level	-0.27	0.21	0.38	0.23
s.d. grade level	-0.19	-0.08	-0.40	0.58
avg. reading score	-1.06	0.97	0.91	-0.23
s.d. reading score	-0.90	-1.95	1.23	-2.20
avg. math score	-0.56	0.26	0.82	0.32
s.d. math score	-1.02	-1.23	0.33	-1.11
ethnicity	-1.03	-0.95	0.20	2.09
mom's education	-0.27	-1.01	-0.21	-0.22
special education	-0.67	-1.12	-0.11	0.68
gifted program	-0.85	0.43	-0.07	-0.52
language	1.13	1.35	0.92	-0.64
afdc	-1.24	0.00	0.13	3.42
avg. age	-0.38	-0.19	0.48	0.66
s.d. age	0.09	0.14	0.00	0.38
avg. educ. exp.	-0.81	-1.22	0.49	-0.71
s.d. educ. exp.	-1.59	-0.80	-0.10	0.94
children in household	0.39	-0.27	-0.40	-0.13
income	0.93	1.47	0.13	2.01
religion	0.01	0.93	-0.97	-0.66
length of residence	-1.29	-1.44	0.54	1.31
dad's work missing	0.39	-1.91	0.70	1.73
food stamps	-2.06	-0.42	-0.35	2.58
mom works	1.29	0.87	0.73	-0.49
birth	0.20	1.26	-0.42	1.34
sex	-0.84	-0.07	-0.17	-1.43

TABLE 1.7. Difference in Means Z-Statistics: Multi-Child Families

the potential matches is displayed in Table 1.7. The initial imbalance in the low/high variable is also present with the multi-child families, but the PMPD still achieves very good overall balance. Compared to the all other designs, the PMPD has lower absolute z-scores for 18 variables, higher z-scores for 8. Again, the gains when PMPD beats the other designs are generally larger than the other way around.

Half-normal quantile-quantile plots for the multi-child families in both experiments, displayed in Figure 1.2, are similar to those for single-child families. Gains in efficiency by a factor of two appear to be achieved by the PMPD over the randomized block design. The stratified random sample performs slightly better than the other alternatives but fails once again to achieve the efficiency of the PMPD.

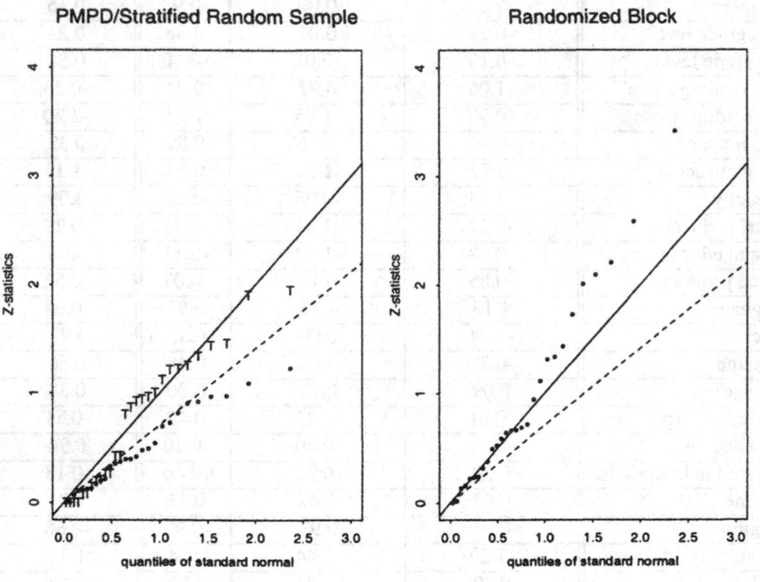

FIGURE 1.2. Half-Normal Plots of Z-Statistics for Multi-Child Families

Although the special test score variables are not quite as well balanced in the PMPD as in the randomized block design for the multi-child families (probably due to correlations between these and the low/high variable), they are still well balanced. Furthermore, the high correlation commonly seen between pre- and post-test scores makes this variable a prime candidate for covariance adjustments within a linear model to take care of the remaining differences between groups. For the single-child families, the PMPD is clearly superior in terms of test score variable balance.

It is worthwhile to note that all of the calculations in the section were performed on an available-case basis to provide statistics comparing balance. They are not directly relevant for drawing causal inference.

7 Imperfect Randomized Experiments

It is important to realize that our randomized experiment does not really randomize the treatment of, for instance, public and private school attendance but rather it randomizes the "encouragement" to attend a private rather than a public school by offering to provide some financial support ($1400) to do so. In some encouragement studies interest may focus on the effect of encouragement itself, but more often when randomized encouragement designs are used, interest focuses on estimating the effect of the treatment being encouraged, here, attending private versus public schools (or participation in the scholarship program). If there were perfect compliance, so that all those encouraged to get the new treatment got it, and all those who were not so encouraged received the standard treatment, then the effect being estimated typically would be attributed to whatever was viewed as the "active" ingredient in the treatment condition. But encouragement designs do not anticipate anything approaching full compliance, and so there is the opportunity to try to estimate different effects for encouragement and the active treatment.

In recent years, there has been substantial progress in the analysis of encouragement designs, based on building bridges between statistical and economic approaches to causal inference. In particular, the widely accepted approach in statistics to formulating causal questions is in terms of "potential outcomes". Although this approach has roots dating back to Neyman and Fisher in the context of perfect randomized experiments (Neyman, 1923; Rubin, 1990), it is generally referred to as Rubin's causal model (Holland, 1986) for work extending the framework to observational studies (Rubin, 1974, 1977) and including modes of inference other than randomization-based, in particular, Bayesian (Rubin, 1978, 1990) . In economics, the technique of "instrumental variables" (IV) due to Haavelmo (1943, 1944) was a main tool of causal inference in the type of nonrandomized studies that dominate economics. Angrist, Imbens, and Rubin (1996) showed how the approaches were completely compatible, thereby clarifying and strengthening each. The result was the interpretation of the IV technology as a way to attack a randomized experiment with noncompliance, such as a randomized encouragement design.

Imbens and Rubin (1997) showed how the Bayesian approach to causal inference in Rubin (1978) could be extended to handle simple randomized experiments with noncompliance, and Hirano, Imbens, Rubin, and Zhou (1999) showed how the approach could be extended to handle fully observed covariates, and applied it to an encouragement design in which doctors were randomly encouraged to give flu shots to at-risk patients.

Our setting is far more complex, because we have missing covariates and multivariate outcomes that are sometimes missing as well. The basic structure for our type of problem was outlined in Barnard, Du, Hill, and Rubin (1998), but our situation is more complex than that because we have a more complicated form of noncompliance – some children attend private school without receiving the monetary encouragement; it is slightly less complicated because we currently have outcomes from only one post-treatment time point. As in Frangakis and Rubin (1999) and Barnard *et al.* (1998), because of the problems described in Section 5.1 we need to make some assumptions about the missing data process and treatment effects for the non-compliers.

The first assumption we make has been called "compound exclusion" by Frangakis and Rubin (1999), when they generalized the exclusion restriction in economics. The way Angrist, Imbens and Rubin define the exclusion restriction is as follows: for those subjects whose behavior cannot be changed by the random assignment in this experiment (i.e., the encouragement to attend private schools), their outcome scores are unaffected by the assignment. That is, for those whose behavior is unaffected by assignment, their outcomes are also unaffected. Thus, under this assumption, the always takers, those who, in the context of this experiment, and defining the treatment as private school attendance, will attend private school whether or not they are encouraged to do so, will have the same outcomes (test grades) in the private school they are attending whether or not they were encouraged. Analogously, those who, in the context of this experiment, will not attend private schools whether or not they are encouraged to do so, will have the same test grades whether or not they are encouraged to attend private school. Actually, this is what Imbens and Rubin (1997) call "weak exclusion" because it says nothing about the compliers in this experiment, whereas the strong exclusion restriction, which is the traditional economic version, adds the assumption that differences in outcomes for assigned and not assigned compliers is due to treatment exposure and *not* assignment to be encouraged or not. The compound exclusion restriction of Frangakis and Rubin (1999) extends the weak exclusion restriction to apply to the missing data pattern of the outcomes as well as the values of the outcomes.

The exclusion restriction focuses attention on the "complier average causal effect" (CACE), which is the average causal effect of assignment for the compliers, rather than the more traditional "intention to treat" effect (ITT), which is the average casual effect of assignment for all subjects. Under exclusion, the average causal effects of assignment for never takers and always takers is zero, so if this assumption is correct, the ITT effect is the weighted average of the CACE and zero.

The second assumption we make has been termed "latent ignorability" of the missing data mechanism by Frangakis and Rubin (1999). Ignorability of the missing data mechanism (Rubin, 1976, Little and Rubin, 1987) basically means that the missingness of the data, given the observed values, is not dependent on missing values themselves or the parameters of the data distribution. Latent ignorability states that ignorability holds if a latent variable were fully observed, here the true compliance status of each subject (complier, never taker, always taker). Notice that we have implicitly made another assumption, namely that there are no defiers, subjects who when encouraged to attend private school will not, but when not encouraged to do so will.

As Imbens and Rubin (1997) and Hirano, Imbens, Rubin, and Zhou (1999) show, none of these assumptions are needed for a valid Bayesian analysis when faced with noncompliance, but they can dramatically simplify the analysis and sharpen posterior inferences. In fact, this is one of the dramatic advantages of the Bayesian approach to this problem: the issue of "identifiability" is put in its proper perspective. It is largely irrelevant to inference if the likelihood function has one mode rather than a small ridge – the important inferential issue is the size of a reasonable interval, e.g. a 90% interval, and not whether or not an ϵ% interval is unique as positive $\epsilon \to 0$.

A final point about our situation, with noncompliance to encouragement and missing outcomes, is that even if the focus of estimation is on the ITT effect and not CACE, one cannot use ad hoc methods to estimate the ITT effect without incurring bias. Under compound exclusion and latent ignorability, Frangakis and Rubin (1999) show that a method of moments estimator analogous to the IV estimator can be used to estimate the CACE and thereby the ITT effect essentially without bias. Of course, our Bayesian analysis does this automatically.

8 Original MPR Analysis Strategy

Before introducing our Bayesian model we first present results from an analysis that combines several existing approaches to each of the three major complications we have discussed. This analysis strategy can be implemented with available software. Missing covariates are handled by limiting the number of covariates to the design variables and the most important predictors, the pre-test scores, and then including in the analysis only individuals for whom these variables are fully observed. Missing outcomes are adjusted for by non-response weights. Instrumental variable models are used to handle the non-compliance. Separate analyses are run for math scores and reading scores (national percentile rankings). Weights are used to make the results for the study participants representative of the population of all eligible single-child families who were screened. The results in this section are obtained using the same analysis strategy that was used in the initial MPR study (Peterson, Myers, Howell, and Mayer, 1999) but now only on the subset of single-child families and separated out by the low/high variable.

Grade at Application	Low		High	
	Reading	Math	Reading	Math
1	-0.97 (170) [0.31]	2.08 (170) [0.88]	4.76 (34) [0.63]	2.59 (34) [0.28]
2	-0.83 (177) [0.40]	2.01 (177) [0.66]	-3.40 (32) [0.51]	2.72 (32) [0.41]
3	3.23 (177) [1.29]	4.95 (177) [1.69]	-8.04 (31) [0.91]	3.98 (31) [0.36]
4	2.65 (116) [0.84]	0.31 (116) [0.08]	27.92 (15) [2.75]	22.67 (15) [1.84]
Overall	0.62 (640) [0.45]	2.03 (640) [1.43]	1.07 (112) [0.26]	0.25 (112) [0.05]

TABLE 1.8. ITT Effect

Table 1.8 presents results from an ITT analysis (so compliance behavior is ignored) broken down by grade and school classification (low/high), therefore the effects represent the gains in test scores attributable to winning a scholarship. Non-bracketed numbers are treatment effect estimates for the appropriate subgroups. Numbers in parentheses are sample sizes. Bracketed numbers are the absolute value of treatment effect t-statistics for a null hypothesis of no treatment effect. Overall, across grades, there appear to be mostly positive effects of a scholarship offer. The only effect that is statistically significant at less than a .05 significance level, however, is for reading scores for children applying in the fourth grade from high-score schools. The corresponding math scores are not quite significant but of a similar direction and near significance. These effects seem quite extreme and certainly not terribly plausible. Their direction can be explained by two facts: (1) for the subset of children with observed pre-test scores and post-test scores there is a positive treatment effect of 16.57, and (2) for the subset of children in this subgroup for whom we observed pre-test scores, the children for whom we don't observe post-test scores had higher pre-test scores than the children for whom we do observe post-test scores (this information was incorporated into the non-response weights). What is particularly noteworthy, however, is that the t-statistic is so large. This points to a problem with using non-response weighting adjustments and complete cases to address such missing data problems: they cannot always reflect our uncertainty about the structure of this missing data particularly when the sample size is as small as it is for this subgroup.

Moreover, the sample sizes for all four subgroups of children applying from schools with high average test scores are quite small and these effects must all be regarded with caution. This sample size issue only worsens in the subsequent two analyses for which the effective sample sizes become even smaller.

Grade at Application	Low		High	
	Reading	Math	Reading	Math
1	-1.33 (124.8) [0.31]	2.86 (124.8) [0.88]	6.71 (26.3) [0.64]	3.65 (26.3) [0.28]
2	-1.05 (139.4) [0.40]	2.52 (139.4) [0.66]	-4.43 (27.4) [0.50]	3.54 (27.4) [0.41]
3	3.86 (138.0) [1.28]	5.91 (138.0) [1.67]	-13.81 (17.7) [0.93]	6.83 (17.7) [.35]
4	3.03 (92.2) [0.84]	0.36 (92.2) [0.08]	27.12 (9.0) [2.75]	22.01 (9.0) [1.79]
Overall	0.77 (494.4) [0.45]	2.52 (494.4) [1.42]	1.53 (80.4) [0.26]	0.36 (80.4) [0.05]

TABLE 1.9. Effect of Program Participation

The results in Table 1.9 were obtained from an analysis in which the treatment was defined as "program participation." That is, a child could only be labeled as having received the treatment if he won a scholarship (which entitled him also to receiving help in finding an appropriate school). Children who did not win scholarships but attended private school were still considered to be not receiving the treatment (therefore they are compliers, not always takers). These effects are similar to those in the preceding table although they are, in general, of larger magnitude. The t-statistics change only incrementally however. In this table the numbers in parentheses represent "effective sample sizes". These numbers correspond to the expected number of compliers for each subgroup; for this treatment definition these numbers simply subtract the expected number of never takers from the sample sizes in Table 1.8. Predictably, these numbers are even smaller than before and hence the corresponding results are even less reliable.

Grade at Application	Low		High	
	Reading	Math	Reading	Math
1	-1.55 (111.7) [0.31]	3.31 (111.7) [0.88]	6.71 (26.3) [0.63]	3.65 (26.3) [0.28]
2	-1.18 (124.2) [0.40]	2.85 (124.2) [0.66]	-5.11 (24.8) [0.51]	4.09 (24.8) [0.41]
3	4.41 (123.8) [1.26]	6.76 (123.8) [1.63]	-16.40 (16.1) [0.91]	8.12 (16.1) [0.36]
4	3.49 (84.3) [0.86]	0.41 (84.3) [0.08]	29.59 (7.8) [2.58]	24.02 (7.8) [1.89]
Overall	0.88 (444.0) [0.45]	2.89 (444.0) [1.42]	1.66 (75.0) [0.26]	0.39 (75.0) [0.05]

TABLE 1.10. Effect of Private School Attendance

The results in Table 1.10 represent the treatment effects for compliers in an analysis which defines the treatment as attendance at a private school. This treatment definition allows for compliers, never takers and always takers. Once again the magnitude of the effects increases in the vast majority of cases, however, the t-statistics across subgroups are only slightly altered, if at all. The effective sample sizes are (in most cases) even smaller for this analysis, with numbers as low as 7.8 for children applying in the 4th grade from schools with high scores.

In sum, the results from these analyses do not provide strong evidence in either direction, though there does seem to be some evidence for positive effects on test scores for a few older subgroups.

9 Notation for our Data Template

An ideal scenario for obtaining valid causal inferences for a binary treatment is the following: (1) the data arise from a randomized experiment with two treatments; (2) the outcome variables are fully observed; (3) there is full compliance with the assigned treatment; and (4) the blocking variables are fully observed; and (5) the background variables are fully observed. Aspect (5) is useful for doing covariate adjustment and subpopulation analyses. For this ideal scenario, there are standard and relatively simple methods for obtaining valid causal inferences. In reality, however, this scenario rarely occurs. Clearly, it does not occur in the SCSF program.

Deviations from the ideal scenario that occur frequently and are present in the SCSF program are the following: (6) there exist missing values in the outcomes; (7) there exist missing values in the background variables; and (8) there is noncompliance with assigned treatment. The standard methods for analyzing the ideal scenario of (1)–(5) generally fail when aspects (6)-(8) are present. Handling these additional complications in a valid and general manner is difficult. Here we present an extremely general data template allowing (6)-(8). When the observed data can be made to conform to this template, we are able to obtain valid causal inferences. Our model will allow us to return to the scenario consisting of (1)–(4).

We now introduce the notation required for the formalization of the probability model corresponding to this template. We assume that for the ith subject, where $i = 1, \ldots, n$, we have the following random variables:

1. Binary indicator of treatment assignment

$$Z_i = \begin{cases} 1 & \text{if subject } i \text{ is assigned to treatment group,} \\ 0 & \text{if subject } i \text{ is assigned to control group.} \end{cases}$$

   Z is the n component vector with i^{th} element Z_i.

2. Binary indicator of treatment receipt

$$D_i = \begin{cases} 1 & \text{if subject } i \text{ received treatment,} \\ 0 & \text{if subject } i \text{ received control.} \end{cases}$$

Because D_i is a post-treatment-assignment variable, it has a potential outcome formulation, $D_i(Z_i)$, where $D_i(0)$ and $D_i(1)$, respectively, refer to the values when assigned control and when assigned treatment.

3. Compliance status

$$C_i = \begin{cases} c & \text{if } D_i(0) = 0 \text{ and } D_i(1) = 1, \\ n & \text{if } D_i(0) = 0 \text{ and } D_i(1) = 0, \\ a & \text{if } D_i(0) = 1 \text{ and } D_i(1) = 1. \end{cases}$$

$C_i = c$ denotes a "complier," a person who will take the treatment if so assigned and will take control if so assigned. $C_i = n$ denotes a "never taker," a person who will not take the treatment no matter the assignment. $C_i = a$ denotes an "always taker," a person who will always take the treatment no matter what the assignment. This template rules out the possibility of "defiers," those who will always do the opposite of what they are assigned, i.e. those i for whom $D_i(0) = 1$ and $D_i(1) = 0$. C denotes the n component vector with i^{th} element C_i.

4. $2P$-component vector of potential outcomes[11], Y_i^{po}, which is composed of two P-length vectors, $Y_i(0)$ and $Y_i(1)$, where

$$\begin{aligned} Y_i(0) &= (Y_{i1}(0), \ldots, Y_{iP}(0)), \text{ and} \\ Y_i(1) &= (Y_{i1}(1), \ldots, Y_{iP}(1)). \end{aligned}$$

Here $Y_{ip}(0)$ is the pth outcome variable corresponding to assignment to the control group for the ith subject; $Y_{ip}(1)$ is the pth outcome variable corresponding to assignment to the treatment group for the ith subject. In other words, for each subject there are P outcome variables, Y_1 through Y_P, and each has two potential values: one corresponding to each of the treatment assignments. $Y(0)$ and $Y(1)$ are used to denote the two $n \times p$ matrices of potential outcomes corresponding to control and treatment assignment respectively.

At times we will refer simply to the P-component vector of outcomes that we *intend* to observe for a person, i.e.,

$$Y_i^{\text{int}} = Y_i(Z_i).$$

---

[11] In general, out template allows for repeated measurements over time. However, currently we have data from one pre-treatment time point and only one post-treatment time point and our notation reflects this simplification.

For convenience, we will henceforth refer to Y_i^{int} as simply Y_i with corresponding elements

$$Y_i = (Y_{i1}, \dots, Y_{iP})$$

In addition, Y represents the $n \times P$ matrix of intended outcomes for all study participants. $Y_{\cdot p}$ is the pth column in this matrix.

5. $2P$-component vector of response patterns for potential outcomes.

$$R_{y_i}(t) = \begin{cases} 1 & \text{if } Z_i = t \text{ and } Y_{ip}(t) \text{ is observed,} \\ & \text{or, if } Z_i \neq t \text{ but } Y_{ip}(t) \text{ would be observed if } Z_i = t, \\ 0 & \text{if } Z_i = t \text{ and } Y_{ip} \text{ is not observed,} \\ & \text{or, if } Z_i \neq t \text{ but } Y_{ip}(t) \text{ would be unobserved if } Z_i = t, \end{cases}$$

These indicators are themselves potential outcomes because we can only observe response indicator $R_{y_i}(t)$ for individual i if $Z_i = t$.

6. P-component outcome response pattern associated with each Y_i

$$R_{y_i} = (R_{y_{i1}}, \dots, R_{y_{iP}}),$$

where

$$R_{y_{ip}} = \begin{cases} 1 & \text{if } Y_{ip} \text{ is observed,} \\ 0 & \text{if } Y_{ip} \text{ is not observed.} \end{cases}$$

R_{y_i} indicates which of the P outcomes are observed and which are missing for subject i. R_y denotes the $n \times P$ matrix of missing outcome indicators for all study participants. $R_{y \cdot p}$ is the pth column in this matrix. We can also think of R_{y_i} as the *intended* portion of $R_{y_i}(0)$ and $R_{y_i}(1)$.

7. K-component vector of fully observed background and design variables

$$W_i = (W_{i1}, \dots, W_{iK}),$$

where W_{ik} is the value of fully observed covariate k for subject i. W is the $n \times K$ matrix of fully observed covariates. $W_{\cdot k}$ is the kth column in this matrix. In this study, application wave, the relative test scores of the school the child attended at time of application (low/high), and grade level are fully observed.

8. Q-component vector of partially observed background and design variables

$$X_i = (X_{i1}, \ldots, X_{iQ}),$$

where X_{iq} is the value of covariate q for subject i. X represents the n by Q matrix of covariates for all study participants. $X_{\cdot q}$ is the qth column in this matrix. In addition, $X^{(\text{cat})}$ refers to the subset of covariates that are categorical and $X^{(\text{cont})}$ refers to the subset of covariates that are continuous.

9. Covariate response pattern associated with X_i

$$Rx_i = (Rx_{i1}, \ldots, Rx_{iQ}),$$

where

$$Rx_i = \begin{cases} 1 & \text{if } X_{iq} \text{ is observed,} \\ 0 & \text{if } X_{iq} \text{ is not observed.} \end{cases}$$

Rx_i indicates which covariates are observed and which covariates are missing out of the Q possible covariates for subject i. Rx denotes the $n \times Q$ matrix of covariate missing data indicators for all study participants. $Rx_{\cdot q}$ is the qth column in this matrix.

This observed data template is extremely general, allowing arbitrary response patterns for the outcomes and covariates.

10 Pattern Mixture Model

Suppose that we have policy relevant covariates that are fully observed, in addition to other covariates, that may be very important for precision of estimation, which are only partially observed. Within the context of a randomized experiment, we can conceive of a sub-experiment within each pattern of missing covariate data, which is also perfectly randomized (just as when we divide a completely randomized experiment into males and females, for instance). That is, indicator variables for pre-treatment missing data patterns can be considered covariates themselves. Consequently, an attractive practical alternative when dealing with missing covariates that are not policy relevant in the above sense is to adopt a pattern mixture approach to the analysis.

Of course if a policy relevant covariate is missing, then this approach is not satisfactory, and that covariate, and not indicators for its missingness, must become part of the model. Fortunately in our setting, the major policy relevant covariates (grade, school test scores), on which decisions regarding viability of new programs may be made, are fully observed. The covariates that are important but missing are individual-level characteristics such as pre-test scores, which we do not consider policy-relevant in the above sense because it is difficult to conceive of a new program targeting subgroups defined by these variables (e.g., it's difficult to imagine that pre-test scores would used as eligibility criteria for a program, whereas school test scores, perhaps as a proxy for school quality, could be used).

In Bayesian approaches, pattern mixture models typically factor the joint distribution of indicators for missing data patterns and data as the marginal distribution of the missing data patterns and the conditional distribution of the data given these patterns. The parameters of the conditional data model are typically under-identified; assumptions regarding missing data mechanisms can help to identify these parameters.

A variety of authors use pattern mixture model approaches to missing data including Rubin (1977, 1978b); Little (1993); Glynn, Laird, and Rubin (1986, 1993); Little (1996). Standard pattern mixture models partition the data with respect to the missingness of the primary variables of interest. In this application we partition the data with respect to only covariate missing data patterns, so the assumptions will differ slightly from the standard usage. One argument that can be made for the pattern mixture approach used in this setting is that it focuses the model on the primary quantities of interest, (functions of) Y or $Y \mid X$. The marginal distribution of X and R_x is ignored.

We describe the model first by stating its structural assumptions, that is, assumptions that can be expressed without reference to a particular distributional family. Then we describe the assumptions of the particular parametric model we assume.

10.1 Structural Assumptions

We now formalize the structural assumptions of our model (some of which were previously introduced in Section 7) and discuss their plausibility for this study.

SUTVA

A standard assumption made in causal analyses of this kind is the Stable Unit Treatment Value Assumption (SUTVA) (Rubin, 1978, 1980, 1990). This assumption implies that one unit's treatment assignment does not affect another unit's outcomes and there are no versions of treatments. Formally, SUTVA is satisfied if $Y_i(Z) = Y_i(Z')$ and $D_i(Z) = D_i(Z')$ if $Z_i = Z_i'$, where Z' is the n-length vector with i^{th} element Z_i'. In this study, for SUTVA to be violated, the fact that one family won a scholarship or did not would have to affect outcomes such as another family's choice to attend private school or their children's test scores. It does not seem a terribly strong assumption to disallow such effects, or, rather, we expect our results to be rather robust to the types and degree of deviations from SUTVA that we might expect in this study.

If we define the treatment as private school attendance, the "no versions of treatments" part of SUTVA is satisfied if the definitions of private school and public school encompass all the varieties of such schools encountered by the study participants. Similarly, if we define the treatment as participation in the scholarship program (winning the money, receiving help in finding a school), treatment homogeneity is satisfied or not depending on how rigidly "participation in the scholarship program" is defined – e.g. is using the money sufficient, or need the families all have received help in finding a new school as well?

Randomization

We assume scholarships have been randomly assigned. This implies

$$\begin{aligned} p(Z \mid Y(1), Y(0), X, W, C, R_y(0), R_y(1), R_x, \theta) &= p(Z \mid W^*, \theta) \\ &= p(Z \mid W^*), \end{aligned}$$

where W^* represents the portion of W that comprises the design variables, and θ is generic notation representing the parameters in any model. We drop the dependence on θ because there are no unknown parameters governing the treatment assignment mechanism. This "assumption" should be trivially satisfied given that MPR administered a lottery to assign scholarships to families and the differential sampling weights for school test score classification (low/high) and application wave are known.

Missing data process assumption – Latent Ignorability
We assume that missingness is ignorable given observed covariates within subgroups defined by compliance status. Here, observed covariates include indicators for missingness of the covariates, R_x, as well. This assumption is defined as "latent ignorability" of the missing data mechanism, formally,

$$p(R_y(0),R_y(1) \mid R_x, Y(1), Y(0), X, W, C, \theta) =$$
$$p(R_y(0),R_y(1) \mid R_x, X^{\text{obs}}, W, C, \theta).$$

where X^{obs} comprises the elements of the covariate data matrix X that are observed. Note that this is a *non-ignorable* missing data mechanism.

Recall that latent ignorability differs from standard ignorability (Rubin, 1978; Little and Rubin, 1987) because it conditions on something that is (at least partially) unobserved or latent, in this case, compliance status, C. This is a more reasonable assumption than standard ignorability because it seems quite likely that the groups of people defined by compliance status would behave differently with regard to whether or not they fill out surveys or show up for post-tests.

Noncompliance process assumption 1 – Compound Exclusion
In order to discriminate among compliers, never takers, and always takers, we need to make an assumption about their behavior. Given that never takers and always takers will participate in the same treatment (control or treatment, respectively) regardless of what they were randomly assigned, it seems plausible to assume that their outcomes and missing data patterns will not be affected by treatment assignment. The compound exclusion restriction, which generalizes the standard exclusion restriction (Angrist, Imbens, and Rubin, 1996; Imbens and Rubin, 1997), reflects this assumption, formally, as

$$p(Y(1), R_y(1) \mid X, R_x, W, C = n) = p(Y(0), R_y(0) \mid X, R_x, W, C = n),$$

for never takers, and,

$$p(Y(1), R_y(1) \mid X, R_x, W, C = a) = p(Y(0), R_y(0) \mid X, R_x, W, C = a),$$

for always takers.

Compound exclusion seems more plausible for never takers than for always takers. Never takers would stay in the public school system no matter whether they won a scholarship or not. Always takers, on the other hand, might be in one private school if they won a scholarship or another if they didn't win a scholarship, particularly since those who won a scholarship had access to resources to help find an appropriate private school. In addition, the scholarship provided the family with $1400 more in resources than was available to the family who didn't win a scholarship and still sent a child to private school; this could in and of itself have had an effect on student outcomes.

Noncompliance process assumption II – Monotonicity

Implicit in the definition of compliance status, C, and as pointed out in Section 9, we exclude the possibility that there exist people who will do the opposite of their assignment. These individuals are referred to in the compliance literature (see, for example, (Angrist, Imbens, and Rubin, 1996)) as "defiers" and have the property that, for individual i,

$$\begin{aligned} D_i(Z_i = 0) &= 1, \text{ and,} \\ D_i(Z_i = 1) &= 0. \end{aligned}$$

The assumption that there exist no defiers for this study is referred to as monotonicity because it implies that for all i, $D_i(Z_i = 1) \geq D_i(Z_i = 0)$ (Imbens and Angrist, 1994). In the SCSF program defiers would be families who would not use a scholarship if they won one, but, would pay to go to private school if they did not win a scholarship. It seems highly implausible that such a group of people exists, therefore the monotonicity assumption appears to be quite reasonable.

10.2 Parametric Model

Our full model needs simultaneously to (1) represent a reasonable approximation to the sampling distribution of the (complete) data, (2) be comprehensive enough to justify our assumptions about the missing data process, (3) incorporate the constraints imposed by the randomization, (4) incorporate the constraints imposed by the exclusion restriction, and (5) incorporate the conditional independence structures imposed by the latent ignorability.

Consider the following factorization of the joint sampling distribution of the potential outcomes and compliance conditional on the covariates and their missing data patterns,

$$\begin{aligned} &p(Y_i(0), Y_i(1), R_{y_i}(0), R_{y_i}(1), C_i \mid W_i, X_i^{\text{obs}}, R_{x_i}, \theta) = \\ &p(C_i \mid W_i, X_i^{\text{obs}}, R_{x_i}, \theta^{(C)}) p(R_{y_i}(0), R_{y_i}(1) \mid W_i, X_i^{\text{obs}}, R_{x_i}, C_i, \theta^{(R)}) \\ &p(Y_i(0), Y_i(1) \mid W_i, X_i^{\text{obs}}, R_{x_i}, C_i, \theta^{(Y)}) \end{aligned}$$

where $\theta = (\theta^{(C)}, \theta^{(R)}, \theta^{(Y)})'$, justified by the preceding assumptions. Note that the response pattern of covariates for each individual is itself a covariate.

The specifications of each of these components are described in the next three sections.

Compliance Status Sub-Model

The specification for the compliance status model comprises a series of conditional probit models defined using indicator variables $C_i(c)$ and $C_i(n)$ for whether individual i is a complier or a never taker, respectively:

$$C_i(n) = 1 \text{ if } C_i(n)^* \equiv g_1(W_i, X_i^{\text{obs}}, R_{x_i})'\beta^{(C,1)} + V_i \leq 0$$
$$C_i(c) = 1 \text{ if } C_i(n)^* > 0 \text{ and } C_i(c)^* \equiv g_0(W_i, X_i^{\text{obs}}, R_{x_i})'\beta^{(C,2)} + U_i \leq 0,$$

where

$$V_i \sim \mathrm{N}(0,1) \text{ and,}$$
$$U_i \sim \mathrm{N}(0,1).$$

The specific models attempt to strike a balance between including all the design variables as well as the variables that were regarded as most important in predicting compliance or having interactions with the treatment effect, and on the other hand trying to maintain parsimony. The results reported in Section 11 use a compliance component model whose link function, g_1, fits, in addition to an intercept: school test scores (low/high); indicators for application wave; propensity scores for subjects applying in the first period and propensity scores for the other waves; indicators for grade of the student; recorded ethnicity (African American or other); an indicator for whether or not the pre-treatment test scores of reading and math were available; and the pre-test scores (reading and math) for the subjects with available scores. The link function g_0 is the same as g_1 with the exception that it excluded the indicators for application wave. This link function, a more parsimonious version of one we employed in earlier models, was more appropriate to fit the relatively small proportion of always-takers.

Because the pre-tests were either jointly observed or jointly missing, one indicator for missingness of pre-test scores is sufficient. The same is true of the post-tests.

The prior distributions for the compliance sub-model are

$$\beta^{(C,1)} \sim \mathrm{N}(\beta_0^{(C,1)}, \{\sigma^{(C,1)}\}^2\mathbf{I}),$$
$$\text{and } \beta^{(C,2)} \sim \mathrm{N}(0, \{\sigma^{(C,2)}\}^2\mathbf{I}),$$

where $(\sigma^{(C,1)})^2$ and $(\sigma^{(C,2)})^2$ are "known" hyperparameters set at ten, and $\beta_0^{(C,1)}$ is a vector of zeros with the exception of the first element which is set equal to $-\Phi^{-1}(1/3) * \{1+\sigma^{(C,1)} \text{ ave}(g'_{1,i} g_{1,i})\}^{\frac{1}{2}}$, where $g_{1,i} = g_1(W_i, X_i^{obs}, Rx_i)$, and ave denotes the average over the students. These priors reflect our a priori ignorance about the probability any individual belongs to each compliance status by setting each of their prior probabilities at 1/3.

Outcome Sub-Model

The specification for the outcome sub-model first posits a latent variable such that

$$Y_i(z)^* \mid W_i, X_i^{obs}, Rx_i, C_i, \theta^{(Y)} \sim$$
$$\text{N}(g_2(W_i, X_i^{obs}, Rx_i, C_i, z)'\beta^{(Y)}, \exp[g_3(X_i^{obs}, Rx_i, C_i, z)'\zeta^{(Y)}]),$$

for $z = 0, 1$, where $\theta^{(Y)} = (\beta^{(Y)}, \zeta^{(Y)})$ and where $Y_i(0)^*$ and $Y_i(1)^*$ are assumed conditionally independent, an assumption which has no effect on inference for super-population parameters (Rubin, 1978). Then

$$Y_i(z) = \begin{cases} 0 & \text{if } Y_i(z)^* \leq 0, \\ 100 & \text{if } Y_i(z)^* \geq 100, \\ Y_i(z)^* & \text{otherwise.} \end{cases}$$

The results reported in Section 11 use an outcome component model whose outcome mean link function, g_2, is linear in, and fits distinct parameters for, the following:

1. For the students of the PMPD design: an intercept; school test scores (low/high); recorded ethnicity; indicators for grade; the propensity score; and an indicator for whether or not the pre-treatment test scores were available, and the pre-test score values for the subjects with available scores.

2. For the students of the other periods: an intercept; school test scores (low/high); recorded ethnicity; indicators for grade; the propensity score; indicators for application wave; an indicator for whether or not the pre-treatment test scores were available, and the pre-test score values for the subjects with available scores.

3. An indicator for whether or not a person is an always-taker.

4. An indicator for whether or not a person is a complier.

5. For compliers assigned treatment: an intercept, one indicator for school test scores (low/high); ethnicity; and indicators for the first three grades (the variable for the fourth grade's treatment effect is a function of the already included variables.)

For the variance of the outcome component, the link function, g_3, includes indicators that saturate the missing data patterns, which are defined by cross-classification of whether or not a person applied in the first wave (i.e., for whom there is a propensity score), and by whether or not the pretreatment test scores were available. This dependence is needed because each pattern conditions on a different set of covariates; i.e., X^{obs} varies from pattern to pattern.

The prior distributions for the outcome sub-model are:

$$\beta^{(\mathrm{Y})} \mid \zeta^{(\mathrm{Y})} \sim \mathrm{N}(0, F(\zeta^{(\mathrm{Y})})\xi \mathbf{I})$$

$$\text{where} \quad F(\zeta^{(\mathrm{Y})}) = \frac{1}{K}\sum_k \exp(\zeta_k),$$

and where $\zeta^{(\mathrm{Y})} = (\zeta_1, \ldots, \zeta_K)$, one component for each of the K (in our case K=4) missing data patterns, and where ξ is an "inflater" which is set at five; and

$$\exp(\zeta_k) \overset{\text{iid}}{\sim} \mathrm{inv}\chi^2(\nu, \sigma^2),$$

where $\mathrm{inv}\chi^2(\nu, \sigma^2)$ refers to the distribution of the inverse of a χ^2 random variable with degrees of freedom ν (set at three) and scale parameter σ^2 (set at 400).

Outcome Response Sub-Model

We also use a probit specification for the sub-model for outcome response, $R_{y_i}(z)$, $z = 0, 1$.

$$R_{y_i}(z) = 1 \text{ if } R_{y_i}(z)^* \equiv g_2(W_i, X_i^{\text{obs}}, R_{x_i}, C_i, z)'\beta^{(\mathrm{R})} + E_i(z) \geq 0,$$

where $R_{y_i}(0)$ and $R_{y_i}(1)$ are assumed conditionally independent (using the same justification as for the potential outcomes) and where

$$E_i(z) \sim \mathrm{N}(0, 1).$$

The link function of the probit model on the outcome response, g_2, is the same as the link function for the mean of the outcome component.

The prior distribution for the outcome response sub-model is

$$\beta^{(\mathrm{R})} \sim \mathrm{N}(0, \{\sigma^{(\mathrm{R})}\}^2 \mathbf{I}),$$

where $\{\sigma^{(\mathrm{R})}\}^2$ is a "known" hyperparameter, set at ten.

11 Results

All of the results below were obtained from the same Bayesian analyses (one for math scores and one for reading scores). Both analyses include latent variables for compliers, never takers and always takers, imposing the exclusion restriction on never takers and always takers. The differences between the results for the first three tables reflect different ways of averaging over the results for these groups, as described in each subsection.

The results are reported by school test scores classification (low/high) and grade – our "policy-relevant" variables – averaging over the other characteristics in the model. Both school test scores classification and grade were thought to have possible interaction effects with treatment assignment. Most of the following estimates are not parameters of the model but functions of parameters, whose posterior distributions are induced by the posterior predictive distributions (multiple imputation) of the compliance statuses. Except when otherwise stated, plain numbers are posterior means and brackets are 95% posterior intervals.

11.1 Test Score Results

In this section we present answers to the three questions posed in Section 4:

1. What is the impact of being offered a scholarship on student outcomes?
2. What is the impact of using a scholarship (participating in the scholarship program) over and above what families and children would do in the absence of the scholarship program?
3. What is the impact of attending a private school on student outcomes?

In all three cases math and reading post-test scores will be used as outcomes. These test scores represent the national percentile rankings within grade. They have been adjusted to correct for the fact that some children were kept behind while others skipped a grade; students transferring to private schools are hypothesized to be more likely to have been kept behind by those schools. The individual-level causal estimates have also been weighted so that the subgroup causal estimates correspond to the effects for all eligible children belonging to that subgroup who attended a screening session. The numbers in parentheses represent either the sample sizes or "effective sample sizes" corresponding to each subgroup, just as described in Section 8, though here the posterior means of the parameters reflecting probabilities for each compliance category were used as estimated probabilities when calculating expected values for each.

ITT results

We examine the impact of being offered a scholarship on post-test scores by estimating the ITT effect as displayed in Table 1.11. We calculate the ITT effect by averaging over the effect in all three compliance groups[12].

Grade at Application	Low		High	
	Reading	Math	Reading	Math
1	2.06 (244) [-1.69, 5.54]	4.89 (244) [1.70, 8.05]	1.31 (46) [-4.74, 6.80]	5.01 (46) [0.07, 9.81]
2	0.20 (244) [-2.85, 3.44]	1.10 (244) [-2.08, 4.21]	-0.71 (45) [-5.99, 4.75]	0.97 (45) [-4.44, 6.55]
3	0.46 (233) [-3.00, 4.13]	3.02 (233) [-0.85, 6.66]	-0.60 (40) [-6.21, 4.81]	2.49 (40) [-3.99, 8.08]
4	2.78 (171) [–1.16, 7.06]	2.65 (171) [-1.50, 6.81]	1.69 (27) [-4.10, 7.81]	2.15 (27) [-3.76, 7.85]
Overall	1.27 (892) [-0.80, 3.42]	2.94 (892) [0.71, 5.15]	0.32 (158) [-4.89, 4.96]	2.73 (158) [-2.01, 7.15]

TABLE 1.11. ITT Effect

These results indicate posterior distributions primarily (i.e. greater than 97.5%) to the right of zero for the treatment effect on mathematics scores for 1st graders, and overall grades from low-score schools. Each indicate an average gain of more than 2.9 percentile points for children who won a scholarship.

Effect of participation in SCSF program

The results displayed in Table 1.12 reflect the effect of participation in the SCSF. They were calculated by measuring the ITT effect for always takers and compliers combined[13]. This analysis defines the SCSF *program* as the "treatment" rather than just private school attendance. This will provide an answer to the second of the questions posed above because the complier

[12]This strategy is an approximation to the most appropriate analysis for this estimand which would relax the exclusion restriction on both the always takers and never takers.

[13]This strategy is an approximation to the most appropriate analysis for this estimand which would relax the exclusion restriction on the always takers.

control group will include children who were able to take advantage of resources beyond those provided by the SCSF program.

Grade at	Low		High	
Application	Reading	Math	Reading	Math
1	2.74 (216.9) [-2.19, 7.32]	6.40 (216.9) [2.24, 10.79]	1.76 (42.6) [-6.46, 9.39]	6.61 (42.6) [0.08, 13.00]
2	0.24 (214.7) [-3.50, 4.20]	1.38 (214.7) [-2.59, 5.29]	-0.97 (41.0) [-8.03, 6.12]	1.31 (41.0) [-5.76, 8.52]
3	0.60 (204.6) [-3.84, 5.18]	3.96 (204.6) [-1.08, 8.91]	-0.84 (36.4) [-8.37, 6.82]	3.46 (36.4) [-5.26, 11.42]
4	3.40 (152.7) [-1.44, 8.34]	3.23 (152.7) [-1.79, 8.34]	2.31 (24.9) [-5.65, 10.29]	2.93 (24.9) [-5.00, 10.20]
Overall	1.63 (788.9) [-1.06, 4.46]	3.74 (788.9) [0.88, 6.56]	0.43 (144.9) [-6.40, 7.11]	3.68 (144.9) [-2.61, 9.74]

TABLE 1.12. Effect of Scholarship Program

We see a similar pattern of effects in this analysis though the posterior means are all larger in absolute value than in the ITT analysis. The intervals are also larger than the ITT intervals which is not surprising given that the estimand now applies to only a subset of the study participants (as reflected by the effective sample sizes in parentheses).

Effect of Private School Attendance

The results in Table 1.13 represent the effect of private school attendance by focusing only on the compliers. This analysis defines the "treatment" as private school attendance. The validity of these results rest, in part, on the assumption that receiving a scholarship and then attending private school is the same treatment as not receiving a scholarship and attending private school.

The effects of private school attendance, displayed in Table 1.13, are quite similar to the scholarship program effects with posterior means that are slightly bigger in absolute value than in the other two analyses. The intervals have also grown reflecting the still smaller effective sample sizes. The effective sample sizes for the subgroup of 4th-graders applying from high-score schools is so small (24.9) as to make these results a bit suspect.

11.2 Composition of compliance status

Table 1.14 gives estimates of the composition of compliance status as a function of school test scores classification and grade. Because the distri-

Grade at Application	Low		High	
	Reading	Math	Reading	Math
1	3.08 (164.2) [-2.56, 8.38]	7.24 (164.2) [2.45, 12.41]	1.89 (32.2) [-7.2, 10.53]	7.14 (32.2) [0.09, 14.36]
2	0.28 (171.8) [-3.94, 4.80]	1.57 (171.8) [-2.98, 6.05]	-1.08 (30.9) [-9.49, 6.75]	1.44 (30.9) [-6.10, 9.68]
3	0.67 (155.9) [-4.28, 5.81]	4.53 (155.9) [-1.22, 9.50]	-0.93 (26.2) [-9.33, 7.25]	3.82 (26.2) [-5.61, 12.58]
4	3.84 (125.2) [-1.76, 9.83]	3.62 (125.2) [-1.95, 9.50]	2.55 (18.4) [-6.15, 11.54]	3.17 (18.4) [-5.29, 11.12]
Overall	1.85 (617.1) [-1.21, 5.13]	4.24 (600.6) [0.99, 7.44]	0.47 (107.7) [-7.20, 7.77]	4.02 (107.7) [-2.83, 10.79]

TABLE 1.13. Effect of Private School Attendance

butions between the two models (mathematics/reading) were comparable in both location and uncertainty (see also *Agreement between the models* below), reported results are from the equal-weight mixture of the distributions of the two models.

Grade at Application	School Test Scores	Never Taker	Complier	Always Taker
1	High	24.6 (5.8)	69.9 (6.6)	5.5 (3.3)
	Low	24.2 (3.5)	67.3 (4.3)	8.5 (2.7)
2	High	24.7 (5.7)	68.7 (6.5)	6.6 (3.4)
	Low	20.0 (3.4)	70.4 (4.3)	9.6 (2.7)
3	High	28.0 (6.3)	65.6 (7.0)	6.4 (3.3)
	Low	23.8 (3.8)	66.9 (4.5)	9.3 (2.6)
4	High	26.3 (6.2)	68.0 (6.8)	5.7 (3.2)
	Low	18.0 (3.7)	73.2 (4.8)	8.8 (3.1)

Posterior standard deviations are in parentheses.

TABLE 1.14. Composition of compliance status

The clearest pattern revealed by Table 1.14 is that, in most cases, high-score schools have more never takers, fewer always takers, and slightly fewer compliers than low-score schools.

11.3 *Impact of missing data*

When the latent compliance groups have differential response (i.e. missing data) behaviors, standard ITT analyses or standard IV analyses are generally not appropriate for estimating, respectively, the ITT or IV estimands. The following table compares response behavior (i) between compliers attending public schools and never-takers, (ii) between compliers attending private schools and always takers, and (iii) between compliers attending private schools and compliers attending public schools.

The observed response behavior on the mathematics and reading was identical within individuals. For this reason, and also because there was satisfactory agreement in the prediction of compliance status between the two models (mathematics/reading, see *Agreement between the models* below), reported results are from the equal-weight mixture of the distributions from the two models. In addition, the posterior distributions of the odds ratios are skewed, so posterior medians and posterior intervals are reported.

For each of the first two comparisons (columns three and four), the groups being compared are attending the same type of school, so any difference in response rate is attributed to the latent compliance status characteristics. For the last comparison (right-most column), any differences are attributed to the treatment. From the table it can be deduced that response is increasing in the following order: never-takers, compliers attending public, compliers attending private, and always-takers. Therefore, the latent compliance behavior appears to be an important predictor of response.

Agreement between the models

Before running the final analyses, we assessed the agreement between the model for mathematics and the model for reading in predicting compliance type (i) at the individual student level, and (ii) as a function of the covariates low/high and grade, aggregating over the students in these classes. Evaluating agreement at such specific levels is important because, although the marginal probability of being a complier is well estimated generally, the two models might have been assigning different probabilities of being a complier to different sets of students.

At the individual level, for each model, and for each student assigned the lottery but whose compliance type was not known, we computed the posterior probability of being a complier. The correlation between the probabilities obtained from the two models was 0.72, and the corresponding correlation for the students assigned control with unknown compliance status was 0.73, indicating a satisfactory level of agreement at the individual level. At

Grade at Application	School Pre-Treat. Scores	Control Complier vs. Never Taker	Treatment Complier vs. Always Taker	Treatment Complier vs. Control Complier
1	High	2.4 [1.2, 5.1]	0.1 [0.0, 2.7]	1.7[0.5, 6.2]
	Low	2.3 [1.2, 4.5]	0.1 [0.0, 0.8]	1.4[0.7, 2.9]
2	High	2.3 [1.2, 5.0]	0.3 [0.0, 4.5]	2.9[0.9, 14.3]
	Low	2.1 [1.1, 4.2]	0.2 [0.0, 1.5]	2.4[1.1, 5.4]
3	High	2.4 [1.2, 6.1]	0.1 [0.0, 1.9]	2.0[0.5, 11.1]
	Low	2.3 [1.2, 4.9]	0.1 [0.0, 1.1]	1.6 [0.7, 3.8]
4	High	2.0 [1.0, 4.2]	0.1 [0.0, 1.7]	2.0 [0.5, 11.1]
	Low	2.1 [1.1, 4.1]	0.1 [0.0, 0.8]	1.6 [0.7, 3.9]

Results are reported combined from mathematics and reading models because they were similar (raw data on response in mathematics and reading were identical).
Numbers are posterior medians and posterior 95% intervals.

TABLE 1.15. Odds ratios comparing response rates among groups.

the level of the cross-classification between grade and low/high the agreement of the posterior distributions of compliance status, summarized by posterior first two moments, was very good.

12 Comparison Between Models

The analyses relying on standard approaches presented in Section 8 (henceforth referred to as MPR analyses) and the Bayesian analyses were performed on the same outcomes and for the same initial subset[14] of children (single-child families from grades one through four). In addition they both attempt to address the same complications which draw the template away from the perfectly controlled randomized experiment. The Bayesian analyses, however, rely on weaker structural (though perhaps slightly stronger parametric) assumptions than the MPR analyses. These strategies, therefore, invite comparison.

Results from the Bayesian analyses lead to somewhat similar, although not altogether consistent, inferences to those indicated by the MPR analyses, largely in the sense that neither analysis shows consistently strong evidence in one direction or another. If we examine the overall results, there is a fair amount of agreement between the approaches (for all three questions asked) for math scores of children applying from low-score schools, across all grades. Both approaches provide evidence for positive gains of two to three and three-quarters percentiles for math scores of children from low-score schools.

There does appear to be a difference with regard to the effect on the other scores. For the effect on reading scores of children who applied from low-score schools, the Bayesian analyses report generally positive, though not very strong, gains across grades, whereas the MPR analyses show both positive and negative mean effects. Another, more specific difference exists on reading and mathematics on 4th graders from high-score schools, between the modest effects reported by the Bayesian analyses and the large effects reported by MPR analyses. Section 8 briefly discusses this difference in terms of the issues involved with non-response weighting. In general, these differences are driven by the fact that these analyses condition on different sources of information. The Bayesian analyses can include

[14]Clearly the exclusion of students with missing pre- and post-test scores as a part of the approach to handling these missing data problems creates a non-randomly smaller sample for the MPR analyses. However both intend their inferences to apply to the same population.

subjects with missing pre-test scores or post-test scores.

One way in which these different sources of information affect inferences concerns the differences between children who took the pre-test and those who didn't. As it turns out, for the people with missing pre-test scores (excluded in the MPR analysis), there is a negative treatment effect for reading post-test scores, which serves to counteract, in part, the positive treatment effects we see in those for whom we observe pre-test scores. Another difference has to do with the amount of smoothing allowed in each model. Currently our Bayesian models are quite parsimonious, so it is possible that we haven't allowed for enough cell to cell variation in treatment effects. The difference between the two model approaches within subgroups defined by grade and type of school is also likely influenced by the relatively smaller sample sizes in these subgroups. Clearly the larger the sample size, the greater chance we have of finding consistent results across the two approaches.

13 Discussion

Future analyses will attempt to learn from these initial models and incorporate additional complexity. We would like to investigate more closely the appropriateness of additivity and smoothing in the model, include both math and reading outcomes in the same model, include more covariates, and test the sensitivity to relaxing each of the exclusion restrictions. With additional years' data we will have to model appropriately the time series nature of the data as well as the more complicated compliance structures that will develop. We also recognize the need to perform more model checks and sensitivity analysis than we have performed to date.

As far as substantive conclusions regarding school choice, both models appear to indicate gains in math scores for children from low-score schools who have either won a scholarship, participated in the scholarship program or attended private school; however, neither analysis strategy leads to convincing and consistent overall evidence in favor of private schools. If it is truly the case the we see greater gains for the children from low-score schools, this information would have policy implications and would provide greater justification for the current Florida school choice initiative which targets more disadvantaged schools. Given the heterogeneity in schools and the noisiness of our outcome measure, it is likely that it is too soon to expect to find sharp differences between the treatment and control groups.

Acknowledgments

David Myers and Paul E. Peterson were co-principal investigators for the evaluation. We wish to thank the School Choice Scholarships Foundation (SCSF) for co-operating fully with this evaluation. This evaluation has been supported by grants from the following foundations: Achelis Foundation, Bodman Foundation, Lynde and Harry Bradley Foundation, Donner Foundation, Milton and Rose D. Friedman Foundation, John M. Olin Foundation, David and Lucile Packard Foundation, Smith Richardson Foundation, and the Spencer Foundation. We are grateful to Kristin Kearns Jordan and other members of the SCSF staff for their co-operation and assistance with data collection. We received helpful advice from Paul Hill, Christopher Jencks, and Donald Rock. Daniel Mayer and Julia Kim, from Mathematica Policy Research, were instrumental in preparing the survey and test score data and answering questions about that data. Additional research assistance was provided by David Campbell and Rachel Deyette; staff assistance was provided by Shelley Weiner. The methodology, analyses of data, reported findings and interpretations of findings are the sole responsibility of the authors and are not subject to the approval of SCSF or of any foundation providing support for this research.

We would also like to acknowledge support for the methodological work from the National Institute of Child Health and Human Development (R01 HD38209), National Institute of Mental Health (R01 MH56639), National Institute for Drug Abuse (R01 DA10184), the H.-C. Yang Memorial Faculty Fund, and National Science Foundation (SBR 9709359 and DMS 9705158).

The introduction to this paper and portions of Sections 3 and 5 were taken from Peterson and Howell (1999). The Design Section is a slightly modified version of portions of Hill, Rubin, and Thomas (1999).

References

Angrist, J. D., Imbens, G. W., and Rubin, D. B. (1996), "Identification of Causal Effects Using Instrumental Variables," *Journal of the American Statistical Association* 91, 444–472.

Ascher, C., Fruchter, N., and Berne, R. (1996), "Hard Lessons: Public Schools and Privatization," Tech. rep., Century Foundation, New York, NY.

Barnard, J., Du, J., Hill, J. L., and Rubin, D. B. (1998), "A Broader Template for Analyzing Broken Randomized Experiments," *Sociological Methods and Research* 27, 285–317.

Bishop, Y. M. M., Fienberg, S. E., and Holland, P. W. (1975), *Discrete Multivariate Analyses: Theory and Practice*, MIT Press.

Bonsteel, A. and Bonilla, C. A. (1997), *A Choice for our Children: Curing the Crisis in America's Schools*, San Francisco, California: Institute for Contemporary Studies.

Brandl, J. E. (1998), *Money and Good Intentions are not Enough, or Why Liberal Democrat Thinks States Need Both Competition and Community*, Washington, D.C.: Brookings Institution Press.

Carnegie Foundation for the Advancement of Teaching (1992), *School Choice: A Special Report*, San Francisco, CA: Jossey-Bass, Inc. Publishers.

Chubb, J. E. and Moe, T. M. (1990), *Politics, Markets and America's Schools*, Washington, D.C.: Brookings Institution Press.

Cobb, C. W. (1992), *Responsive Schools, Renewed Communities*, San Francisco, California: Institute for Contemporary Studies.

Cochran, W. G. and Rubin, D. B. (1973), "Controlling Bias in Observational Studies: A Review," *Sankhya* 35, 417–446.

Coleman, J. S., Hoffer, T., and Kilgore, S. (1982), *High School Achievement*, New York: NY: Basic Books.

Cookson, P. W. (1994), *School Choice: The Struggle for the Sould of American Education*, New Haven, CT: Yale University Press.

Coulson, A. J. (forthcoming), "Market Education: The Unknown History."

D'Agostino, Ralph B., J. and Rubin, D. B. (1999), "Estimating and Using Propensity Scores With Incomplete Data," pending publication in JASA.

Derek, N. (1997), "The Effects of Catholic Secondary Schooling on Educational Achievement," *Journal of Labor Economics* 15, 1, 98–123.

Frangakis, C. E. and Rubin, D. B. (1999), "Addressing Complications of Intention-to-Treat Analysis in the Combined Presence of All-or-None Treatment-Noncompliance and Subsequent Missing Outcomes," *Biometrika* 86, 365–380.

Fuller, B. and Elmore, R. F. (1996), *Who Chooses? Who Loses? Culture, Institutions, and the Unequal Effects of School Choice*, New York: Teachers College Press.

Gelfand, A. E. and Smith, A. F. M. (1990), "Sampling-based Approaches to Calculating Marginal Densities," *Journal of the American Statistical Association* 85, 398–409.

Glynn, R. J., Laird, N. M., and Rubin, D. B. (1986), "Mixture Modeling Versus Selection Modeling for Nonignorable Nonresponse," in *Drawing Inferences from Self-Selected Samples*, ed. H. Wainer, pp. 115–142, Springer-Verlag.

Glynn, R. J., Laird, N. M., and Rubin, D. B. (1993), "Multiple Imputation in Mixture Models for Nonignorable Nonresponse With Follow-ups," *Journal of the American Statistical Association* 88, 984–993.

Goldberger, A. S. and Cain, G. G. (1982), "The Causal Analysis of Cognitive Outcomes in the Coleman, Hoffer, and Kilgore Report," *Sociology of Education* 55, 103–22.

Gu, X. S. and Rosenbaum, P. R. (1993), "Comparison of Multivariate Matching Methods: Structures, Distances, and Algorithms," *Journal of Computational and Graphical Statistics* 2, 405–420.

Gutmann, A. (1987), *Democratic Education*, Princeton, NJ: Princeton University Press.

Haavelmo, T. (1943), "The Statistical Implications of a System of Simultaneous Equations," *Econometrica* 11, 1–12.

Haavelmo, T. (1944), "The Probability Approach in Econometrics," *Econometrica* 12, 1–115, (Supplement).

Hill, J. L., Rubin, D. B., and Thomas, N. (1999), "The Design of the New York School Choice Scholarship Program Evaluation," in *Donald Campbell's Legacy*, ed. L. Bickman, Sage Publications.

Hirano, K., Imbens, G. W., Rubin, D. B., and Zhou, A. (1999), "Estimating the Effect of an Influenza Vaccine in an Encouragement Design," to appear in *Biostatistics*.

Holland, P. (1986), "Statistics and Causal Inference," *Journal of the American Statistical Association* 81, 396, 945–970.

Imbens, G. W. and Angrist, J. D. (1994), "Identification and Estimation of Local Average Treatment Effects," *Econometrica* 62, 467–476.

Imbens, G. W. and Rubin, D. B. (1997), "Bayesian Inference for Causal Effects in Randomized Experiments with Noncompliance," *The Annals of Statistics* 25, 305–327.

Levin, H. M. (1998), "Educational Vouchers: Effectiveness, Choice, and Costs," *Journal of Policy Analysis and Management* 17, 3, 373–392.

Little, R. J. A. (1993), "Pattern-mixture models for multivariate incomplete data," *Journal of the American Statistical Association* 88, 125–134.

Little, R. J. A. (1996), "Pattern-mixture models for multivariate incomplete data with covariates," *Biometrics* 52, 98–111.

Little, R. J. A. and Rubin, D. B. (1987), *Statistical Analysis With Missing Data*, New York: John Wiley & Sons.

Meng, X.-L. and Rubin, D. B. (1993), "Maximum Likelihood Estimation Via the ECM Algorithm: A General Framework," *Biometrika* 80, 267–278.

Metropolis, N., Rosenbluth, A. W., Rosenbluth, M. N., Teller, A. H., and Teller, E. (1953), "Equations of state calculations by fast computing machines," *Chemical Physics* 21, 1087–1091.

Mosteller, F. (1995), "The Tennessee Study of Class Size in the Early School Grades," in *The Future of Children*, vol. 5, pp. 113–27.

Neyman, J. (1923), "On the Application of Probablity Theory to Agricultural Experiments Essay on Principles. Section 9," translated in *Statistical Science* 5, 465–480, 1990.

Peterson, P. E. and Hassel, B. C., eds. (1998), *Learning from School Choice*, Washington, D.C.: Brookings Institution Press.

Peterson, P. E. and Howell, W. G. (1999), "What Happens to Low-Income New York Students When They Move from Public to Private Schools," in *City Schools: Lessons from New York*, eds. D. Ravitch and J. Viteritti, Johns Hopkins University Press, forthcoming.

Peterson, P. E., Myers, D. E., Howell, W. G., and Mayer, D. P. (1999), "The Effects of School Choice in New York City," in *Earning and Learning; How Schools Matter*, eds. S. E. Mayer and P. E. Peterson, Brookings Institution Press.

Rasell, E. and Rothstein, R., eds. (1993), *School Choice: Examining the Evidence*, Washington, D.C.: Economic Policy Institute.

Roseman, L. (1998), "Reducing Bias in the Estimate of the Difference in Survival in Observational Studies Using Subclassification on the Propensity Score," Ph.D. thesis, Harvard University.

Rosenbaum, P. R. and Rubin, D. B. (1983), "The central role of the propensity score in observational studies for causal effects," *Biometrika* 70, 1, 41–55.

Rosenbaum, P. R. and Rubin, D. B. (1984), "Reducing Bias in Observational Studies Using Subclassification on the Propensity Score," *Journal of the American Statistical Association* 79, 516–524.

Rosenbaum, P. R. and Rubin, D. B. (1985), "Constructing a control group using multivariate matched sampling methods that incorporate the propensity score," *The American Statistician* 39, 33–38.

Rubin, D. B. (1973), "The Use of Matched Sampling and Regression Adjustment to Remove Bias in Observational Studies," *Biometrics* 29, 185–203.

Rubin, D. B. (1974), "Estimating Causal Effects of Treatments in Randomized and Non-Randomized Studies," *Journal of Educational Psychology* 66, 688–701.

Rubin, D. B. (1977), "Assignment to Treatment Groups on the Basis of a Covariate," *Journal of Educational Statistics* 2, 1–26.

Rubin, D. B. (1978a), "Bayesian Inference for Causal Effects: The role of randomization," *The Annals of Statistics* 6, 34–58.

Rubin, D. B. (1978b), "Multiple Imputations in Sample Surveys: A Phenomenological Bayesian Approach to Nonresponse (C/R: P29-34)," in *ASA Proceedings of Survey Research Methods Section*, pp. 20– 28.

Rubin, D. B. (1979), "Using Multivariate Matched Sampling and Regression Adjustment to Control Bias in Observational Studies," *Journal of the American Statistical Association* 74, 318–328.

Rubin, D. B. (1980), "Comments on "Randomization Analysis of Experimental Da ta: The Fisher Randomization Test"," *Journal of the American Statistical Association* 75, 591–593.

Rubin, D. B. (1990), "Comment: Neyman (1923) and Causal Inference in Experiments and Observational Studies," *Statistical Science* 5, 472–480.

Rubin, D. B. and Thomas, N. (1992), "Characterizing the Effect of Matching Using Linear Propensity Score Methods With Normal Distributions," *Biometrika* 79, 797–809.

Rubin, D. B. and Thomas, N. (1996), "Matching using estimated propensity scores: Relating theory to practice," *Biometrics* 52, 249–264.

Schafer, J. L. (1997), *Analysis of Incomplete Multivariate Data*, New York: Chapman and Hall.

Wilms, D. J. (1985), "Catholic School Effect on Academic Achievement. New Evidence from the High School and Beyond Follow-Up Study," *Sociology of Education* 58, 98–114.

Appendix A Computations

Computations of the posterior distribution of the missing compliance statuses C^{mis} and parameters were based on a Gibbs sampler (Gelfand and Smith, 1990). The Gibbs sampler we used draws, in this order: the missing compliance statuses C^{mis}; the latent variables $C_i(n)^*$ and $C_i(c)^*$ for the current set of never takers, compliers, and always-takers; the possibly latent variables $Y_i^* \equiv Y_i(Z_i)^*$ for the outcome model; all latent variables $R_{y_i}(Z_i)^*$ for the response model; the parameters for the compliance model, $\beta^{(c,1)}, \beta^{(c,2)}$; the response model parameters $\beta^{(R)}$; and the mean and variance outcome parameters $\beta^{(Y)}$ and $\zeta^{(Y)}$ respectively. For all steps drawing is done cyclically and with conditioning that ensures convergence of the Gibbs Sampler to the posterior distribution. The first step must exclude $C_i(n)^*$ and $C_i(c)^*$ from the conditioning in order for the Gibbs sampler to converge to the posterior distribution. Moreover, at this step, the conditioning on Y_i^* and $R_{yi}(Z_i)^*$ can be replaced, respectively, by $Y_i * R_{yi}(Z_i)$ and $R_{yi}(Z_i)$ for algorithmic efficiency. In the following we let $H_i \equiv (W_i, X_i^{obs}, R_{xi})$ and ϕ denote all of the model parameters. The distributions involved in the Gibbs sampler are as follows.

1. The conditional distribution required for C_i^{mis} at this step is

$$p(C_i | Y_i, H_i, D_i, Z_i, R_{yi}, \phi).$$

   This distribution is obtained from the joint $p(C_i, Y_i, D_i, R_{yi} | Z_i, H_i, \phi)$. For example, a subject with $Z_i = D_i = 0$ can be a complier or a never-taker, and the conditional Bernoulli distribution of C_i is proportional to

$$\{l(c, Z_i, H_i, Y_i, R_{yi}, \phi)\}^{I(C_i=c)} \{l(n, Z_i, H_i, Y_i, R_{yi}, \phi)\}^{I(C_i=n)},$$

   where we define

$$\begin{aligned} l(c_0, z_0, h_0, y_0, r_0, \phi) = {} & p(C_i = c_0 | H_i = h_0, \phi) \\ & \times \{p(Y_i = y_0 | C_i = c_0, H_i = h_0, z = z_0, \phi)\}^{r_0} \\ & \times p(R_{yi}(Z_i) = r_0 | C_i = c_0, H_i = h_0, \phi). \end{aligned}$$

   Therefore, the conditional probability of the subject being a complier is

$$l(c, Z_i, H_i, Y_i, R_{yi}, \phi)\{l(c, Z_i, H_i, Y_i, R_{yi}, \phi) + l(n, Z_i, H_i, Y_i, R_{yi}, \phi)\}^{-1}.$$

   The drawing of C_i for subjects with $Z_i = D_i = 1$ is done in a similar way. Note that the drawing of the compliance at this step uses information on the response behavior (R_y).

2. The drawing of $C_i(n)^*$ is from $p(C_i(n)^* | H_i, C_i, \phi)$. This distribution is the same as the defining model $p(C_i(n)^* | H_i, \phi)$ but truncated either to the left or to the right of zero depending on C_i. The drawing of the truncated normal is done using its inverse distribution function, which is readily calculable. For subjects that have been imputed as always-takers or compliers at the previous step, drawing of $C_i(c)^*$ is done in a similar way.

3. The drawing of Y_i^* is from $p(Y_i^*|H_i, C_i, Y_i, z = Z_i, \phi)$: when Y_i is in $(0, 100)$, $Y_i^* = Y_i$; when $Y_i = 0$ then Y_i^* is drawn from the tail of the defining normal distribution $p(Y_i^*|H_i, C_i, z = Z_i, \phi)$ left of 0, and the method of simulation is as with the compliance latent normals; there was no observation equal to 100.

4. The drawing of $R_{y_i}(Z_i)^*$ is from $p(R_{y_i}(Z_i)^*|H_i, C_i, R_{y_i}, z = Z_i, \phi)$. This distribution is the same as the defining model $p(R_{y_i}(Z_i)^*|H_i, C_i, z = Z_i, \phi)$ except that it is truncated to the right or left of zero depending on R_{y_i}. Drawing is as with the compliance and outcome latent normals.

5. Drawing of the coefficients $\beta^{(c,1)}$ is from $p(\beta^{(c,1)}|\{ \text{ all } C_i(n)^*, H_i\})$, which is a Bayesian linear regression based on the defining likelihood and prior. Drawing of the coefficients $\beta^{(c,2)}$ is from the distribution $p(\beta^{(c,2)}|\{ \text{ all } C_i(c)^*, H_i : C_i = a \text{ or } c\})$, and drawing of the coefficient $\beta^{(R)}$ is from $p(\beta^{(R)}| \text{ all } R_{y_i}(Z_i)^*, H_i, Z_i, C_i)$, both of which are Bayesian linear regressions.

6. The drawing of the parameters of the outcome model is further divided in two steps. In one, with $\zeta^{(Y)}$ conditioned at the values from the previous cycle, $\beta^{(Y)}$ is drawn from $p(\beta^{(Y)}|\{ \text{ all } Y_i^*, H_i, Z_i, C_i : R_{y_i} = 1\}, \zeta^{(Y)})$, which is a weighted normal linear regression with known weights. With the mean parameters $\beta^{(Y)}$ conditioned at the drawn value, there is still no known direct way of drawing from the distribution of $\zeta^{(Y)}$. Nevertheless, because its distribution is easily calculable up to proportionality, the Metropolis-Hastings algorithm (Metropolis, Rosenbluth, Rosenbluth, Teller, and Teller, 1953) was used. Because the dimension of the parameters is large, it is important to obtain a good jumping density. By defining, $\widetilde{Y}_i = |Y_i^* - g_2(H_i, C_i, Z_i)\beta^{(Y)}|$ for the observed outcomes we have that

$$\frac{(\widetilde{Y}_i)^2}{\exp(g_3(H_i, C_i, Z_i)\zeta^{(Y)})} \sim \chi_1^2, \quad \text{so}$$
$$E(2\log(\widetilde{Y}_i)) = E(\log(\chi_1^2)) + g_2(H_i, C_i, Z_i)\zeta^{(Y)},$$

where χ_1^2 is a chi-squared random variable with one degree of freedom and the distribution and expectation above are conditional on all variables except Y_i^*, and on all parameters including $\zeta^{(Y)}$. Using the regression estimates from the last relation, we obtained the two moments for a normal jumping density for $\zeta^{(Y)}$. Because the jumping density does not use the values of $\zeta^{(Y)}$ from the previous cycle, the asymmetric version of Metropolis-Hastings was used.

Initial values for the missing compliance statuses were drawn based on the moment estimates given assignment arm and school attended. The parameters were initialized to generalized linear model estimates given the initialized compliance statuses. Subsequently, the models were run each for an initial burnout series of 5000 iterations. We assessed convergence with ad hoc methods. Then a main series of an additional 5000 iterations was run for each model, upon which the results are based.

Appendix B

Variable	Description
Baseline variables (pre-lottery)	
Application wave	Indicator for each of five waves
Won a scholarship?	No/Yes
Low/high-score school	Indicator for each category
Child's birth location	U.S./Puerto Rico/Other
Grade level of child when applying	Kindergarten through 4th grade
Female guardian's ethnicity	Puerto Rican, Dominican, Other Hispanic/Black,African American/Other
Female guardian's education	Some high school/High school graduate or GED/Some college/Graduated from a 4 year college/More than a 4 year degree
Child participated in special education in the last year?	No/yes
Child participated in gifted programs in the last year	No/yes
continued on next page	

continued from previous page

Variable	Description
Main language spoken in home	English/Other
Family participates in AFDC	Yes/No
Family participates in Food-stamp Program	Yes/No
Female guardian's work status	Fulltime/Part-time/Not working but looking/Not working not looking
Education expectations for child	Some high school will not graduate/Graduate from high school/Some college/Graduate from 4-year college/More than a 4-year college degree
Number of children under 18 in household	
Female guardian's birth location	United States/Other
Female guardian's length of residence at current address	More than 2 years/1-2 years/3-11 months/Less than 3 months
Data on father's work status missing?	No/Yes
Female guardian's religion	Other/Catholic
Sex	Male/Female
Income	0-$4999/$5000-7999/.../More than $50,000
Age of the child on 4/1/97 in years	
Pre-test reading score (percentile)	
Pre-test math score (percentile)	
Pre-test reading score (normal curve equivalent)	
Pre-test math score (normal curve equivalent)	

continued on next page

continued from previous page

Variable	Description
Attendance at private school during previous year	No/Yes
Survey respondent one of child's primary caretakers what portion of the time during the past year?	None/Some/All
Time student has attended day care/school outside the US?	None/Some
Where send child to school next year (if no scholarship)?	Public/Religious Private/Secular Private
How many times during the school year have you spoken to someone from this child's school about "problems with this child's' behavior at school"?	None/1 or 2/3 or 4/More Than 4
How many times during the school year have you spoken to someone from this child's school about "this child's attendance"	None/1 or 2/3 or 4/More than 4
How many times have during the school year have you spoken to someone from this child's school about "placing this child in special classes or programs"	None/1 or 2/3 or 4/More than 4
Variables recorded one year after the lottery	
Post-test reading score (percentile)	
Post-test math score (percentile)	
Post-test reading score (normal curve equivalent)	
Post-test math score (normal curve equivalent)	

TABLE B.1. Description of Variables

Appendix C Finer Strata

The model of Section 10.2 can be used to estimate the effect of the program on finer strata that may be of interest. For example, to estimate effects stratified by ethnicity, models for compliance, outcomes, and response analogous to those in Section 10.2 were estimated with parameters for recorded ethnicity (dichotomized as African American (AA) or other). As in Section 11, the models induce a posterior distribution for the causal effect of the program for each child, and children are subsequently stratified by the variables: grade (1-4), type of child's originating school's past scores (high/low), and ethnicity. The results for this stratification are reported in Tables C.2, C.3, C.4, for the estimands of ITT, effect of scholarship, and effect of attendance of private versus public school respectively.

The results, as in those of Section 11, support evidence for effect on mathematics for certain subgroups of children. Here, the effects on mathematics are largest for African American children (with tighter intervals around the estimates in the low-score schools which constitute on average 85% of this ethnic sub-sample), smallest for non African Americans originating from high past-score schools, and in the middle range for the remaining students.

Grade at Application	Ethnicity	Low		High	
		Reading	Math	Reading	Math
1	AA	2.75 [-1.65, 6.58]	6.40 [2.59, 10.12]	1.84 [-4.93, 7.77]	6.25 [0.70, 12.03]
	other	1.43 [-2.36, 5.53]	3.56 [0.25, 7.21]	0.60 [-4.95, 5.89]	3.30 [-1.77, 8.10]
2	AA	0.72 [-2.89, 4.50]	2.28 [-1.20, 6.14]	-0.25 [-6.77, 6.20]	2.22 [-3.88, 8.38]
	other	-0.21 [-3.22, 3.23]	0.18 [-3.31, 3.77]	-1.09 [-6.16, 4.03]	-0.17 [-5.51, 4.99]
3	AA	0.99 [-2.94, 4.90]	4.43 [0.03, 8.58]	-0.06 [-6.72, 6.46]	4.29 [-3.10, 11.02]
	other	-0.03 [-3.55, 3.78]	1.73 [-2.40, 5.56]	-0.91 [-6.13, 4.14]	1.33 [-4.40, 6.49]
4	AA	3.54 [-0.94, 8.44]	4.03 [-0.54, 9.03]	2.45 [-4.10, 10.06]	3.66 [-2.88, 10.37]
	other	2.05 [-2.09, 6.24]	1.29 [-3.06, 5.83]	1.12 [-4.64, 6.48]	0.88 [-5.07, 6.46]

TABLE C.2. ITT Effect.

Grade at Application	Ethnicity	Low		High	
		Reading	Math	Reading	Math
1	AA	3.36 [-1.98, 7.89]	7.70 [3.18, 12.28]	2.27 [-6.14, 10.09]	7.62 [0.86, 14.71]
	other	2.06 [-3.34, 7.92]	5.02 [0.35, 10.08]	0.92 [-7.53, 9.09]	5.04 [-2.51, 11.88]
2	AA	0.82 [-3.23, 5.06]	2.61 [-1.39, 7.00]	-0.32 [-8.27, 7.42]	2.70 [-4.56, 10.35]
	other	-0.29 [-4.33, 4.30]	0.25 [-4.43, 5.06]	-1.67 [-9.03, 6.04]	-0.24 [-7.73, 7.64]
3	AA	1.20 [-3.69, 5.85]	5.30 [0.03, 10.36]	-0.07 [-8.25, 7.92]	5.19 [-3.75, 13.12]
	other	-0.06 [-5.11, 5.40]	2.51 [-3.34, 8.22]	-1.44 [-9.60, 6.32]	2.11 [-6.68, 10.06]
4	AA	4.00 [-1.08, 9.42]	4.53 [-0.60, 10.15]	2.90 [-4.96, 11.49]	4.29 [-3.45, 12.21]
	other	2.72 [-2.74, 7.98]	1.71 [-3.86, 7.57]	1.68 [-7.12, 9.85]	1.32 [-7.41, 9.22]

TABLE C.3. Effect of Scholarship Program.

Grade at Application	Ethnicity	Low		High	
		Reading	Math	Reading	Math
1	AA	3.75 [-2.15, 8.91]	8.66 [3.52, 13.78]	2.44 [-6.83, 10.87]	8.18 [0.93, 15.46]
	other	2.34 [-3.71, 9.59]	5.72 [0.42, 11.79]	1.00 [-8.37, 10.02]	5.51 [-2.66, 12.70]
2	AA	0.92 [-3.59, 5.68]	2.91 [-1.55, 7.88]	-0.35 [-9.03, 9.02]	2.92 [-4.90, 11.17]
	other	-0.34 [-5.10, 4.94]	0.29 [-5.03, 5.70]	-1.90 [-10.78, 6.73]	-0.23 [-8.25, 8.69]
3	AA	1.34 [-3.96, 6.72]	6.00 [0.04, 11.87]	-0.07 [-8.76, 8.73]	5.61 [-3.86, 14.67]
	other	-0.09 [-6.47, 6.30]	2.91 [-3.96, 9.73]	-1.65 [-10.83, 7.82]	2.39 [-7.32, 11.67]
4	AA	4.44 [-1.20, 10.91]	5.01 [-0.67, 11.37]	3.13 [-5.43, 12.31]	4.55 [-3.69, 13.03]
	other	3.12 [-3.13, 9.10]	1.96 [-4.46, 8.70]	1.88 [-7.95, 10.78]	1.46 [-8.05, 10.46]

TABLE C.4. Effect of Private School Attendance.

Discussion

Stephen Ansolabehere Department of Political Science, MIT

"School Choice in New York City" engages one of the most important policy debates of the 1990s and probably of the coming decade. Primary and secondary education is by far the largest category of local public expenditure in the United States. It is one of the largest categories of public employment in the United States. And, through PTOs and other organizations, it is one of the main ways that people become involved in politics and public affairs in the United States. Needless to say, any attempt to change the way that we provide this public good sparks intense ideological and political reactions from citizens, politicians, public employees unions (especially teachers), and even from usually objective academics.

Political scientists, economists, and other social scientists typically come at the problem with survey and observational data. What do parents think of public education? How well do different schools perform? The picture arising from these data is murky, often contradictory. For example, a majority of respondents to sample surveys state that they are satisfied with the education that their own children receive, but a large majority are unsatisfied with the education provided to children in the country as a whole. Similarly, states often implement various policies to "experiment" with reform. California, for instance, recently adopted an initiative that sets maximum class sizes for the state's public schools. Changes in the laws, though, are undertaken precisely because the problems have become so severe that perhaps any reform will lead to improvement: they reflect high public demand for reform.

Field experiments such as the one undertaken in New York City public schools are rare. Rarer still is the magnitude of the experiment attempted. Social science field experiments tend to be small in magnitude. The lessons from experimental studies such as this one are therefore especially valuable.

So what did I learn? As with most good papers in statistics that I've read, I learned a lot of statistics: Some dos and don'ts and even a few things to try in my own empirical work. The final analyses presented here, though, did not lead me to change my beliefs much about school choice. There are two issues. Should the basis for financing schools be changed so that failing schools close? Which vehicle for education - public, private, or parochial school - performs best, and should thus be encouraged? This paper concerns only the latter of these issues, and its documents, especially in Tables 1.11, 1.12, and 1.13, that the private school attendance and scholarships lead to noticeably better reading and test scores.

I did not know, however, what to make of these effects, because the ultimate calculation that must be made is whether this program is superior to the alternatives, including changing nothing about public education. Several calculations need to be made using the estimated effects in order for the statistics presented to inform the public debate. First, what does the magnitude of the effects mean in terms of abilities? Is a four to six point difference in tests scores the difference between literacy and illiteracy? a grade-level of reading or math? Harry Potter versus Camus? Second, how does a rise of 4 or 5 points in reading and math scores affect students' lives? Can anything be said about the economic value of this improved level of performance? For example, how well do these scores predict high school graduation or college admission, which economists have shown permanent effects on income? Relatedly, the value of the study might be linked to data used to evaluate the relative performance of educational programs in different countries, such as the statistics offered by the OECD. These comparisons may be drawn directly with further information about the Iowa Test Scores and how they relate to levels of education and achievement and to other benchmarks in education.

I am not prepared to make such assessments, as I am merely a consumer of information in this area, and certainly not an advocate. In order to draw conclusions about the policy reforms that lie behind this social experiment, consumers of social science data need to know how to assess the gains documented in experiments such as this. Social scientists typically do not do a very good job assessing the policy gains or social benefits associated with their research. It is admittedly hard to do, because such projections involve predictions that are far out of sample: What if all schools in New York City were on a voucher system? Is the social gain driven by the 5 percent increase in math test scores, or does the value lie elsewhere?

A third sort of calculation regards the comparison of the scholarship program examined in these experiments with other possible reforms. A 5 point gain in test scores is achieved with a \$1400 scholarship, which induces students to attend private schools. How does that compare with other uses of the funds, such as hiring private tutors or lowering class sizes? At \$1400 a student, the school district could hire one tutor for each scholarship student for one-hour a week at a rate of \$40 an hour (assuming a 180 day school year). Or, for every 42 scholarship students the school district could hire an additional teacher (assuming $60,000 cost per new teacher), thereby lowering class sizes. In other words, if this experiment were imposed system wide in New York City, the cost would be approximately the same as cutting class sizes in half. Would the additional tutoring produce even bigger improvements in test scores? Would the reduction in class sizes produce similar or even larger effects, as proponents of the California reforms claim? I do not know the answers to these questions. They are best answered through social experiments such as this one. To my knowledge such experiments have not been attempted. Given the results and comparisons offered here, further experiments with other means of improving test scores are warranted. The effects of private school attendance do not arise just by chance. However, the cost of this program compared to potential alternatives, especially of reducing class sizes (which might be the reason that private schools perform better), leads me to believe that it is worth conducting experiments to see if similar or even greater gains can be made through other, potentially less disruptive means.

These data may illuminate a further feature of the debate over school choice. Detractors of school reform often allege that the gains from private schooling are largely artificial and are the result of creaming, which adversely affects the students who remain in public schools. The very best students select into the private system. They enjoy modest academic gains, but their (and their parents') departure from the public school system leaves the public school system significantly worse off. The New York school experiment shows some gain to private schooling, as revealed by the Complier Causal effects. The data gathered here seem appropriate for estimating whether the students who choose to not participate (non-compliers) are especially difficult or bad students. Do they have much worse test scores than students assigned to the control group? Is the difference between their scores and the control groups much greater than the improvement posted by those who entered private schools? Comparing Tables 1.8 and 1.13, my guess is that the non-compliers are not appreciably worse students than the compliers, but this may be ascertained only through more careful consideration of the data.

Setting political science and public policy issues aside, one troubling design issue nagged at me as I worked through this article. Experiment effects (such as the Hawthorne effect) may easily arise in a study such as this one. By virtue of being studied, knowing what the study is about, and knowing what is at stake in the study, the students, families, and schools might behave differently than they would if this program were introduced for the entire city and they were not under scrutiny. The schools, in particular, have strong incentives to react to the experiment. The public schools and the teachers' unions stand to lose revenues and political power if the public school students in the control group perform poorly and the city, as a result, elects to adopt school choice for the entire system. Likewise, the private schools participating in the study could eventually gain more revenues and more high quality students if the participants posted superior performance and the city adopted the program district-wide.

Clinical trials in medicine and many social science experiments, especially in psychology, use blind or double-blind designs to guard against contamination of the estimated causal effects by human reactions to experiments. Often, the incentives to react to the experiment itself are not as strong as they are in the New York school choice experiment. It would be extremely difficult to incorporate such design features into a study such as this one, but it is not impossible. For example, experimenters could assign some private school students to other private schools, informing the recipient private school that these students are actually public school students participating in the school choice experiment. Likewise, some public school students could be assigned to other public schools. The differential performance of these students would give the researcher some leverage on experiment effects. Such control groups might be too costly, and, alas, the New York study did not employ such assignments.

I wish to end this comment, then, with a challenge. Statistical research has produced new methods for dealing with missing data, non-response, and non-compliance in experiments and observational studies. These techniques are implemented at the data analysis stage, and many of them are showcased in "School Choice in New York City." But applied researchers do not have good data analytic models for Hawthorne effects and other experiment effects. What, if anything, should be done in the data analysis to address experiment effects?

Discussion

Brian W. Junker, Carnegie Mellon University
Alix I. Gitelman, Oregon State University

1 Introduction

1.1 Educational Reform and School Choice

At least since the landmark study *A Nation at Risk* (National Center for Excellence in Education, 1983) catapulted the plight of American education and the need for educational reform into the national spotlight, educational research and educational policy analysis—especially for grades K through 12—have experienced boom times. The lessons of *A Nation at Risk* have been echoed in recent international comparisons of student performance and curricular design in mathematics and science (e.g. Valverde and Schmidt, 1997). Stories of the decay of the American public education system generally—and urban school districts in particular—are legion, from advocates of public education (e.g. Chapter 1 of Ascher, Fruchter and Berne, 1996) to its antagonists (e.g. Lieberman, 1993). Reform efforts revolve around standards and accountability (e.g., Ravitch, 1995; CCSSO, 1995; Achieve, Inc., 1999), as well as research in the institutional design of schools (e.g. Glennan, 1998) and mechanisms of institutional change (e.g. Fullan, 1999; Bryk, Sebring, Kerbow, Rollow, and Easton, 1998).

A recurring theme in the educational policy debate, going back at least to Friedman (1962), has been the notion of injecting market forces into the "education industry". This notion generally goes by the name *school choice* and has been studied or partially implemented in federal and state programs for over thirty years (Fuller, Elmore and Orfield, 1996, p. 2). School choice programs exist in a variety of limited forms, ranging from magnet schools accessible by lottery within a public school system, to state-sanctioned charter school programs, to vouchers for economically stressed families who wish to opt out of public schooling, to tax deductions and tax credits for parents who send their children to non-public schools (Peterson, 1998, pp. 7–8).

Proponents of the expansion of school choice, such as Peterson and Hassel (1998), argue that eliminating centralized power to assign students to schools will make parents effective market actors, thereby increasing parental satisfaction with their children's education; that market forces will weaken the hold that bureaucracies and interest groups such as teachers' unions have on schools, as well as strengthen the educational programs in schools that respond to these forces; and that student achievement ultimately will improve as a consequence of these choices (Fuller, Elmore and Orfield, 1996, pp. 11–12; Peterson, 1998, p. 6). Other claimed benefits include increased teacher satisfaction, reduced student mobility, and higher college matriculation rates (Peterson, 1998, pp. 17–23). More broadly, Gilles (1998) rejects the legal/philosophical view of *liberal statism*, which emphasizes the interest of the state in ensuring that children become good liberal citizens, over the interests of parents in individualized approaches to raising their children to become virtuous and flourishing men and women, in favor of *liberal parentalism*, which essentially reverses these priorities (cf. Gilles, 1998, p. 398).

Those who take a more cautious view of school choice, such as Fuller and Elmore (1996), worry that parents' tendency to exercise choice can vary according to their affluence and ethnicity, reinforcing social-class inequality; question whether choice really will lead to stronger accountability, more discriminating and involved parents, or the rise of novel and more effective forms of schooling; and wonder which of these might lead to greater student achievement (Fuller, Elmore and Oldfield, 1996, p. 12). Opponents of choice and privatization of education generally subscribe to some form of the liberal statist stance rejected by Gilles (1998; see for example Ascher, Fruchter and Berne, 1996, p. 9), arguing that broadly available choice would further balkanize communities, erode the education of citizens for a democratic society, and undermine the assimilation of socially isolated groups, such as new immigrants. Moreover, even if the theoretical argument for school choice is strong, Peterson (1998, p. 9) for example concedes that transforming a longstanding public institution into a market system is likely to be difficult in practice and should be approached with caution.

1.2 Case Study: The New York School Choice Scholarship Foundation Program

The most compelling practical arguments in favor of experimenting with school choice usually revolve around voucher programs for poor families in the cachement area of a dysfunctional urban school system (Peterson, 1998, pp. 10–23). Few would rise to defend educational systems that abysmally fail their constituents, and it is hard not to be moved—from a personal as well as a social policy point of view—by the documented desire of many inner-city parents to find a way out of the poverty-and-poor-education trap for their children. Although several urban voucher experiments have been conducted, none have provided definitive answers to relatively straightforward questions comparing achievement between students in a choice program and students in a control group, largely because political constraints, and/or selection effects associated with voluntary participation, in these largely observational studies, undermine causal interpretations of the results (e.g. Peterson, 1998, pp. 20–21). The New York School Choice Scholarships Foundation (SCSF) Program, whose design, implementation, and evaluation involved a collaboration between statisticians Barnard, Frangakis, Hill, and Rubin, and educational policy researchers Paul E. Peterson and David Meyers, provides a rare opportunity to conduct a randomized controlled experiment that attempts to address some of these concerns.

Barnard, Frangakis, Hill and Rubin (1999) present an extensive account of the SCSF experiment, in which eligible families were selected by lottery, i.e. randomly assigned, to receive scholarship/voucher money and other assistance for sending their child(ren) to a private school of their choice. Their case study illustrates some of the complications involved in educational research, even in working with a "clean" experiment like the SCSF study. Education is delivered to institutionally-structured groups of students—classrooms, grades, schools, districts—in a social mode in which selection effects and compliance and confounding issues abound. It is difficult to ensure that students who have been targeted for an educational intervention actually receive a fairly uncontaminated version of it, independently of other targeted students; difficult to attribute achievement gains to the intended intervention, as opposed to interactions with subjects' background variables or other unintended side effects of the intervention; and difficult to ensure that those not targeted do not receive some form of the intervention. For structural and ethical reasons, educational studies usually must be organized as observational studies rather than the classic randomized controlled experiments that are the gold standard of causal analysis in biostatistics and other areas. While not all of these problems arise in Barnard et al.'s (1999) work—in particular institutional grouping effects seem to be minimized—subject non-compliance and non-response, and the possibility of unmeasured confounders and/or side effects of the way in which the school choice "treatment" was administered, force the analysis away from a relatively straightforward randomized controlled experiment approach suggested by the lottery design, towards a more complex observational study approach. Barnard et al.'s key technical contributions lie in accounting for these problems, especially in the areas of subject non-compliance and non-response.

Their case study also clearly illustrates the benefits of statisticians' deep involvement in the the conception, design and implementation phases of a complex program evaluation study like the SCSF study, guided by the desire to minimize design complications and maximize data quality for the intended ultimate inferential analysis. This final analysis was cast in a Bayesian framework, not so much from the attraction of subjectivist ways of reasoning about data, but because the "complete probability model" that provides the context for formally Bayesian methods, in which all variables—parameters and data—are endowed with a single joint distribution, is a powerful way to keep track of the complex interaction of observed and missing outcome data and covariates, and to think about and account for various program evaluation and design decisions along the way. Practical limits on time and resources, and the natural evolution of the final model as the study evolves through design and implementation, make it impossible—and perhaps undesirable—for every decision along the way to be formally Bayes with respect to the final model, but of course a formally Bayesian approach to the final inferential analyses follows naturally from the complete probability model that Barnard et al. have built up for the SCSF program evaluation.

In the remainder of our discussion we will consider first some aspects of the experimental design, data collection and matching, and then some aspects of the final inferential model used to reach causal conclusions, in Barnard et al.'s case study. The work of Barnard et al. stands as an exemplar of the convergence of a variety of methodologies for the analysis of observational studies and broken experiments, developed by Don Rubin and his colleagues and students over the past two decades and more. Although our comments below are directed more toward supplementary analyses and connections with other approaches that we would like to have seen, we applaud the vast effort and resourcefulness required to design, implement, and undertake a reasonably correct analysis of, a study such as this. Finally we will return to the broader school choice debate sketched in this introduction: what does the SCSF experiment tell us about the desirability of school choice? Here, our conclusions are mixed: we feel that this case study illustrates exemplary practical methodology chasing a flawed substantive question. We would like to encourage the application of similar methodology to more focused questions in educational research.

2 Design, Data Collection and Matching

2.1 *Intake, Assignment and Followup*

Subjects entered the New York School Choice Scholarship Program (NYS CSP) in five intake waves over a period of three months; in all there were over 20,000 entrants. The School Choice Scholarship Foundation (SCSF) had funding to award 1,300 scholarships, each of which provided $1,400 annually per child to cover tuition, books, uniforms, and other expenses for the child to attend a private school. Remaining tuition, which ranged roughly from $1,500 to $2,500 for the mostly parochial schools chosen, and any other expenses, had to come from families' pockets, or other sources such as additional scholarships that might be provided by the chosen school (SCSF, 1999; Peterson, Myers and Howell, 1998). The SCSF targeted *a priori* 85% of its scholarship awards for students from "low-score" public schools, that is, schools with average test scores lower than the City-wide median; the remaining 15% were to be awarded to students from the complementary "high-score" schools.

In the first intake wave, all entrants were invited to screening sessions where their eligibility was determined. Then, all eligible entrants participated in the lottery, with chances of winning adjusted to achieve the 85%/15% target allocation of scholarships to student from low- and high-score schools. Because of the effort involved in screening the unexpectedly large number of entrants, in the second through fifth waves, an initial pre-screening selection of entrants was added, thereby reducing the number of entrants allowed to the screening sessions. In addition, this pre-screening in waves two through five selected entrants in the correct proportions from low-score and high-score schools. In the first wave, then, the lottery selection mechanism was relied upon to select winners in the correct proportion from low-score and high-score schools. Furthermore, for every lottery winner in wave 1, there were roughly five control families, while this ratio was 1:1 in the subsequent waves due to the pre-screening process (see Barnard et al., Table 1.3). All participants—that is, all who entered the lottery—were followed up at the end of the first year of the study.

In the control arm, an 80% follow-up response rate was achieved. In the treatment group, of the 75% who actually used the scholarship to attend a private school, the follow-up rate was 89%. Of the 25% of the lottery winners who declined the scholarship—mostly for lack of funds to make up the difference between $1,400 and the actual costs of private schooling—there was only a 66% follow-up response rate (Barnard et al., 1999; Peterson, Myers and Howell, 1998).

Strong incentives to followup—in the form of cash rewards and a "second-chance" lottery, were used to raise the followup rate in the control arm; still, there appear to be at least economic differences, apparently not captured by the matching processes described below, between the non-compliers in the treatment arm and the control arm subjects. We would like to understand how these differences might affect the study outcome of student achievement after one year.

2.2 *Matching*

Barnard et al. adroitly exploited the shift in the intake process described above to conduct a sub-study comparing a "propensity matched pairs design" (PMPD) to select matched controls from the pool of lottery non-winners in the first wave, with the randomized block design in effect for subsequent four waves, and with a stratified random sampling approach (stratifying by low/high school test score indicator) to selecting a control group in wave one that would have been implemented in absence of the PMPD technology. The construction of the control group via PMPD proceeded by best-first Mahalanobis matching within 0.10 SD calipers of the estimated linear propensity score, effectively logit P[win lottery|covariates], subject to exact match on certain special discrete variables. Since the lottery was independent of all covariates except for the low/high school test score indicator, the expected value of the propensity score should be constant in the other covariates—there is no systematic bias—so that (random) Mahalanobis matching would be expected to predominate. However, estimated propensity score matching can also reduce sample-specific randomly generated bias, and Barnard et al. demonstrate its improved efficiency for differences in covariate means, over simple random sampling, in this sub-study, e.g. their Table 1.6.

It is difficult to tell from Barnard et al.'s account exactly which covariates were included in the implemented PMPD procedure, but we believe it was all variables in, e.g., Table 1.6: that is, the same set of variables on which the efficiencies of matching in the three control-group designs were compared. Thus, while the balance is demonstrably wonderful for PMPD on the variables included in the matching procedure, it is difficult to say what is happening with *unobserved* covariates, such as parental motivation, that may plausibly affect the outcome of student achievement. We know of no guarantees about balance in expectation or otherwise for unobserved covariates—as are available for simple and stratified random sampling—for propensity score methods. To this end, a more convincing balance demonstration would hold out some observed covariates from the matching process and consider the balance on these variables obtained by matching; however the general uneasiness about unobserved covariates remains. This, combined with the complexity of communicating the PMPD approach to non-experts, may make the simpler but less efficient stratified random sampling (perhaps stratifying on several discrete variables in addition to low/high school score indicator) and randomized block approaches to control group construction preferable in some circumstances.

Another feature of Barnard et al.'s propensity score matching procedure is that indicators of missing data are captured as covariates and can be modeled into the propensity score estimate. Since missingness, e.g. in mother's education and father's employment status, may be relevant to the student achievement outcome, including it in the propensity scores for matching seems reasonable. The technology required to estimate propensity scores in the presence of missing data is more complex, involving an application of the ECM algorithm to the general location model (D'Agostino and Rubin, 1999; Schafer, 1995) but worth the effort in terms of addressing a potentially important matching issue.

Given the effort that went into matching in the first wave, we would also like to have seen an explicit matched pair analysis. That is, since the propensity scores were used to find good individual matches for each of the lottery winners in the first wave, it would seem sensible to conduct a second sub-study analyzing student achievement in matched pairs from the first wave. This would represent a more compelling use of the PMPD matches, improving on a technique that goes back at least to Aikin (1942) in studying the effects of different forms of schooling on outcomes such as achievement, college matriculation rates, etc.

2.3 *Generalizability*

Finally, we would like to have seen a discussion of the generalizability of the results of this study beyond the applicant pool. While 20,000 applicants is a phenomenal number, again expressing the interest of urban families in finding ways out of the poverty-poor-education cycle for their children, it is still true that these subjects' parents and guardians had to be motivated enough to seek out and apply for the SDSF program—i.e. there is a potential self selection problem at intake that no amount of design and statistical machinery after intake can address. What fraction of all possible applicant families is 20,000? Did these 20,000 families differ from other possible applicants in ways that might affect our predictions if the voucher program were made available to all on an income- or means-tested basis?

3 The Inferential Model

In Section 3.1 we try to summarize the main technical points of the inferential model employed by Barnard et al., which seemed useful to us as we considered the results of their analyses. In Section 3.2 we comment on the main student achievement results of the SCSF study, viewed through both the lens of a conventional linear regression based instrumental variables model and the lens of Barnard et al.'s more elaborate Bayesian potential outcomes model. We focus on Barnard et al.'s analysis of families with a single eligible student.

3.1 A Potential Outcomes Model for Broken Experiments and Encouragement Designs

Causal models in statistics are models designed to enable causal inferences of similar validity to those drawn from classic controlled randomized experiments, in nonrandomized or broken experiments and encouragement designs in which we are forced to accomodate subject self-selection into treatment groups or subject non-compliance with the assigned or encouraged treatment. In addition, statistical models for social experiments must often deal with substantial subject non-response on background and outcome variables.

The inferential model of Barnard et al. is intended to deal with both of these problems; it is a potential-outcomes version of the the path diagram of Figure 1.1, expressed in a Bayesian or complete probability model framework. This is an instance of an "encouragement design" model (Holland, 1988), in which winning the lottery or not, Z, affects the likelihood of exposure to treatment, $D(Z)$, e.g. private school attendance, but the treatment $D(Z)$ is ultimately selected by the subject and not the experimenter. These in turn affect students' post-test scores $Y(D,Z)$ and missingness $R_Y(D,Z)$ of the post-test scores. In addition, subject covariates[1] X intended to be observed in the study—including subject background covariates, good/bad school score indicator for the public school previ-

[1]To ease the notational burden we have subsumed Barnard et al.'s fully-observed covariates W into their partially-observed covariates X.

ously attended by subjects, and subjects' pretest scores—indicators R_X for missing X's, and other covariates not intended to be recorded for the study (and therefore largely unobserved), U, may affect both self selection of treatment $D(Z)$ and outcomes $Y(D, Z)$ and $R_Y(D, Z)$. This framework accomodates subject non-response by allowing for explicit modeling of R_Y and R_X and their influence on D and Y.

Their model accomodates subjects' self-selection of treatment by explicitly accounting for one hypothesized member of the class of unrecorded variables U, the subject's compliance status C. This variable encodes whether the subject is a "complier" who would attend private school ($D = 1$) if and only if he/she were a lottery winner ($Z = 1$); a "defier" who would always do the opposite of the lottery outcome ($D = 1 - Z$); an "always taker" who would attend private school regardless of the lottery outcome ($D \equiv 1$); or a "never taker" who would attend public school regardless of the lottery outcome ($D \equiv 0$). Since each subject either does or does not win the lottery, and either does or does not go to private school, we only get to observe one of the two potential treatment exposures $D(0)$ and $D(1)$, and one of the four potential outcome values $Y(0,0), Y(0,1), Y(1,0), Y(1,1)$, per subject. Compliance status C may or may not be observable, depending on the definition of D (see Section 3.2 below) but the modeling framework allows us to impute its likely values in any case.

The lottery outcome Z is randomized within low/high school score indicator, so that observed and unobserved covariates are uninformative about treatment assignment/encouragement. Treatment exposure D is treated as an outcome covarying with achievment outcome Y that is informative about compliance status C. Y is modeled as depending on treatment assignment Z directly—so that the notation $Y(D, Z)$ is simplified to $Y(Z)$ in what follows—and on exposure D only indirectly through observed or inferred compliance status C (Barnard et al., Section 10.2).

Additional basic assumptions of this class of models include "monotonicity" of $D(Z)$, so that there are no defiers, and the stable unit-value treatment assumption (SUTVA), which asserts that each subject selects a treatment (private school or not), and responds to the treatment, independently of every other subject. SUTVA is potentially somewhat controversial in educational work in which treatments are administered to groups of students, e.g. classes, that are allowed or even encouraged to interact (and therefore an epidemic or contamination model along the lines of Halloran and Struchiner, 1995, would seem more reasonable, in principle; see also Gitelman, 1999). SUTVA could be violated in the SCSF study if partic-

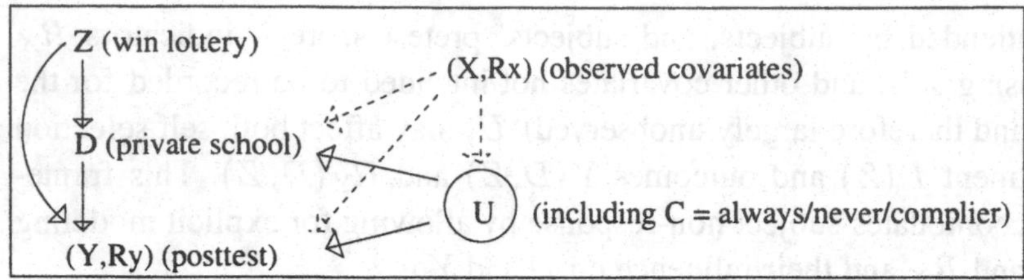

FIGURE 1.1. Simplified path diagram for the final inferential model of Barnard et al. (1999), described in Section 3.1. In each of Barnard et al.'s three analyses, Z is treatment assignment, that is, winning the lottery. D, treatment exposure, varies in Barnard et al.'s analyses; see Section 3.2 for details. Y and R_Y are post-test scores and indicators of missing post-test scores, respectively. X and R_X are background covariates and design variables intended to be collected in the study, and indicators of missingness for them. U represents unmeasured/latent confounders and other covariates, including subject compliance type C. Arrows indicate structure of conditional distributions in the model specification, i.e. putative causal directions and conditional independence assumptions, as usual. Dashed lines are intended to improve readability of the figure only.

ipating families decide together whether to send their children to private school, or if SCSF students form study groups within a school, etc. However, like Barnard et al., we see no obvious potential for violations of SUTVA strong enough to undermine the SCSF study conclusions.

Barnard et al. make two further assumptions that are helpful in taming the complexity of the model. First, they assume that the missing data mechanism R_Y for Y satisfies "latent ignorability," that is, within levels of the observable covariates (X, R_X) the potential outcomes $R_Y(1)$ and $R_Y(0)$ are assumed to be informative about the potential outcomes $Y(1)$ and $Y(0)$ only through the latent compliance status[2] C. This is a specific behavioral model relating missing outcome, compliance status, and outcome itself.

Second, they assume a "compound exclusion restriction", that is, for always takers and never takers, it is assumed that the joint distribution of $(Y(1), R_Y(1))$ is the same as for $(Y(0), R_Y(0))$, within levels of observable covariates[3] (X, R_X). Technically this allows for a simpler complete

[2] Roughly, the R's and Y's are conditionally independent given (X, R_X) and C. See Barnard et al., p. 42, for details.

[3] Again, see Barnard et al., p 42, for details.

description of the probability model; however as Barnard et al. point out, it is only plausible to the extent that an increase in family resources of $1,400 among always-takers would not affect student achievement outcome. As in Imbens and Rubin (1997), we imagine that this restriction could be weakened in the Barnard et al. model, at the expense of wider interval estimates and consequently less power to distinguish outcomes in the various study conditions.

The latent ignorability and compound exclusion restriction assumptions allow for factoring the model into three components: a compliance sub-model, an outcome sub-model and an outcome response sub-model. The outcome response sub-model is a linear probit model for the response pattern (i.e., response versus non-response). The two other sub-models are also linear models—a compound linear probit model for the compliance status, and a normal linear model with heterogeneous variances for the outcomes. Following current computational practice for complex Bayesian models, this model is estimated using a Markov Chain Monte Carlo approach.

Let us stop and take our breath. While the model just described is plausible and sensible, it contains many assumptions, both technical and substantive. Both the latent ignorability and compound exclusion restriction assumptions involve assertions about subject behavior that, while reasonable as first approximations, are open to debate and impose restrictions on the form of the data we should see. The linear normal and probit models that form the "plumbing" of Barnard et al.'s setup, as well as prior distribution assumptions not reviewed here, entail specific technical restrictions on the distribution of the observed or future data. Causal models in statistics are very sensitive to underlying assumptions (e.g. Gitelman, 1999), and so it would be interesting and informative to examine the sensitivity of the *causal* conclusions (which after all have enormous policy implications) to, e.g., the latent compliance modeling assiumptions.

While we were therefore disappointed not to see any model selection or model validation in Barnard et al.'s, paper, we understand that various model checking, including posterior predictive checks of specific modeling assumptions as well as comparisons with an independently developed model, are in the works, and we look forward to their future publication. It would also be informative to know how the estimates of causal effects obtained here fair in relation to bounds such as those of Manski (1990) and Robins (1989) or Balke and Pearl (1994), each of which rely on fewer assumptions.

3.2 Achievement Results

Barnard et al. consider three different analyses of the effects of the SCSF study conditions on student achievement, corresponding to different policy-related questions, using each of two different inferential models that follow the general structure of the path diagram in Figure 1.1.

First, the three analyses are presented under a conventional linear regression based model in which missing covariates are handled by variable-wise and case-wise deletion, and missing outcomes are handled by reweighting the cases in which outcomes are observed to reflect the original subject pool; non-compliance is handled with instrumental variable (IV) models treating $Z =$ "winning the lottery" as the instrument. Second, the analyses are presented using the Bayesian potential outcomes model summarized in Section 3.1 above. The potential outcomes model allows us to include cases with missing data (and exploit the missingness itself as additional information about the effects of the SCSF study conditions), and replaces the instrumental variables approach with an explicit model for compliance status C, including behavioral postulates relating non-response (missing data), compliance status and outcome as encoded by the latent ignorability and compound exclusion restriction assumptions.

To analyze the student achievement effect of merely being offered a $1,400 scholarship, Z is taken to be "winning the lottery" and D is defined to be private school attendance (with or without SCSF aid) as in Figure 1.1, and effects of Z on Y are calculated for compliers, always-takers and never-takers combined, that is, comparing all lottery winners with all lottery non-winners, regardless of actual school attended. This is sometimes called an intention-to-treat (ITT) analysis.

To analyze the student achievement effect of attending private school, only the effects for compliers are considered in the potential outcomes model, and an IV estimate—whose estimand is a kind of compliers-only effect; see Angrist, Imbens and Rubin (1996)—is calculated.

To analyze the student achievement effect of participating in the SCSF program itself, effects for for compliers and always-takers only (that is, excluding never-takers) are combined in the potential-outcomes model. A similar result is obtained in the instrumental variables model by redefining D to be accepting the $1,400 and using SCSF school placement services, since the IV estimand is a compliers-only effect where compliance now means accepting the $1,400 and placement services.

Analyses under both inferential models—instrumental variables vs. potential outcomes—show that, as expected, complier effects and complier-plus-always-taker effects are generally larger than ITT effects, which ignore compliance behavior completely[4]

However, the pattern of significant and trend effects across grade level (1–4), subject (reading or math) and low/high school score indicator were the same *within* each inferential model, as we move from ITT effects to complier-plus-always-taker effects to complier-only effects, but somewhat different *between* inferential models. For this reason we will not distinguish between ITT, complier and complier-plus-always-taker effects as we compare results under the instrumental variables and potential outcomes models.

Both models weakly suggest small positive treatment effects for overall math achievement for students coming from low-scoring schools; and essentially no overall treatment effects for reading (with the exception of a large positive effect found under the instrumental variable model for reading achievement for students in grade 4 from high-scoring schools, to which we will return in a moment).

The strongest effects under the potential outcomes model are for math achievement in grade 1 among students coming from either low- or high-scoring schools; under the instrumental variables model the students from low-scoring schools show only a trend effect for math in grade 3, and those from high-scoring schools show only a trend effect for math in grade 4.

The strongest effects under the instrumental variables linear model are for reading and math in grade 4, among students coming from high-scoring schools; we agree with Barnard et al. that these latter effects do not seem terribly plausible.

Beyond this, the effects estimated under both models are rather weak, and somewhat contradictory.

For example, where the instrumental variables model shows a large effect for grade 4 reading among students coming from high-scoring schools; the potential outcomes model does not even show a trend effect for the same study condition. In general, the potential outcomes model shows more positive point estimates than the instrumental variables model, but the standard errors are so large for most of these estimates under both models that they cannot even be called trend effects.

[4]This is to be expected in the potential-outcomes model anyway, from the compound exclusion restriction. Weakening this restriction, as discussed in Section 3.1, might bring the ITT effects more in line with the complier effects, as well as widening interval estimates. A similar intepretation of the effect of the usual exclusion restriction in an instrumental variables model is available in Angrist, Imbens, and Rubin (1996).

We agree with Barnard et al. that part of the problem may be small sample sizes for both inferential models, and the weakness of weighting as a way to deal with outcome non-response in the instrumental variables model. They also mention that students with missing pretest scores—which is a source of information available to the Bayesian potential outcomes model but not the instrumental variables model—show a negative treatment effect for reading posttest scores; depending on the observed or imputed compliance status for these subjects, this might also guide a finer re-evaluation of the latent ignorability and compound exclusion restriction assumptions. More broadly, these suggestions of disagreement between the models raise again the need for model selection and model validation, as well as comparisons with relevant bounds on the causal effects, as discussed at the end of Section 3.1.

4 School Choice and Student Achievement

Despite the exemplary collaborative and technical methodology displayed by Barnard et al.'s case study, the ultimate substantive results might be thought to be disappointing, from at least two perspectives. First, the effort spent on developing and estimating the Bayesian potential outcomes model did not buy much beyond simpler intention-to-treat (ITT) analyses, which policy analysts may be most interested in: it is easier to make policy about whether vouchers are available to students, than whether they must use them, for example. And second—whether because the 2,000 student sample available for study was too finely sliced by the models, or because educational treatment effects just don't show up strongly after a one-year intervention—the achievement effects found under any of the analyses are generally difficult to get excited about. Indeed, Peterson, Myers and Howell (1998) write, of earlier analyses of the SCSF study data, "Overall, the [achievement] differences between all those ... who used their scholarships ... and the control group ... were small—around two percentile points in both [math and reading]."

Let us suppose that with larger sample sizes, or after the entire multi-year SCSF study is completed, we would find significant achievement gains for those who attended private school under the SCSF program. What should we make of this? We would like to consider two answers, suggesting that omnibus public/private school achievement comparisons are perhaps irrelevant to some of the issues underlying the school choice debate, and are insufficient to settle other issues.

First, let us ask whether achievement comparison is the right yardstick for the school choice question. After their assessment of the SCSF achievement effects quoted above, Peterson, Myers and Howell (1998) go on to say that, nevertheless, "Parents of scholarship users are much more satisfied with their children's education." Indeed over 20,000 applicants–presumably largely unaware of previous studies of achievement and school choice–were interested in applying to the SCSF program, so math and reading achievement gains may not have been their primary concern. And much of the literature on school choice surveyed in Section 1 of this discussion–from Lieberman (1993) and Gilles (1998) to Ascher, Fruchter and Berne (1996)–is couched in a rhetorical debate that is broader than achievement. While the achievement evidence provided by the SCSF study—as well for example as other empirical studies in the anthologies of Fuller, Elmore and Oldfield (1996) and Peterson and Hassel (1998)–may help inform the discussion, fundamental political issues of whose interests and rights should predominate in deciding where to send children to school may ultimately outweigh achievement comparisons.

Second, let us ask a question for which achievement comparison is the right yardstick: what should we do to improve students' achievement the most? As an answer to this educational research question, the SCSF study by itself is not very interesting. Given that we eventually find evidence of a positive effect of private schooling on achievement, a better set of questions would be analogous to Bloom's (1984) "2-sigma" paper: What is it that makes private schools better? Most observers (see for example the assumptions underlying the analyses of Rangazas, 1997) agree that under any likely broadening of families' school choice options, the public school system will not disappear. How could we, or *could* we, modify public school institutions so that they become better, perhaps so much so that choice might be offered—satisfying the liberal parentalist argument identified by Gilles (1998)—but considered unnecessary by most eligible families—maintaining the opportunity for social education underlying the liberal statist view.

There is a research and development tradition that tries to address these questions, going back at least to the Eight-Year Study (Aiken, 1942), which included an informal paired comparisons study examining college matriculation rates and other measures on students attending various forms of experimental and conventional public and private high schools. More recently, Bryk, Lee and Holland (1993) developed qualitative case studies, as well as statistical analyses of national data bases, to examine the distinctive features of Catholic high schools that provide supportive social environments and promote academic achievement for a broad cross section of students, drawing lessons for the renewal of America's educational institutions generally. The New American Schools program (e.g., Glennan, 1998) represents a large research initiative on the institutional design of schools, and Fullan (1998) and Bryk, Sebring, Kerbow, Rollow, and Easton (1998) explore internal and external mechanisms of institutional change in public schools. Much of current educational reform efforts are driven by the standards and accountability movement (e.g., Achieve, Inc., 1999; CCSSO, 1995; Ravitch, 1995), and are implemented through professional development, management reform, and quality monitoring efforts such as the Institute for Learning (LRDC, 1999).

These research and development efforts could benefit from greater involvement of statisticians armed with the sophisticated and practical methodology illustrated by Barnard et al.'s case study. One of the few existing examples of this is the Tennessee STAR study (e.g. Mosteller, 1995), a randomized controlled longitudinal study of specific manipulations within a school system that affect student achievement, which has established substantial long-term effects for smaller class sizes in public schools in grades 1–3. What makes such studies successful is precisely statisticians' deep involvement in their conception, design and implementation phases, and the sophisticated and appropriate technical machinery that is available today to draw strong inferences from the messy data that even well designed educational studies produce.

Acknowledgments

On leave at the Learning Research and Development Center (LRDC), University of Pittsburgh. Junker's work was supported in part by NSF grants #SES-99.07447 and #DMS-97.05032, and by the generous hospitality of LRDC. Junker gratefully acknowledges conversations with Lauren Resnick and Laura D'Amico, LRDC, and Peter Rangazas, Indiana University/Purdue University at Indianapolis. However, the views expressed, and any errors, in this discussion are our responsibility and not theirs.

References

Achieve, Inc. (1999). Home Page. Obtained from the World Wide Web at address `http://www.achieve.org`, September 1999.

Aikin, W. M. (1942). *The story of the eight-year study, with conclusions and recommendations.* Volume 1 in: Commission on the Relation of School and College, Progressive Education Association (ed.) (1942). *Adventure in American Education.* New York: Harper & Brothers.

Angrist, J. D., Imbens, G. and Rubin, D. B. (1996). *Identification of causal effects using instrumental variables. Journal of the American Statistical Association, 91,* 444–472.

Ascher, C., Fructer, N. and Berne, R. (1996). *Hard lessons: public schools and privatization.* New York: Twentieth Century Fund Press.

Balke, A. and Pearl, J. (1994). Counterfactual probabilities, computational methods, bounds and applications, in *Proceedings of the Tenth Conference on Uncertainty in Artificial Intelligence*, R. Lopez de Mantaras and D. Poole, eds. Morgan Kaufman.

Barnard, J., Frangakis, C., Hill, J., and Rubin, D. B. (1999). School choice in New York City: A Bayesian analysis of an imperfect randomized experiment. Invited Case Study, Fifth Workshop on Bayesian Statistics in Science and Technology. Pittsburgh, PA, September 24–25, 1999.

Bloom, B. S. (1984). The 2-sigma problem: the search for methods of group instruction as effective as one-to-one tutoring. *Educational Researcher, 13,* 4–16.

Bryk, A. S., Lee, V. E. and Holland, P. B. (1993). *Catholic schools and the common good.* Cambridge MA: Harvard University Press.

Bryk, A. S., Sebring, P. B., Kerbow, D., Rollow, S., and Easton, J. Q. (1998). *Charting Chicago school reform: democratic localism as a lever for change.* Boulder, CO: Westview Press.

Council of Chief State School Officers (CCSSO). (1995). *State responsibility for student opportunity: commitment and issues.* A statement of the Council of Chief State School Officers. Washington DC: Author. Obtained from the World Wide Web at address `http://www.ccsso.org/oppol.html`, August 1999.

D'Agostino, R. J., and Rubin, D. B. (1999). Estimating and using propensity scores with incomplete data. To appear, *Journal of the American Statistical Association.*

Friedman, M. (1962). *Capitalism and Freedom.* Chicago IL: University of Chicago Press.

Fullan, M. (1999). *Change forces: the sequel (Educational Change and Development Series).* Philadelphia: Falmer Press.

Fuller, B., Elmore, R. F. and Orfield, G. (1996). Policy-making in the dark. Chapter 1, pp. 1–21 in Fuller, B. and Elmore, R. F. with Orfield, G. (eds.) (1996). *Who chooses? Who loses? Culture, institutions, and the unequal effects of school choice.* New York: Teachers College Press.

Fuller, B. and Elmore, R. F. with Orfield, G. (eds.) (1996). *Who chooses? Who loses? Culture, institutions, and the unequal effects of school choice.* New York: Teachers College Press.

Gilles, S. G. (1998). Why parents should choose. Chapter 15, pp. 395–407 in Peterson, P. E. and Hassel, B. C. (eds.) (1998). *Learning from school choice.* Washington DC: Brookings Institution Press.

Gitelman, A.I. (1999). Treatment integrity concerns in comparative education studies. Unpublished Ph.D. Dissertation, Carnegie Mellon University, Pittsburgh PA.

Glennan, T. K. Jr. (1998). *New American Schools after six years.* Santa Monica CA: RAND Corporation.

Halloran, M.E. and Struchiner, C.J. (1995). Causal inference in infectious diseases, *Epidemiology, 6, 2* 142–151.

Holland, P.W. (1988). Causal inference, path analysis, and recursive structural equation models, *Sociological Methodology, 18*, 449–484.

Imbens G. W., and Rubin D. B. (1997). Bayesian inference for causal effects in randomized experiments with noncompliance. *Annals of Statistics, 25*, 305–327.

Learning Research and Development Center (LRDC) (1999). *Outreach and implementation.* Obtained from the World Wide Web at address `http://www.lrdc.pitt.edu/research/oai99.htm`, September 1999.

Lieberman, M. (1993). *Public education: an autopsy.* Cambridge MA: Harvard University Press.

Manski, C. (1990). Nonparametric bounds on treatment effects, *American Economic Review, Papers and Proceedings, 80*, 319–323.

Mosteller, F. (1995). The Tennessee study of class size in the early school grades. in *The Future of Children, Vol. 5, No. 2, Summer/Fall 1995.* Obtained from the World Wide Web at address `http://www.futureofchildren.org/cri/08cri.htm`, September 1999.

National Commission on Excellence in Education. (1983). *A Nation at Risk: The Imperative for Educational Reform.* Washington DC: author.

Peterson, P. E. (1998). School choice: a report card. Chapter 1, pp. 3–32 in Peterson, P. E. and Hassel, B. C. (eds.) (1998). *Learning from school choice.* Washington DC: Brookings Institution Press.

Peterson, P. E. and Hassel, B. C. (eds.) (1998). *Learning from school choice.* Washington DC: Brookings Institution Press.

Peterson, P. E., Myers, D. E and Howell, W. G. (1998). *An Evaluation of the New York City School Choice Scholarships Program: The First Year.* Cambridge MA and Washington DC: Harvard Program on Educational Policy and Governance and Mathematica Policy Research. Obtained from the World Wide Web at address `http://data.fas.harvard.edu/pepg/NewYork-First.htm`, September, 1999.

Rangazas, P. (1997). Competition and private school vouchers. *Education Economics, 5,* 245–263.

Ravitch, D. (1995). *National standards in American education: a citizen's guide. Updated with a new introduction.* Washington DC: Brookings Institution Press.

Robins, J.M. (1989). The analysis of randomized and non-randomized AIDS treatment trials using a new approach to causal inference in longitudinal studies, *Health Service Research Methodology: A Focus on AIDS*, L. Sechrest, H. Freeman and A. Mulley, eds. NCHSR, U.S. Public Health Service.

Schafer, J. L. (1997). *Analysis of incomplete multivariate data.* New York: Chapman and Hall.

School Choice Scholarships Foundation (SCSF) Inc. (1999). Home Page. Obtained from the World Wide Web at address `http://nygroup.com/scs/`, September, 1999.

Valverde, G. A. and Schmidt, W. H. (1997). Refocusing U.S. Math and Science Education. *Issues in Science and Technology Online, Winter 1997* Obtained from the World Wide Web at address `http://www.nap.edu/issues/14.2/schmid.htm`, November, 1999.

Rejoinder

Ansolabehere discussion

Ansolabehere raises many interesting questions and points the way towards useful extensions of our work. He details the need for calibrating our results (i.e., what are the implications of a two point gain in test scores), for comparing effects found here with those attributable to competing reforms[1], and for addressing the more difficult question regarding the system-wide changes that would occur were school choice to be implemented as policy. We agree strongly with Ansolabehere's statements (which presumably, however, refer to only a subset of these issues) that "they are best answered through social experiments such as this one" and "given the results and comparisons offered here, further experiments with other means of improving test score are warranted." The questions that Ansolabehere raises are all *causal* in nature and deserve the same amount of thoughtfulness as the ones we addressed in this paper. For instance, we believe it would border on irresponsibility to simply take the results of this experiment and plug them into a linear regression obtained from some other dataset relating test scores and other types of outcomes of interest.

With regard to creaming, Ansolabehere correctly points out a limitation in our presentation. It would have been strengthened by a more thorough comparison of the differences between compliance groups, which will become easier to perform once we move to a selection model approach. We were, however, confused by the notion of comparing the difference between the scores of "non-compliers" and the members of the control group with "the improvement posted by those who entered private schools" because this reflects a comparison between a causal effect and a descriptive difference.

Regarding Hawthorne effects and other complications (e.g., expectancy effects, Rosenthal and Jacobson (1968)), it's true that in social science experiments there are more issues than in laboratory experiments or even clinical trials, which leave these studies more subject to departures from important assumptions such as SUTVA. We cannot assign children to particular schools in the same way a laboratory researcher assigns mice to a

[1] As a point of clarification in this debate, the financing issue is a bit more complicated than Ansolabehere suggests. These reforms, when implemented as public policy, tend to transfer existing money, on a per pupil basis, from one school to another, so in some sense it requires no additional funding. However, this issue is complicated by the economies of scale required to run schools efficiently.

drug dosage. The best we can do is to take advantage of naturally occurring randomized experiments such as this one and learn as much as we can from them. We are hopeful that Hawthorne effects are at a minimum in this particular study for three reasons: (1) children in schools (as well as teachers in those schools) are so used to being tested, evaluated and scrutinized that the added measure of scrutiny created by this study is likely to be small, (2) the ITBS was chosen as a test measure because it is not currently the standardized test currently administered by the public school system in New York City, therefore minimizing the chance of teachers "teaching to the test", and (3) the children in this study were such a small percentage of the overall school population in the city and scattered across so many schools that these effects, once again, should be smaller than otherwise. Ansolabehere's concern about these types of influences, however, and his claim that care should be taken in addressing them are surely valid.

Junker and Gitelman discussion

Junker and Gitelman (henceforth JG) have created a thorough and thought-provoking summary of our paper. In addition they have positioned our study within an expanded summary of the school choice debate, for which we are grateful.

At the end of Section 2.1, JG, while discussing the follow-up problems in the study, express concern about "economic differences, apparently not captured by the matching processes ... between the non-compliers in the treatment arm and the control arm subjects." We were confused about this comment. We would expect to see differences between "non-compliers" (we're assuming this refers to both never takers and always takers) and the full control group; that is why we use covariates to discriminate between these groups. These differences would not be affected, on average, by the matching procedure.

JG state in Section 2.2 of their discussion that they were unclear about which variables were included in the propensity score model, and say that they would have preferred that the comparisons we present include some variables that were not included in this model. These are valid points. In fact though, we describe the variables included in this model in Section 6.12 of our paper, and indeed five of the variables presented in the subsequent across-design comparisons (child's age, reading pre-test scores, indicator of whether mom works, educational expectations, and length of residence) were not among those in that model. The balance in these is better than for the PMPD design for three out of the five when compared to the stratified random sample, two out of five when compared to the simple random sample, and three out of five when compared to the randomized block design of waves two through five.

With regard to JG's concern that we don't know what is happening to the balance of the unobserved covariates, we offer the following intuition. If the unobserved variables are correlated with the variables we are now balancing better, then these unobserved variables will have better balance, on average (across samples), as well as a result. If, on the other hand, these unobserved covariates are uncorrelated with the variables we are balancing, then balancing should not affect the balance of the unobserved variables. Focusing on the balance of these observed variables could lead to worse balance in the unobserved variables *for a given sample* but this will only affect efficiency (and then only if those unobserved variables play a more prominent role in the response surface then our more highly balanced observed variables), not bias (which has to do with the average imbalance over repeated samples). Of course this intuition relies on normality, but results using ellipsoidal symmetry and empirical distributions in (Rubin and Thomas, 2000) support the claim that this intuition applies more broadly.

Matching was done primarily to select, given limited resources, an appropriate subset of students to follow up. For analysis, a method validly addresses the matching design as long as it uses the variables that formed the basis of the matching (see definition of ignorable assignment in Rubin (1978)). The analysis, therefore, can be either: (i) at the pair level, which is what JG call explicit; or (ii) regression-assisted, which is a more general analysis framework and includes JG's suggestion as a special case. With full compliance and fully observed outcomes, JG's suggestion has an intuitive appeal. With noncompliance and missing outcomes, however, JG's proposal cannot be similarly implemented, at least not for the estimands we target, because a pair of observations with incomplete compliance status or outcomes or both is simply too crude to be used as a unit of analysis. Our method is valid because it uses the propensity score that determined the matches, so that assignment is ignorable (Rubin, 1978) in our data, and is also practical to implement together with the additional structure of our data.

JG would have liked to have seen a discussion of generalizability, which is a fair point. Peterson, Myers, Howell, and Mayer (1999) address this issue by comparing characteristics of lottery applicants to those of all eligible families in NYC. They find that the lottery applicants, on average, tend to have slightly higher rates of dependence on welfare and social security, but tend to be better educated and more likely to be employed. In addition, the applicants are more likely to be African American and to have English as their primary language. One problem with this comparison, however, is that the information on eligible families is based on 1990 census data (background information on applicants was recorded in 1997) so some of these differences may be due to the different time periods (for instance 1990 was a recession year). It is difficult to predict whether those who applied to (and took advantage of) the scholarship program in this private initiative would be representative of those who would do so were a similar program to be implemented as policy.

We agree with JG (Section 3.1) that it would be useful to explore the sensitivity of our conclusion to some of the latent compliance modeling assumptions (although it should be noted that the general sensitivity to underlying assumptions found in "causal models" that they use to motivate this concern are certainly less of an issue in this scenario because of the underlying randomization). More generally, they express a desire to have seen greater effort spent on model selection and model validation. We agree with this sentiment and plan to work on these issues as we continue to pursue this research.

JG suggest comparisons to the work on bounds for causal effects. We don't find work on bounds interesting because they do not represent bounds on the treatment effect, they represent bounds on an estimate of this effect which don't incorporate posterior uncertainty. Consequently, we prefer sensitivity analyses, which we have yet to perform.

Section 4 of the JG discussion describes their disappointment in the substantive results of our case study. There are several points here.

1. Why was it worth going beyond ITT? There seems to be a common misconception among many researchers that ITT measures the effect of an initiative were it to be implemented as policy. Not so. It only would do so if the initiative being examined were the same as that which would be implemented as policy, the types of people who participated in the program were exactly the same, the compliance rates were the same, there were no "experiment effects", and so on. Our analysis allows us to measure both the ITT effect and the presumably more "scientific effect" of this particular initiative on the program participants. Ignoring the issues of differences in types of people who might participate in the public policy incarnation of this same program, we can use the information from our analysis to examine the "policy implication" of this program under a range of dif-

ferent assumptions about compliance rates. The ITT effect muddles these two effects.

2. The results are weak and therefore "disappointing." JG seem to be implying that interesting results are necessarily synonymous with large results. Given the expectations of many that the results here would be quite large (such expectations even leading some scholars to question why we felt compelled to pursue this question at all), there is an argument to be made that weak results may actually be of greater policy interest than strong results in this context.
3. Using achievement scores as sole result. It would of course be interesting to examine other measure of the "success" of this program. We chose not to do so in the interest of space.
4. Getting inside the black box. It is not the theory for why private schools could produce better results that is lacking, it is the empirical evidence. Discovering evidence of treatment effects in this study should provide the justification for further pursuit of support for the various theories that already exist regarding the causal mechanisms by which these effects were achieved. Arguably, if we never saw such effects we would not want to waste resources further pursuing these issues.

Overall

All of the discussants make worthwhile points about additions, but none suggest where we could have cut our already novella-length paper to accommodate these. Moreover, both discussions seem to vastly underestimate the difficulty in answering the questions in which they are interested. Academicians have been performing educational research for decades without generating anything that resembles consensus regarding "what works best." We have at least provided a framework within which we believe these questions could be addressed, one at a time.

The principal investigators of this study (Paul Peterson of Harvard University and David Myers of Mathematica Policy Research) did not set out to create a study that would answer all questions about school choice. Rather they saw the opportunity, in this initiative sponsored by the SCSF, to take advantage of a naturally occurring randomized experiment to provide some solid evidence about the relative efficacy of private vs public school experiences – a question that had never before been addressed by a well-designed randomized study. We believe that a careful analysis of this study has taken us one step towards a better understanding of some of the key issues involved in the school choice debate and discovery of profitable directions to further explore.

References

Peterson, P. E., Myers, D. E., Howell, W. G., and Mayer, D. P. (1999), "The Effects of School Choice in New York City," in *Earning and Learning; How Schools Matter*, eds. S. E. Mayer and P. E. Peterson, Brookings Institution Press.

Rosenthal, R. and Jacobson, L. (1968), *Pygmalion in the Classroom: teacher expectation and pupils' intellectual development*, Holt, Rinehart and Winston.

Rubin, D. B. (1978), "Bayesian Inference for Causal Effects: The role of randomization," *The Annals of Statistics* 6, 34–58.

Rubin, D. B. and Thomas, N. (2000), "Combining Propensity Score Matching with Additional Adjustments for Prognostic Covariates," *JASA*.

Adaptive Bayesian Designs for Dose-Ranging Drug Trials

Donald A. Berry, Peter Müller, Andy P. Grieve, Michael Smith, Tom Parke, Richard Blazek, Neil Mitchard, Michael Krams

ABSTRACT In the standard type of phase II efficacy trial, patients are assigned to a dose from among those being considered (usually 4 to 12 in number). Assignment is random, usually with equal numbers of patients assigned to each dose. Based on the results of the trial, a decision is made to either enter phase III in the drug's development, stop the drug's development, or conduct another phase II trial. Such a design is inefficient, in terms of both time and resources. We have developed an innovative class of designs that we are introducing into practice. In this case study we describe the designs, address difficulties in implementing them in actual clinical trials, and relay our experience with using them.

The first stage allows for a wide range and a large number of doses, including placebo. The purpose of this stage is to assess dose-response in an informative and efficient way. Assignment to dose is sequential, in the following sense. As patients are treated, they are followed and their responses are communicated to a central database. Doses are assigned to subsequent patients so as to obtain maximal and rapid information about the dose-response curve. We impute missing final responses using their predictive distributions given current responses and given assigned doses.

As information accrues about dose-response from the dose-finding stage, if this information is suggestive that the drug is effective then the assignment procedure shifts to a confirmatory stage ("pivotal" phase III). Two doses will be identified based on the dose-response information and patients will be randomized to these two doses and placebo in a balanced fashion. The shift will be seamless and not recognizable by physicians and others involved in the trial (except for members of the trial's data and safety monitoring committee).

The timing of the shift from dose-finding to pivotal is critical. Whether to shift will be based on a Bayesian decision analysis using forward simulation and dynamic programming. A decision to shift will depend on the available information about dose-response, the costs of entering additional patients, and the requirements of the FDA and other regulatory agencies concerning information needed for eventual marketing approval of the drug. Although

the design and the determination of the sample size of the pivotal stage is Bayesian, the decision analysis recognizes the need to provide regulatory agencies with a frequentist analysis of the trial results.
Our efficient dose assignment scheme more accurately identifies effective drugs and it more accurately identifies ineffective drugs. Moreover, efficient dose assignment can significantly shorten a drug's clinical development. First, the number of patients in a sequential trial will usually be substantially smaller than when using standard designs. This has important economical and ethical implications. Second, a seamless transition between the dose-finding and confirmatory stages eliminates the time required to set up a second trial.

1 Introduction

1.1 Stroke and Neuroprotective Agents

If you hold this text in your right hand and read it aloud, parts of the left side of your brain would be very active. This activity would consume considerable amounts of energy. Arterial blood provides a constant source of fresh energy to the brain in the form of glucose and oxygen. Now imagine that there is a sudden interruption in the arterial flow to these parts of your brain. Brain cells would start to die within minutes. You would lose control and power in your right arm and the text would drop. You would want to say something but would be unable to speak, and you may be unable to understand what other people are saying to you. These symptoms would develop very rapidly. Each year in the United States some 600,000 patients experience such symptoms. A doctor would diagnose "stroke" or "brain attack." In industrialized countries, stroke is the leading cause of disability, and it is the third leading cause of death, behind only heart disease and cancer. Stroke survivors can show either full or partial recovery. At about three months after the event most patients have reached their best recovery and at most small improvements occur after three months. The overall economic burden of stroke is enormous – approximately $41 billion in the U.S. in 1997.

Until recently, there were few therapeutic options for stroke patients. This is now changing: both patients and medical teams have learned to deal with a brain attack as a medical emergency. The motto is "Time is brain." There are two therapeutic approaches that have been shown to improve outcome (Hill, 1998):

1. Stroke units: Specialized semi-intensive care units, where blood pres-

sure, blood sugar and other important factors relevant for outcome can be properly controlled and where patients receive physiotherapy.

2. Thrombolytic therapy: If a patient suffering from an ischemic stroke is admitted to a stroke unit within 3 hours of onset of symptoms, tPA (tissue plasminogen activator) can help to dissolve the blood clot that is occluding the cerebral artery. tPA can thus help to reinstall perfusion of the affected parts of the brain (that is, get the blood flowing through them again). However, tPA has risks: given to the wrong patient, it can lead to serious complications, such as secondary hemorrhage.

A third therapeutic concept involving neuroprotective agents is under development but is still experimental (Lees, 1998). The example considered throughout this report is the development of a particular neuroprotective agent. The logic of neuroprotective therapy in acute stroke is based on the pathophysiology of brain attacks. The following is a brief review of that pathophysiology (see also Dirnagl, 1999). Brain cells cannot function properly without a constant supply of oxygen and glucose. Stopping blood flow halts this supply. An interruption for longer than a few minutes will result in the death of affected brain cells. Once cells start to die, the fine equilibrium of intra- and extracellular ions spirals out of control. Neurotransmitters excite surrounding brain cells. This makes them hyperactive in an environment where the energy supply is already at dangerously low levels, which leads to further cell death. Complex inflammatory processes are triggered and complicate the situation further. The area of the brain that receives no or minimal blood flow is called the infarct core. Cells in the infarct core are destined to die. The penumbra is the area of the brain that surrounds the core and that receives some blood flow, but at a level insufficient to allow brain cells to function properly. Cells in the penumbra are able to survive. After a time they may recover some or all of their function. Think of a stroke as a rapidly expanding sphere inside a brain. The interior of the sphere (the infarct core), but not its surface (penumbra), is condemned to death. Now imagine that it is possible to intervene and interrupt the expansion of this sphere. Fewer brain cells would die. The goal of neuroprotective therapy in stroke is to stop or slow the expansion of the infarct core. To be effective it may have to be administered very soon after the onset of stroke.

To test whether a particular compound has neuroprotective properties, such a compound is first tested in animal stroke models. These tests are controlled for important factors such as:

(i) Type of stroke (ischemic/hemorrhagic, cortical/subcortical).

(ii) Stroke severity (localisation - proximal or distal occlusion of artery; duration over which a cerebral artery is occluded).

(iii) Time delay between onset of stroke and administration of therapy (the shorter the delay, the more efficient the therapy).

(iv) Administration of study drug (dose and duration over which drug is given).

If a reduction of infarct volume is observed in the treated group, this is taken as evidence for a compound to have neuroprotective properties. Whether these properties translate into a clinical benefit in human stroke patients, however, can only be seen by conducting clinical trials involving human stroke patients. Generally in drug development, the first clinical trials are in human volunteers and the goal is to assess the agent's safety. If it is proven safe then it is used in patients who have suffered stroke. The goal is to determine whether the agent is effective, and of course to continue assessing its safety.

Assessing neuroprotective drug efficacy in human stroke patients is not easy. An important question is: What is a relevant efficacy endpoint? Infarct volume is the endpoint often used in preclinical studies. In principle it is possible to assess infarct volume also in human stroke patients, by using sophisticated imaging techniques (diffusion/perfusion weighted magnetic resonance imaging). But it is as yet unclear how well reduction of infarct volume and clinical recovery correlate. The problem is that the neurological deficit depends on a) the stroke locale, b) the extent of the brain damage, and c) the stroke's etiology. A large cortical lesion in the primary motor cortex can produce a neurological deficit similar to a very small lesion along the corticospinal tract. And a rather large lesion in certain parts of the frontal lobes may produce only minor deficits, or least deficits that are difficult to measure. As regards etiology, approximately 80% of strokes are called "ischemic," often involving the obstruction of a cerebral artery. And about 15% of strokes are "hemorrhagic:" if an arterial vessel bursts, blood will be pushed with high pressure into the delicate brain tissue, destroying the fine neuronal structures. Sometimes a hemorrhagic stroke can be a secondary complication of an ischemic stroke. This can happen when the lack of blood perfusion mollifies the brain tissue. When the occluded artery becomes cleared, blood may shoot with high pressure into the area of the brain that had been damaged, and potentially resulting in secondary hemorrhage.

In practice, most stroke trials take the principal endpoint to be the patient's condition 90 days after the stroke. The patient's condition is measured by one or more outcome rating scales. There is no general agreement regarding the appropriate way to assess recovery in stroke patients. Instruments are difficult to standardize across different types of strokes, different cultures, and different expectations. Some stroke scales for assessing recovery are dichotomous, with patients classified as "dependent" or "independent." Other scales have a finer rating of neurological deficit. We will use one of the latter, the Scandinavian Stroke Scale (SSS) (Scandinavian Stroke Study Group 1985). The SSS has nine components, each with possible scores ranging from 0 to some maximum number of points. The maxima are shown in parentheses in the following list: consciousness (6), eye movement (4), upper extremity (6), hand (6), lower extremity (6), orientation (6), aphasia (10), facial (2), walking (12). Higher score means better performance. The maximum total SSS is 58.

Inclusion/exclusion criteria in stroke trials are very important, although exactly what they should be is not clear. Patients with mild stroke are easier to recruit into trials, in part because there are more of them. Such patients tend to improve even without therapy. On the other hand, perhaps they would improve even faster with neurological therapy. To help us understand the response over time of untreated patients we used an extensive longitudinal database over time for in excess of 1000 acute stroke patients, the Copenhagen Stroke Database (Jorgensen et al. 1994). Perhaps the most impressive aspect of this database is that it demonstrates the difficulty of developing drugs for treating stroke: the variability in response is enormous, even considering patients with similar baseline characteristics, including baseline SSS. Information from this database led us to focus on moderate to severe stroke in the trial described in this report. In addition, we use this database to simulate trials, as described later in this report.

Among the most important aspects of drug development programs is selecting the correct dose. Too many programs have focused too little attention on this aspect of drug development. There have been a number of failures in recent trials investigating neuroprotective agents in acute stroke and these may be due to an imperfect understanding of the dose-response relationship and a consequent incorrect selection of dose.

One of the reasons for the inadequate understanding about dose-response in development programs of neuroprotective agents is that proper learning requires trials with a very large number of patients – at least with traditional designs. Given the time and resource constraints, some programs may have extrapolated dose-response relationships from preclinical programs, sometimes assuming that from the perspective of efficacy, "more is better."

Our approach described in the following section considers only single-shot IV therapy to be given as soon as possible after the stroke. We do not consider duration of therapy as an independent factor. A more comprehensive model would allow for learning about this dimension of dose.

We have suggested that acute stroke is a difficult indication for which to develop therapies. The design we propose in this report improves on the efficiency of drug development programs by applying Bayesian thinking and adaptive design. This methodology makes learning about novel therapies more efficient. It will help in the treatment of stroke patients by reducing the numbers of patients exposed to non-efficacious doses or drugs during clinical development of a new compound and by bringing efficacious therapies to the marketplace earlier. The approach is applicable to a wide range of drugs and indications, with stroke being but an example.

1.2 *Critical aspects of standard dose-selection trials*

In a standard dose-selection trial, patients are assigned randomly to a set of pre-defined doses, typically between 3 and 5 in number. A standard trial is balanced in the sense that equal or nearly equal numbers of patients are assigned to the various doses.

Such a trial may be inefficient. If the test drug is effective then the dose-response curve has a positive slope for some interval of doses. The sloping part of the curve may be located at doses greater than or less than those considered in the trial. In either case some of the observations at the opposite end from where the slope occurs may be wasted. Moreover, if the sloping part of the curve occurs within the range of doses considered and is steep relative to the interval between doses then observations at both ends of the range may be largely wasted. The case in which the drug is ineffective is similar to the one in which a positive slope occurs at doses larger than those considered in the trial; namely, many of the observations at lower doses may be wasted.

To illustrate some of these points, consider the sample of dose-response curves shown in 2.1, (a)-(f). The true dose-response curves are logistic. Four active doses (labeled 1, 2, 3, 4) plus placebo (dose 0) are assigned in a balanced fashion so the sample sizes are equal. The true values of the response curves at these doses are shown as solid circles – these are the true mean responses for the various treatment arms in the trial. The vertical lines attached to the circles represent the set of likely values for the observed mean response at the corresponding dose. For example, this interval may be the one containing the middle 95% of the probability.

In Figure 2.1(a) the sloping part of the dose-response curve is between doses 0 and 1, with all four active doses having essentially the same underlying mean response. In retrospect, many (but not all) of the patients assigned to doses 2, 3 and 4 would have been more informative about the slope of the dose-response curve (and also about the ED95, the smallest dose at which 95% of the maximal response is achieved) had they been assigned to doses between 0 and 1.

The accuracy with which a trial can determine the ED95 is limited by the separation between doses. In Figure 2.1(a), for example, the ED95 will likely be identified as being between doses 0 and 1. Such accuracy may be sufficient for some compounds, but for drugs that may be less safe at higher doses and for drugs that are costly to produce, it may be essential to identify the optimal dose with greater accuracy.

In Figure 2.1(b) the slope of the dose-response curve is the same as that in Figure 2.1(a) but the entire curve is shifted to the right. If the dose-response curve happens to be the one in Figure 2.1(b) then in retrospect it would have been more informative to have assigned patients to doses between 2 and 3 and also additional patients at doses 3 and 4. In addition, sampling variability present in the observed responses at doses 3 and 4 will leave open the possibility that the asymptote of the curve has not been reached. For example, the observed means at doses 3 and 4 may be as indicated by the open circles. But regardless of the observations at these two doses, the value of the asymptote and the ED95 will be poorly known, and in retrospect assigning patients to doses higher than 4 (safety permitting) would have been more informative.

On this same point, the conclusions drawn from trials in which the curves are as shown in Figures 2.1(c) and 2.1(d) will be similar. For example, the five observed mean responses may be as shown by the open circles, and these are identical in the two figures. In 2.1(c) but not in 2.1(d) the asymptote will have been reached (unbeknownst to the investigators). Moreover, if the responses at dose 4 in comparison to placebo are not clinically important then in 2.1(c) the drug should be abandoned while in 2.1(d) doses higher than 4 should be explored, safety permitting. The observed results of the trial will not help in deciding which strategy is better.

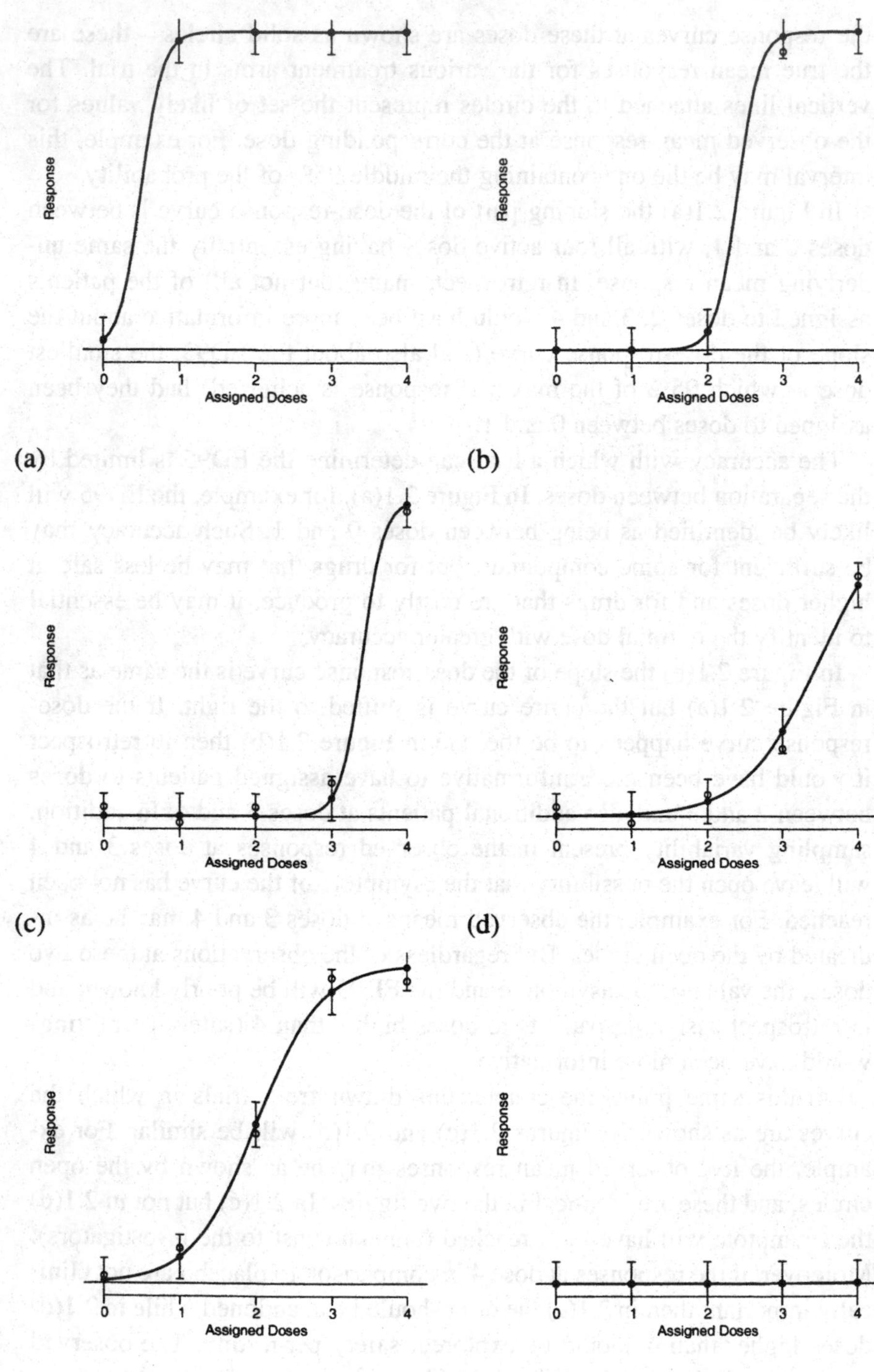

FIGURE 2.1.

In Figure 2.1(e) the doses assigned happen to be ideally placed in the sense that they encompass the gamut of responses and they are spread out along the sloping part of the curve. But even here, because of uncertainty concerning the true slope of the curve between doses 3 and 4, the ED95 is not very well identified.

In Figure 2.1(f) there is no response over the range of doses considered. Assuming that the dose-response curve is known to be monotonic, information from patients assigned to doses 1, 2 and 3 will be largely wasted and would have been more informative had they been placed at either end. Either the drug is ineffective or its effect occurs at doses higher than 4. Explorations beyond dose 4 are essential (if safe) to decide which possibility is the correct one.

Another drawback to standard types of designs is the use of fixed sample sizes. Even if the doses considered are judged in retrospect to be appropriate, the variability in response across patients assigned to the same dose may be less or greater than anticipated. In the former case the sample size chosen may be unnecessarily large and in the latter case it may be too small to accomplish the study's goals. Figure 2.2 addresses this issue. In both 2.2(a) and 2.2(b) the dose-response curve is the same as in Figure 2.1(e). In Figure 2.2(a) the measurement standard deviation is half that of Figure 2.1(e) and in Figure 2.2(b) it is twice that of Figure 2.1(e). The standard errors at each dose will then be half as large and twice as large, respectively, as anticipated.

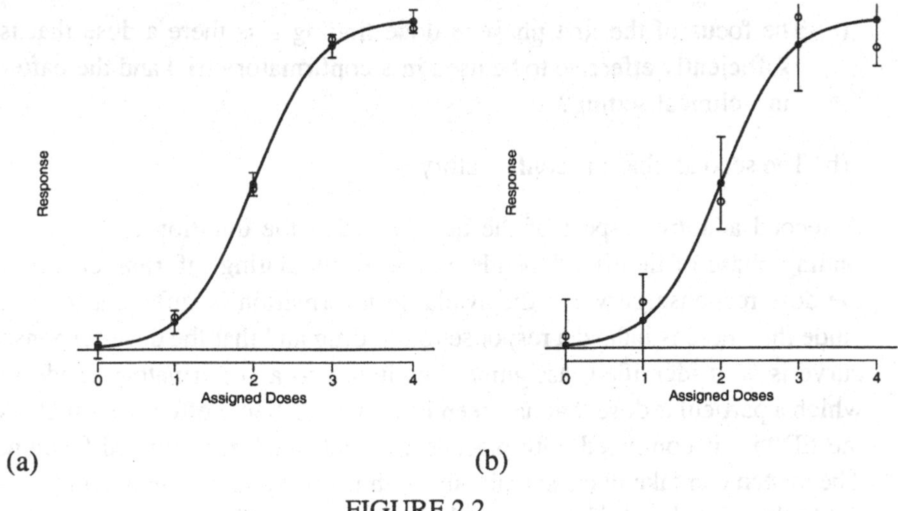

(a) (b)

FIGURE 2.2.

Not surprisingly, a retrospective view of conventional dose-finding trials suggests that assigning to doses in a different fashion from that actually used would have been more informative, or that a different sample size would have been more appropriate, or both.

At least two types of improvements to balanced designs with pre-selected doses are possible:

(i) Increase the number of doses considered. This can facilitate identifying critical aspects of the dose-response curve, but if the numbers of patients assigned at each dose is the same then the cost in terms of sample size may be prohibitively large.

(ii) Avoid large sample numbers of observations at doses where the form of the curve can be reasonably well estimated from a smaller number by borrowing the information available from observations at nearby doses.

Both improvements are accomplished using the procedure we describe in this report. The result is that more accurate information is available about critical aspects of dose-response with, generally, smaller sample sizes. Moreover, when the total sample size is larger, this larger size is necessary to identify important aspects of the dose-response curve.

In this report we propose an adaptive scheme in which the doses assigned are allowed to depend on information from patients treated previously. The design is in two phases:

(a) The focus of the first phase is dose finding – is there a dose that is sufficiently effective to be used in a confirmatory trial and thereafter in a clinical setting?

(b) The second phase is confirmatory.

A second adaptive aspect of the design is that the duration of the dose-finding phase of the trial depends on the accumulating information about the dose-response curve: If the available information is sufficient to conclude that there is indeed a response to the drug and that the dose-response curve is well identified, assignment switches to a confirmatory mode in which a particular dose that has been identified as being effective – such as the ED95 – is compared with placebo in a balanced, randomized fashion. The switch can take place seamlessly with no delay in patient accrual.

On the other hand, if the available evidence is sufficient to suggest that there is no dose-response or that the maximal effect is too small to make the drug worthwhile clinically, then dose finding will cease and the procedure will recommend terminating the trial.

The design is adaptive in both assigning dose and deciding when to stop dose finding. And it uses Bayesian updating and Bayesian decision analysis. However, the regulatory approval process is largely frequentist. Therefore, the decision analysis involves predicting statistical significance (in the usual frequentist sense) of the confirmatory trial, and its sample size is determined accordingly.

There is a regulatory tradition of requiring two trials, both with statistical significant results. The design we propose embodies two trials, one the dose-finding phase and the other the confirmatory phase. Statistical significance in the first phase can be for the slope of the dose-response curve or for the comparison of the ED95 with placebo, assuming a particular probability model. The current version of our program does not require statistical significance in the first phase (adjusted for the sequential nature of the design) before converting to the second phase. However, this option is being incorporated into the program so as to meet this requirement if it is imposed by regulatory agencies.

2 Design Process

The proposed design for dose selection is in four parts, which by the nature of sequential, adaptive designs are not necessarily ordered in time. The process is schematically represented in Figure 2.3, the 11 components of which will be described below and related to the four parts of the dose selection process.

2.1 Allocating patients to optimal dose

On a particular day during the trial, a number of patients, typically fewer than 5, will be admitted to the trial [1 in Figure 2.3]. Consider the first patient. Based on the available information our knowledge concerning the ED95 will be represented by a posterior distribution with some variance. Assigning a particular dose to the next patient and observing the response will effect a change in the posterior distribution of the ED95, and hence in its variance. We can predict how the chosen dose and the corresponding response of the patient will change the consequent posterior variance. The most informative dose [11] – say $\tilde{z}$ – is the one with the smallest average variance of the ED95, where the average is with respect to the predictive probabilities of the various future observations, given the dose assigned. Before assigning the dose, however, some amount of randomization is employed [2]. First, if $\tilde{z}$ is not placebo then placebo is assigned with some minimal probability – such as 15%. Second, consider the set of doses for which the expected response is within 10% of that of $\tilde{z}$. The dose assigned is selected randomly from within this set. Subsequent patients admitted on a particular day – up to 5 – are considered as above, but conditioning on the previously assigned doses and averaging with respect to the predictive distributions of the responses at these doses.

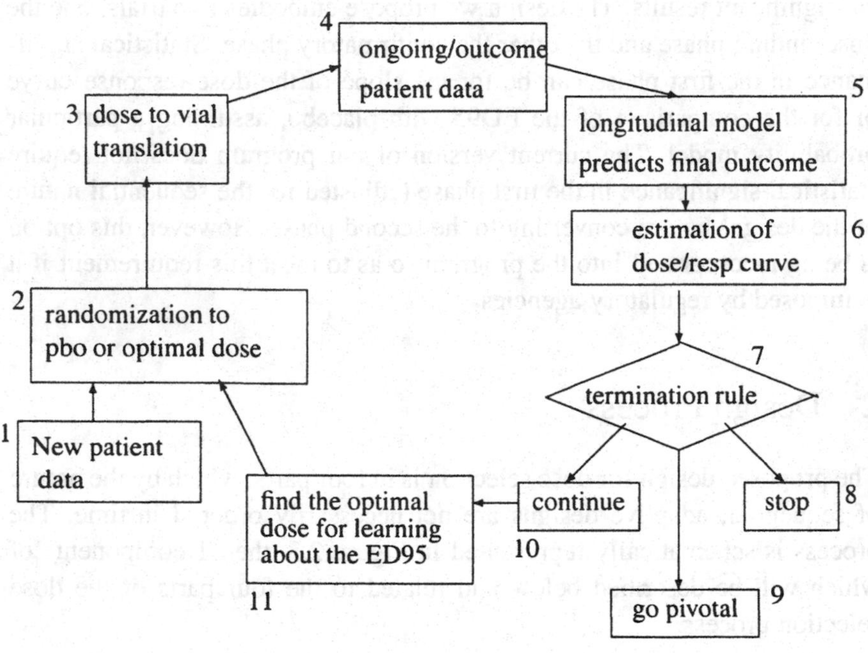

FIGURE 2.3.

2.2 *Predicting final outcome: Longitudinal model and user interface*

At any particular time, patients having entered the trial will provide information concerning their responses [4 in Figure 2.3]. Through the use of a longitudinal, predictive model to be described in Section 7, we impute the final response of the patient [5 in Figure 2.3]. This model allows an early update of knowledge about the dose-response relationship, which in turn allows for better choices concerning the doses to be allocated to future patients. In view of the critical importance of delivering neuroprotective therapy quickly, such updating requires rapid data processing. A rapid communications interface between the statistical system and the centers where the study is carried out has been developed. Details of the interface are given in Section 8.

2.3 *Estimating dose-response*

We use a dose-response model based on Normal Dynamic Linear Models (NDLM) as described, for example, in West and Harrison (1997). This is essentially a piecewise linear model. It provides the necessary flexibility to encompass both monotonic and non-monotonic dose-response relationships. An additional advantage of the NDLM is the existence of analytical results for the determination of the posterior distribution of the dose-response curve, and also for the ED95. Together with data from those patients who have completed the study, the imputed values described in the previous section are used to update the relevant posterior distributions. Details of the model and the updating procedure are given in Section 6.

2.4 *Confirmatory decision*

On each day of the dose-finding phase the algorithm implements a decision rule [7 in Figure 2.3] that addresses the question of whether the process should continue in dose-finding [10 in Figure 2.3], end because of futility [8 in Figure 2.3], or shift to a confirmatory phase [9 in Figure 2.3]. These decisions are made based on the predictive probabilities of failure to show a benefit and successfully showing a benefit. (The decision to stop the trial is not made in the absence of human intervention. A Data Safety Monitoring Board reviews any decision concerning the conduct of the trial.) The shift from dose-finding to confirmatory may be seamless in that the phase of the trial may not be recognizable by physicians and others involved. If a shift is made in a seamless fashion then the follow-up information from patients treated most recently in the trial will not be complete.

The timing of the shift from dose-finding to confirmatory is important. Whether to shift is based on a Bayesian decision analysis. A component of this analysis is an assessment of the probability of showing an eventual benefit in a confirmatory trial at the current ED95. The decision to shift will depend on the available information about dose-response. The design of the confirmatory study will be based on the predictive power calculations, but whether to shift to the confirmatory phase of drug development is based on Bayesian calculations of the probability of showing a benefit over placebo.

3 The System

3.1 Notation

Let y_i denote the final response (at week 13) for the i-th patient, measured as difference from baseline SSS (Scandinavian Stroke Score). Let $y_{ij}, j = 1, \ldots, M-1$, denote responses in weeks 1 through $M-1 = 12$. We use y without an index to stand for week 13 response. Vectors $y(i) = (y_1, ..., y_i)$ stand for data for the i-th patient and for the first i patients, respectively. We will use D to denote all data collected in the study. Let x_i denote a vector of covariates for the i-th patient. In the current implementation the only covariate used is the baseline SSS score. But x_i could include other relevant patient-specific information, such as age, time between onset of stroke and start of treatment, etc. We use z to refer to a treatment dose, z_i to refer to the dose assigned to patient i, and $\{Z_j; j = 0, \ldots, J\}$ to refer to the set of allowable doses, where placebo $Z_0 = 0$. We use $Y_{jk}, j = 0, \ldots, J$, to denote the response of the k-th patient assigned to dose Z_j. In Sections 5 and 6 we describe a sampling model for y_i and y_{ij}. In earlier sections the probability model is general. Our descriptions of dose assignment (Section 4) and stopping time (Section 5) apply for any underlying probability model. We will need generic notation to refer to the dose/response curve for y_i and the longitudinal data model for y_{ij}. We use $f(z, \theta)$ to denote the dose/response curve $E(y|\theta)$ as a function of dose z, parametrized by an unknown parameter vector θ, $E(y_i|\theta) = f(z_i, \theta)$. Given the mean curve $f(z, \theta)$, we assume normal errors:

$$y_i = f(z_i, \theta) + \epsilon_i \text{ with } \epsilon_i \sim N(0, \sigma^2). \tag{2.1}$$

Let $df(z, \theta) = f(z, \theta) - f(0, \theta)$ denote the advantage over placebo. For a given parameter vector θ let $\hat{z}_\theta$ denote the dose with maximum expected response, i.e., $f(\hat{z}_\theta, \theta) = \max_j\{f(Z_j, \theta)\}$. Denote with $z95_\theta$ the ED95 of the unknown dose response curve, defined as the minimum dose Z_j with mean improvement greater or equal than 95% of the maximum possible improvement over placebo:

$$z95_\theta = \min\{Z_j : \; df(Z_j, \theta) \geq 0.95 \; df(\hat{z}_\theta, \theta)\}$$

We use $f(z)$ for the posterior expected dose/response curve,

$f(z) = E[f(z, \theta)|D]$, and $z95$ for the ED95 of the posterior mean curve $f(z)$.

Also, we shall use $\overline{z95}$ to denote the posterior mean of $z95_\theta$, i.e.,

$E(z95_\theta \mid D)$. In view of the nonlinear nature of the ED95, $\overline{z95} \neq z95$. For technical reasons, in the program we sometimes consider conditional posterior means of $z95_\theta$, keeping some aspects of the model fixed. We indicate when this is the case.

3.2 *Dose-finding study*

The first phase of the trial is for dose-finding. Details of the dose allocation are discussed in Section 4. We will use D_{1t} to denote the data up to week t of the dose-finding phase, and n_{1t} to denote the number of patients enrolled by week t. Similarly D_1 denotes the data from the complete dose-finding phase, and n_1 denotes the total number of patients enrolled in the dose-finding phase.

3.3 *Confirmatory phase*

In the confirmatory phase, patients are randomized equally to placebo ($z = 0$), recommended treatment dose z^*, and possibly a third dose z^{**} chosen as alternative to z^*. As a recommended treatment dose we use the ED95 dose, $z^* = \overline{z95}$. As alternative dose z^{**} we use the ED50, $z^{**} = \overline{z50}$. We will use D_2 to denote the data from the confirmatory phase, and n_2 to denote the number of patients enrolled in this phase at each of the three doses (i.e., the total sample size of the confirmatory phase is $3\,n_2$).

To choose n_2 we use predictive power. Let df^* denote the advantage $df(z^*, \theta)$ of treatment over placebo at the recommended dose z^*. Let (m, s) denote posterior mean, $m = E(df^*|D_1)$, and standard deviation, $s^2 = Var(df^*|D_1)$, at the end of the dose-finding phase. Details of this estimation depend on the underlying probability model (See Section 6 for a discussion of the actual probability model used). Let $\hat{\sigma}^2$ denote the posterior mean $E(\sigma^2|D_1)$ conditional on all current data. Fixing σ^2 at $\hat{\sigma}^2$ and approximating $p(df^*|D_1) \approx N(m, s)$, we can find the minimum sample size n_2 that gives a desired predictive power (90% in our implementation).

4 Adaptive Dosing: The Allocator

During the dose-finding phase of the trial we must choose a dose for each new patient. As indicated in the introduction, we consider a decision-theoretic approach. To introduce notation and to clarify the context, we briefly review the general set-up of a Bayesian decision problem. Decision making under uncertainty is choosing an action d to maximize expected utility

$$U(d) = \int u(d, \theta, y) \, dp_d(\theta, y).$$

Here, $u(d, \theta, y)$ is the utility function modeling preferences of consequences and $p_d(\theta, y)$ is a probability distribution of parameter θ and observation y, which may depend on the chosen action d. Typically $p_d(\theta, y)$ is specified as a prior $p(\theta)$ on the parameters and a sampling model $p_d(y|\theta)$. Utility $u(d, \theta, y)$ must be specified only for a specific realization (θ, y) of the experiment. To compute expectation $U(d)$ means averaging over the quantities that are unknown at the time of decision making. Often some of the data, say y_0, is already known at the time of decision making. Assume $y = (y_0, y_1)$. Then $p(\theta)$ and $p_d(y|\theta)$ are replaced by $p(\theta|y_0)$ and $p_d(y|\theta, y_0)$. See Chaloner and Verdinelli (1995) and Verdinelli (1992) for reviews of Bayesian approaches to decision problems traditionally known as optimal design. Spiegelhalter, Freedman and Parmar (1994), Berry (1993) and Berry and Stangl (1996) discuss general issues related to the use of Bayesian optimal design methods in medical decision problems.

4.1 *Dose allocation as a decision problem*

Central to our proposal is a utility function that quantifies relative preferences over alternative outcomes. The proposed utility function is related to learning about the unknown dose/response curve. Learning is formalized as minimizing the posterior variance for some key parameter $g(\theta)$ of the dose/response curve. In the current implementation we choose as key parameter $g(\theta)$ the mean response $f(z95_\theta, \theta)$ at the ED95. The posterior variance of $g(\theta)$ includes uncertainty in the unknown ED95 dose, as well as the unknown response at that dose.

The proposed approach is myoptic in the sense that when we consider the optimal dose for the next patient we proceed as if he or she was the last patient to be recruited into the trial. Assume we currently have N patients enrolled in the trial. Let $\tilde{y}_k = y_{N+k}$, $\tilde{x}_k = x_{N+k}$ and $\tilde{z}_k = z_{N+k}$ denote the response, covariate and assigned dose for the next K patients, $k = 1, \ldots, Kx$. Let $\tilde{y}(k) = (\tilde{y}_1, \ldots, \tilde{y}_k)$, $\tilde{z}(k) = (\tilde{z}_1, \ldots, \tilde{z}_k)$, and $\tilde{x}(k) = (\tilde{x}_1, \ldots, \tilde{x}_k)$, denote responses, assigned doses and covariates up to the k-th new patient. Let D_N denote the observed data for the first N patients, and let $\tilde{D}$ denote the still missing final responses for already enrolled patients. We define the utility function for choosing the dose $\tilde{z}_k$ as

$$u_k[\tilde{z}(k), \tilde{y}(k), \tilde{x}(k), \tilde{D}, D_N] = -Var[g(\theta) \mid D_N, \tilde{D}, \tilde{y}(k), \tilde{x}(k), \tilde{z}(k)].$$

Of course we have to decide upon $\tilde{z}_k$ before observing $\tilde{D}$, $\tilde{y}(k)$ and $\tilde{x}(k)$. Thus we choose the dose $\tilde{z}_k$ by maximizing the utility $u_k(\cdot)$ in expectation, averaging with respect to $\tilde{D}$, $\tilde{y}(k)$ and $\tilde{x}(k-1)$. The relevant distributions for $\tilde{D}$ and $\tilde{y}(k)$ are the posterior predictive distributions given the current data. For the covariates $\tilde{x}_1, \ldots \tilde{x}_{k-1}$ we use the empiricial distribution $p(x)$ from the Copenhagen Stroke Study data base, assuming independence, i.e.,

$$p[\tilde{x}(k-1)] = \prod_{h=1}^{k-1} p(\tilde{x}_h).$$

And $\tilde{x}_k$ is fixed at "typical" covariate values x^o, i.e., we find the optimal dose for an average next patient. For $\tilde{z}_1, \ldots, \tilde{z}_{k-1}$ we substitute the optimal values found by optimizing the expected utilities $U_1(\cdot), \ldots, U_{k-1}(\cdot)$.

$$U_k[\tilde{z}_k, \tilde{z}(k-1), \tilde{x}_k, D_N] = \int u_k[\tilde{z}(k), \tilde{y}(k), \tilde{x}(k), \tilde{D}, D_N] \times \\ \times p(\tilde{D} \mid D)\, p[\tilde{y}(k) \mid D, \tilde{z}(k)]\, p[\tilde{x}(k-1)]\, d\tilde{D}\, d\tilde{y}(k)\, d\tilde{x}(k-1). \quad (2.2)$$

Maximizing $U_k(\cdot)$ over $\tilde{z}_k$ we find the optimal action.

4.2 *Evaluating Expected Utility*

Critical for a successful implementation of the proposed decision theoretic dose allocation is the availability of analytical or efficient numerical integration to evaluate the integrals in (2.2). Key to our implementation strategy is to rewrite the expected utility integral (2.2) as an integral with respect to the posterior distribution $p(\theta \mid D)$

$$U_k[\tilde{z}(k), \tilde{x}_k, D_N] = \int u_k[\tilde{z}(k), \tilde{y}(k), \tilde{x}(k), \tilde{D}, D_N] \times \\ \times p(\tilde{D} \mid \theta, D)\, p[\tilde{y}(k) \mid \theta, \tilde{x}(k), \tilde{z}(k)]\, p[\tilde{x}(k-1)] \\ d\tilde{D}\, d\tilde{y}(k)\, d\tilde{x}(k-1)\, p(\theta \mid D)\, d\theta. \quad (2.3)$$

Most models allow efficient random variate generation from the posterior distribution $p(\theta \mid D)$ using Markov chain Monte Carlo (MCMC) simulation. See, for example, Tierney (1994) or Gilks et al. (1996) for a summary of MCMC methods. Details of implementing the appropriate MCMC scheme depend on the specific probability model $p(D \mid \theta)$. In Section 6 we will discuss how we implement simulations from the posterior distribution in our model. But we do not need these details in the following discussion. We assume only that by appropriate simulation techniques it is possible to generate an (approximate) posterior Monte Carlo sample $\Theta = \{\theta^1, \ldots, \theta^T\}$ with $\theta^t \sim p(\theta \mid D)$. Using the Monte Carlo sample Θ we can evaluate expected utilities by replacing (2.3) with a corresponding Monte Carlo sample average. For each θ^t we simulate covariates $\tilde{x}_h^t \sim p(x)$, $h = 1, \ldots, k-1$, responses $\tilde{y}_h^t \sim p(y_h|\tilde{z}_h, \tilde{x}_h^t, \theta^t)$ and missing responses $\tilde{D}^t$ of current patients $\tilde{D} \sim p(\tilde{D} \mid \theta^t)$, assuming that these models are all available for efficient random variate generation. For each simulated experiment we then evaluate observed utility $u_k^t = u_k(\tilde{z}(k), \tilde{y}(k)^t, \tilde{x}(k)^t, \tilde{D}^t, D)$, and replace expected utility by a Monte Carlo average

$$\hat{U}_k(\tilde{z}(k), \tilde{x}_k, D) = \frac{1}{M} \sum_{t=1}^{M} u_k^t$$

Evaluating expected utility $\hat{U}_k$ for a grid of possible choices $\tilde{z}_k$ we find the optimal dose as the dose with maximum $\hat{U}_k(\cdot)$. To reduce numerical error in the approximation of U_k by $\hat{U}_k$ we use common random numbers, i.e., whenever possible we use the same Monte Carlo sample Θ and random variates $\tilde{x}_h^t$, $\tilde{y}_h^t$, $\tilde{D}^t$ when we evaluate $\hat{U}_k(\tilde{z}_k)$ for alternative choices of $\tilde{z}_k$.

Figure 2.4 shows a typical expected utility curve. Of course, the expected curve changes from week to week. Assume at one time it is optimal to allocate to high doses. As more patients are allocated in that part of the dose range it will usually become more advantageous to allocate new patients in other parts of the dose range to learn about, for example, the response at placebo. While it is difficult to intuitively understand the expected utility curve and the dose assignment at a given time, typical patterns of dose allocations over the course of a clinical trial do seem intuitively meaningful. Figure 2.12 (Section 9) shows an example of dose assignments in a simulated trial.

4.3 Additional Randomization: The Recommender

Maximizing $\hat{U}_k(\cdot)$ delivers the optimal doses $\tilde{z}_k^*$ be assigned to the next patients, $k = 1, \ldots, K$. Assume z_k^* equals Z_{j^*} in the list of allowable doses. Before actually assigning a dose to a new patient we use an additional randomization. First, we guard against the possibility of a drift in the patient population over the course of the trial by fixing a minimum proportion p_0 of patients assigned to placebo. Second, because of safety

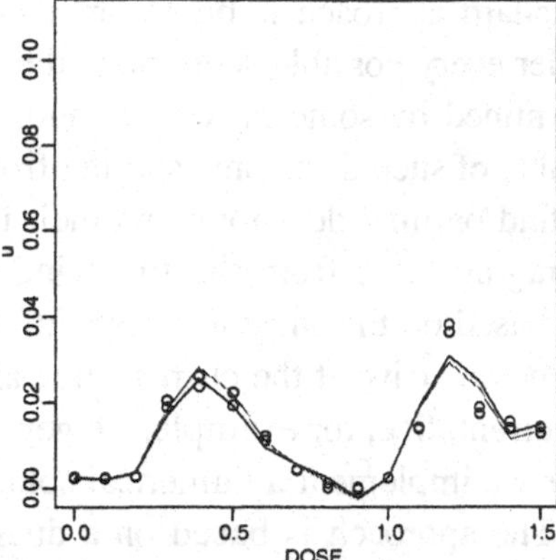

FIGURE 2.4. Approximate expected utilities $\hat{U}_k$ plotted against $\tilde{z}_k$. The circles show the Monte Carlo estimates of expected utilities at the allowable doses. The curve shows a smooth fit through the Monte Carlo estimates. For comparison a second curve (almost indistinguishable from the other curve) shows a smooth fit through Monte Carlo estimates of expected utilities using half the Monte Carlo sample size.

concerns we want to avoid unnecessarily high doses. To achieve these two aims we allocate with probabilities p_0 at placebo and split the remaining probability $(1 - p_0)$ uniformly over all doses Z_j within a neighborhood of Z_{j^*}, defined as the set of all doses less than or equal and having estimated mean response within 10% of the estimated mean response at Z_{j^*}.

5 Optimal Stopping: The Terminator

The dose-finding phase of the trial involves two important decision problems, dose assignment and termination. We discussed an adaptive dose allocation scheme in Section 4. In the present section we discuss the second problem: optimal stopping for the dose-finding phase. At each period t of the trial, say once a week, we have to decide (d_t) whether to terminate the trial and abandon the drug $(d_t = A0)$, continue with the dose-finding phase $(d_t = A1)$, or terminate the dose-finding phase and switch to confirmatory mode $(d_t = A2)$.

Making an optimal decision is complicated. The decision to continue (A1) requires evaluating the expected utility of this action which in turn requires knowing the utility of next week's optimal termination decision for all possible interim results, i.e., we need to solve a sequential decision problem. A standard approach is backward induction: Consider the optimal solution under every possible scenario at the latest possible termination time T (determined by some maximum number of patients in the trial), and record a table of such decisions, starting from T. Then, proceeding backwards, we find optimal decisions and their utilities at the penultimate week by averaging those from the following week using the predictive probabilities based on the current week's information. Proceeding backwards in this way we arrive at the current time and the decisions that are optimal for the current. See, for example, Berger (1985, chapter 7).

For the *terminator* we implement a numerical solution of the backward induction problem. The approach is based on a dual strategy of using a reduced action space to constrain the number of scenarios which we need to consider in backward induction, and forward simulation to evaluate expected utility under all relevant scenarios. Central to our approach is a formulation of the problem as a decision problem with a probability model describing all relevant uncertainties and a utility function which describes the relative preferences of possible outcomes.

Alternative Bayesian approaches to optimal sequential design in phase I drug trials are discussed in, among other references, Thall, Simon and Estey (1995), who define stopping criteria based on posterior probabilities of clinically meaningful events. Thall and Russell (1998) define ad-hoc sequential procedures based on monitoring posterior probabilities and evaluate their frequentist properties. Vlachos and Gelfand (1998) follow a similar strategy. Whitehead and Brunier (1995) and Whitehead and Williamson (1998) use a Bayesian m-step look-ahead procedure to find an optimal dose for the next m patients in a dose-finding study.

5.1 The Terminator

Before we discuss the underlying decision-theoretic framework, we present the form of the final implementation and attempt to make it intuitively appealing. Let $z^* = \overline{z95}$ denote the ED95 dose. Let $(m_t; s_t)$ denote the posterior mean and standard deviation of the advantage over placebo at the ED95 dose conditional on the data available at time t, i.e., $m_t = E[df(z^*, \theta) \mid D_{1t}]$ and $s_t^2 = Var[df(z^*, \theta) \mid D_{1t}]$. The proposed stopping rule is a function of (m_t, s_t). We define cut-offs for (m_t, s_t) that partition the space into three subsets, corresponding to decisions $A0$, $A1$ and $A2$. Figure 2.5 provides an example partition. The pattern is intuitive. For very small and very large m the terminator recommends $A0$ (abandon the drug) and $A2$ (continue to the confirmatory phase), respectively. For intermediate values the terminator recommends continuing the dose-finding trial. And for smaller s there is less ambiguity about whether to stop.

s[t] 1.40	1.50	1.60	1.70	1.80	1.90	2.01	2.11	2.21	2.31	2.41	2.51	2.61	m[t]
0	0	0	0	0	0	0	0	0	0	0	0	0	-0.36
0	0	0	0	0	0	0	0	0	0	0	0	0	0.10
0	0	0	0	0	0	0	0	0	0	0	0	0	0.57
0	0	0	0	0	0	0	0	0	0	0	0	0	1.03
1	1	1	1	1	0	0	0	0	0	0	0	0	1.49
1	1	1	1	1	1	1	1	1	1	1	0	0	1.95
1	1	1	1	1	1	1	1	1	1	1	1	1	2.41
1	1	1	1	1	1	1	1	1	1	1	1	1	2.88
1	1	1	1	1	1	1	1	1	1	1	1	1	3.34
1	1	1	1	1	1	1	1	1	1	1	1	1	3.80
1	1	1	1	1	1	1	1	1	1	1	1	1	4.26
1	1	1	1	1	1	1	1	1	1	1	1	1	4.72
2	2	1	1	1	1	1	1	1	1	1	1	1	5.18
2	2	2	2	2	2	2	2	2	1	1	1	1	5.65
2	2	2	2	2	2	2	2	2	2	1	1	1	6.11
2	2	2	2	2	2	2	2	2	2	2	1	1	6.57
2	2	2	2	2	2	2	2	2	2	2	2	1	7.03
2	2	2	2	2	2	2	2	2	2	2	2	2	7.49
2	2	2	2	2	2	2	2	2	2	2	2	2	7.96
2	2	2	2	2	2	2	2	2	2	2	2	2	8.42

FIGURE 2.5. Example terminator. The optimal decision for each pair (m, s) is either $A0$, $A1$, or $A2$, indicated by entries 0, 1, and 2, respectively.

To decide whether to terminate at time t, we compute (m_t, s_t) and use the entry shown in Figure 2.5. For any two scenarios with equal summary (m_t, s_t) we use the same decision and ignore any additional information beyond (m_t, s_t) which might be contained in the data. Thus the decision depends on the current data only indirectly through (m_t, s_t) which we use to look up the decision table. The main reason for this constraint is to simplify the resulting computations.

5.2 *A decision-theoretic stopping rule*

To find the boundaries in the (m_t, s_t) table we use a decision-theoretic argument. Specifying a utility function for each outcome of the experiment allows us to find the worth of a given sequential strategy. Maximizing this utility function in expectation defines the optimal termination decision.

The Utility Function

We start the process by defining the utilities for all possible outcomes assuming that the whole experiment is known. Expected utilities of decisions can then be found by finding expectations over all unknowns. For a given realization the utility includes a sampling cost for the number of patients recruited into the trial, and a payoff for successfully developing and marketing the drug. Of course that payoff is included only if the drug is approved for marketing by the regulatory authorities. And it depends on the magnitude of the drug's benefit in comparison with placebo as observed in the two phases of the trial taken together – a better drug has a greater payoff.

Utility under $d_t = A2$. If we were to make decision $d_t = A2$ then the resulting utility will include the sampling cost (negative) for the confirmatory phase of the study plus the payoff if the confirmatory study eventually concludes a significant treatment effect.

Let B denote the event in which the null hypothesis of no treatment effect is rejected in the confirmatory phase at some fixed significance level α. Let $\bar{y}_0$ and $\bar{y}^*$ denote the sample average of the responses under placebo and treatment (z^*), respectively, in the confirmatory phase. B is the event

$$B = \{(\bar{y}^* - \bar{y}_0)/\sqrt{2\,\hat{\sigma}^2/n_2} > q_\alpha\}$$

where q_α is the $(1-\alpha)$ standard normal quantile. As in Section 3.3, n_2 is the sample size in the confirmatory study, and $\hat{\sigma}^2$ is the posterior mean of the measurement variance.

Let c_1 denote the sampling cost per patient, and let c_2 denote the payoff for a successful drug. Assuming that this payoff is proportional to the size of the effect we specify c_2 as payoff per point advantage over placebo. Let $m_2 = E(df^* \mid D_1, D_2)$.

$$u(d_t = A2, D_1, D_2) = \begin{cases} -3\; n_2\; c_1 & \text{if not } B \\ -3\; n_2\; c_1 + c_2\; m_2 & \text{if } B \end{cases}$$

The available data enters into the definition of $u(\cdot)$ implicitly through n_2, which depends on D_1, m_2, which is a statistic depending on D_2, and B_2, which is an event in D_2. The sampling cost includes only those patients in the confirmatory study and omits the sampling cost for the first n_1 patients in the dose-finding phase. Including the latter would add the same term to the utilities of all three alternative actions ($A0$, $A1$ and $A2$) and would thus not change the decision problem. Of course, decision d_t needs to be made before observing D_2. The relevant expected utility is an average over D_2:

$$U(d_2 = A2, D_1) = \int u(d_t = A2, D_1, D_2)\, dp(D_2 \mid D_1). \tag{2.4}$$

As indicated in Section 3.3, n_2 is a function of (m, s). It remains to compute $P(B|D_1)$ and $E(m_2|D_1, B)$. Both can be computed analytically if we use the following approximation: $p(df^* \mid D_1) \approx N(m, s)$.

Utility under $d_t = A0$. If we were to decide $d_t = A0$ then the trial is over and thus $U(d_t = A0) = 0$, again omitting the sampling cost for the first n_1 patients.

Utility under $d_t = A1$. Finally, if we were to decide $d_t = A1$, then the utility depends on what we will decide in the next period $t + 1$. Let n^+ denote the number of patients enrolled in week $(t + 1)$ and let D^+ denote the data collected in week $(t + 1)$.

$$u(d_t = A1, D_{1t}, D^+) = -n^+ \cdot c_1 + U^*_{t+1}(D_{1t} \cup D^+),$$

where $U^*_{t+1}(D_{1,t+1})$ is the expected utility under the optimal action at time $(t + 1)$. By taking the expectation with respect to D^+ we obtain

$$U(d_t = A1, D_{1t}) = \int u(d_t = A1, D_{1t}, D^+)\, dp(D^+ \mid D_{1t}). \tag{2.5}$$

We are assuming that n^+ is fixed or known. Extension to random n^+ is straightforward. Evaluation of the integral (2.5) is discussed below in the subsection *Constrained backward induction.*

The optimal decision $d^*_t(D_{1t})$ at time t and its value $U^*_t(\cdot)$ can, in principle, be derived as

$$U^*_t(D_{1t}) = U(d_t = d^*_t, D_{1t}) = \max_{d_t \in \{A0, A1, A2\}} \{U(d_t, D_{1t})\}. \tag{2.6}$$

There are at least two impediments to implementing (2.6). First, the solution requires backward induction in the sense that the definition of $U(d_t = A1, \ldots)$ entails solving (6) for period $t + 1$. Second, the definitions of $U(d_t = A2, \ldots)$ and $U(d_t = A1, \ldots)$ involve integrals that are typically analytically intractable. In the following two subsections we outline an implementation strategy based on constrained backward induction and forward simulation. See Müller, Berry, Grieve, Smith and Krams (1999) for a detailed discussion.

Constrained Backward Induction

To allow a practical solution to the backward induction problem we constrain decisions d_t to depend on the current data only indirectly through (m_t, s_t), with (m_t, s_t) given for a finite grid, i.e., we use $d^*(m_t, s_t)$. This amounts to constraining the action space. Instead of allowing decisions to depend on the full information set D_{1t} they are allowed to depend on the current data only indirectly through (m_t, s_t). Even without this constraint, expression (2.4) for $U(d = A2, D_1)$ depends on D_{1t} only indirectly through (m, s). We write $U(d = A2, m, s)$ to emphasize this. However, $U(d_t = A1, D_{1t})$ depends on D_{1t} beyond the statistic (m, s). See the subsection *Forward Simulation* for the definition and evaluation of $U(d_t = A1, m, s)$. We will use $U^*(m, s) = U(d_t^*(m, s), m, s)$ for the expected utility when making an optimal decision.

Computation proceeds as follows. Starting with an initial guess for $d^*(m, s)$ and $U^*(m, s)$ we update the tables using (2.4) through (2.6). We continue updating until the table remains unchanged over a complete cycle of updates.

Forward Simulation

There remains the problem of evaluating expected utility in (2.5). We use forward simulation (Carlin, Kadane and Gelfand, 1998). The idea is to simulate a trial including up to the maximum number of possible patients in the dose-finding phase. Then, many such trials, say T, are simulated and the results averaged.

Figure 2.6 illustrates how these forward simulations are used to compute posterior integrals. Assume we need a posterior integral conditional on data D_{1t}. We first compute the corresponding summary statistic (m_t, s_t), and then approximate the desired posterior integral by a sample average over all simulated trials whose trajectories pass through (m_t, s_t). If (m_t, s_t) were in fact a sufficient statistic for the unknown parameter vector then the sample average would provide a (simulation) consistent and unbiased estimate of the desired expectation. Carlin et al. (1998) use forward simulation in such a set-up. In general, some approximation is involved.

Let D_{1t} and n_{1t} denote the data and number of enrolled patients at the time of making the decision d_t. Let $\tilde{D}$ denote the future data, including the remaining $n_1 - n_{1t}$ patients, as well as still missing final measurements on current patients who have been recruited during the last 13 weeks. The T trials for the forward simulation are generated by simulating from the posterior predictive distribution $p(\tilde{D}|D_{1t})$. Denote the simulated values as $\tilde{D}^{(j)}$. For each week of each simulation we record the moments

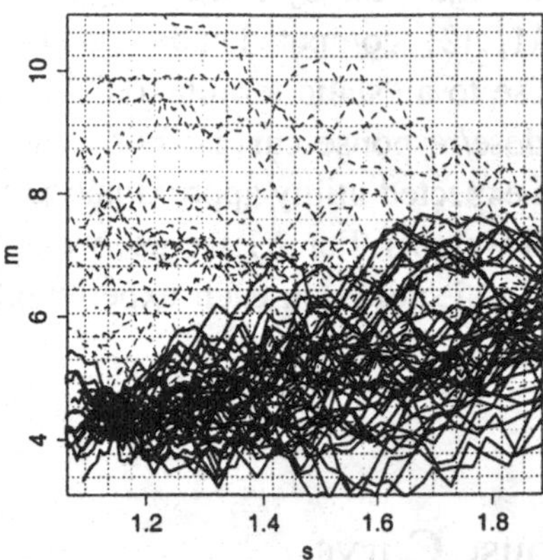

FIGURE 2.6. Forward simulation. The figure shows trajectories in the $(m_t; s_t)$ space for some simulated trials. See the text for how these simulations are used to compute a posterior integral.

$(m_t^{(j)}, s_t^{(j)})$.

Assume in the constrained backward induction we have current estimates $d^*(m, s)$ and $U^*(m, s)$. Assume we are updating a specific cell (m, s) and want to find the optimal decision for that cell. Expected utility $U(d = A2, m, s)$ is easily computed using (2.4), and $U(d = A0, m, s) = 0$ is fixed. To compute integral (2.5) for $U(d = A1, \dots)$ we consider the subset A of simulations which pass through cell (m, s) at some time, i.e., $j \in A$ if $(m_{tj}, s_{tj}) = (m, s)$ in some week t_j. Let T' denote the number of indices in A. Approximate (2.5) by the sample average

$$\hat{U}(d = A1, m, s) = -n^+ \, c_1 + \frac{1}{T'} \sum_{j \in A} U^*(m_{t_j+1}^{(j)}, s_{t_j+1}^{(j)}),$$

using the current estimates for $U^*(m, s)$. The approximation for $U(d = A1, ..)$ is the same for all D_{1t} with the same summary statistic (m, s). Thus we write $U(d = A1, m, s)$. Compare with $U(d = A2, m, s)$ and $U(d = A0, m, s)$ to find the optimal decision $d^*(m, s)$ and its value $U^*(m, s)$. Repeating the same process for all cells (m, s) updates the currently imputed values for $d^*(m, s)$ and $U^*(m, s)$. We repeat updating until none of these values change after a cycle of updates.

5.3 *From stopping rule to terminator*

The process described in the previous two sections is computationally intensive. It is impractical to repeat this computation every week, each time with slightly different data D_{1t}. Instead, we build a single decision table such as the one in Figure 2.5 by combining estimated utilities $\hat{U}(d_t = a, m, s), a \in \{A0, A1, A2\}$, computed under a set of "typical" dose/response curves. This gives rise to a "static terminator." In the current implementation we use four dose/response curves. For each cell in the (m, s) we compute the average expected utility approximation under the four curves. For A2 we use equal weights. For A1 we use weights proportional to the number of simulated trajectories in each experiment that passed through the specific (m, s) cell.

6 Dose/Response Curve

6.1 *Probability model*

The choice of the probability model for $f(z)$ is guided by the following considerations. First, the model should allow for posterior inferences that are analytically tractable to facilitate efficient computation of expected utilities. Second, we want a flexible model which supports a priori a wide range of dose/response curves. Although a monotonic curve with asymptotes near 0 and for large doses is likely, the model should allow for non-monotonicity and other irregular features. This is particularly important since death is a possible consequence of stroke. The drug may have a different dose-effect as regards rehabilitating patients and mortality.

Based on these considerations we chose a normal dynamic linear model (NDLM). West and Harrison (1997) give a formal definition and discussion of NDLMs. Before we describe details of the model, we outline some important features. Denote by $Z_j, j = 0, \ldots, J$, the range of allowable doses, and by $\theta_j = f(Z_j, \theta), j = 0, \ldots, J$, the vector of mean responses at the allowable doses. The underlying idea is to formalize a model which locally, for z close to Z_j, fits a straight line $\theta_j + (z - Z_j)\ \delta_j$, with level θ_j and slope δ_j. This is illustrated in Figure 2.7. When moving from dose Z_{j-1} to Z_j the parameters $\alpha_j = (\theta_j, \delta_j)$ change by adding a (small) so-called evolution noise e_j and adjusting the level $\theta_j = \theta_{j-1} + \delta_{j-1}$. Let Y_{jk}, $k = 1, \ldots, \nu_j$, denote the k-th response observed at dose Z_j. Therefore, $Y_j = (Y_{jk},\ k = 1, \ldots, \nu_j)$ is the vector of responses y_i of all patients with assigned dose $z_i = Z_j$. The resulting model is

$$Y_{jk} = \theta_j + \epsilon_{jk}, \qquad j = 1, \ldots, n,\ k = 1, \ldots, \nu_j$$

and

$$\begin{pmatrix} \theta_j \\ \delta_j \end{pmatrix} = \begin{pmatrix} \theta_{j-1} + \delta_{j-1} \\ \delta_{j-1} \end{pmatrix} + e_j, \tag{2.7}$$

with independent errors $\epsilon_{jk} \sim N(0, V\sigma^2)$ and $e_j \sim N(w_j,\ W_j\sigma^2)$. The first equation describes the distribution of Y_{jk} conditional on the state parameters and is referred to as the "observation equation." The second equation formalizes the change of α_j between doses and is referred to as the "evolution equation". For a given specification of $\{V, W_j,\ j = 1, \ldots, n\}$ and prior

$$p(\alpha_0) = N(m_0, C_0), p(\sigma^{-2}) = Gamma(n_0/2, S_0/2)$$

with given hyperparameters m_0, C_0, S_0, and n_0, there exists a straightforward recursive algorithm to compute posterior distributions $p(\alpha_j|Y_0, \ldots, Y_j)$ and therefore any other desired posterior inference. It can be shown that $p(\alpha_j|Y_0, \ldots, Y_j)$ is bivariate normal $N(m_j, C_j)$ with some moments m_j, C_j. For later reference: the predictive distributions $p(Y_j|Y_0, \ldots, Y_{j-1})$ are normal distributions with moments f_j, Q_j, and the posterior distributions $p(\alpha_j| Y_0, \ldots, Y_J)$ are bivariate $N(m_j^*, C_j^*)$. West and Harrison (1997) give the recursive equations to compute $m_j, C_j, f_j, Q_j, m_j^*, C_j^*$ and other posterior moments. An algorithm known as Forward Filtering Backward Sampling (FFBS) allows efficient random variate generation from the full posterior distribution. It is described in Frühwirt-Schnatter (1994) and Carter and Kohn (1994).

Specifying the evolution variances W_j is not easy. As an alternative to specifying W_j *a priori*, West and Harrison (1997) propose to define W_j as a scalar multiple of the posterior variance C_j, i.e., $W_j = (1-r)/r\ C_j$ for some scalar $r \in (0, 1)$. This corresponds to thinking of the evolution noise e_j as discounting some of the current information on α_j as represented by the posterior variance C_j. The scalar factor r is the "discount factor". A large discount factor implies a small variance W_j. This in turn implies that a small change e_j between adjacent doses is likely and there is greater smoothing. On the other hand, a smaller r implies larger e_j and therefore allows for larger changes between adjacent doses.

6.2 *Prior specification for the NDLM*

A minor shortcoming of the NDLM in the present application is that the prior specification with the hyperparameters m_0, C_0 and r does not naturally allow us to fix arbitrary desired prior moments for $\theta_j = f(Z_j)$. Only $E(\theta_0)$ and $Var(\theta_0)$ are fixed as m_0 and C_0. Prior expectation and

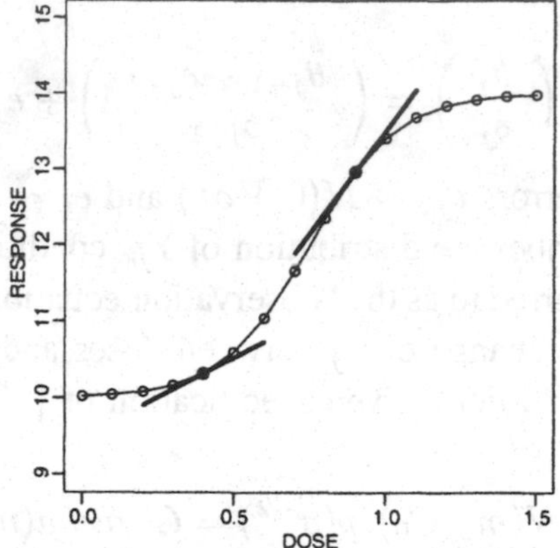

FIGURE 2.7. The NDLM fits a smooth curve to the data by defining for each dose $j, j = 0, \ldots, J$, a locally straight line, parametrized by α_j. The plot shows the local lines fit at doses $z_j = 0.4$ and $z_{jo} = 0.9$. The solid circle marks the point and a solid line segment indicates the fitted line. Between dose levels the slope of the line changes by adding an evolution noise e_j.

variance for θ_j, $j > 0$, are then implied by the evolution equation. To increase the number of prior parameters and allow for essentially arbitrary prior moments $E(\theta_j)$ and $Var(\theta_j)$ we augment the model by introducing dummy observations $\tilde{Y}_j$, $j = 1, \ldots, J$, with associated observation variance $\tilde{\sigma}_j^2$. When going through the FFBS scheme for posterior inference in the NDLM we proceed then as if $\tilde{Y}_j$ were data sampled from $\tilde{Y}_j \sim N(\theta_j, \tilde{\sigma}_j^2)$. By appropriate choice of $\tilde{Y}_j$ and $\tilde{\sigma}_j^2$ we can achieve any prior moments for θ_j, subject only to technical constraints (for example, the marginal prior variances $Var(\theta_j)$ can not be larger than those implied without the dummy observations).

7 Longitudinal Model

For many patients with missing final score y_i we have earlier scores, y_{ij} available, where $j < M$. From historical data (Copenhagen Stroke Database) we expect y_{ij} to be a good predictor of y_i, even as early as week 4 ($j = 4$). To allow formal imputation we extend the probability model with a longitudinal data model for $(y_{ij}, j = 1, \ldots, 13)$.

$$y_{ij} | y_i \sim N(m_j + a_j\, y_i, s_j^2)$$

We assume that given a patient's current status, the patient's future status does not depend further on the patient's previous history. That is, given y_{ij}, y_i and any earlier response $y'_{ij}, j' < j$, are conditionally independent. This implies

$$p(y_{ij}|y_{i,j+1},\dots,y_{iM}) = p(y_{ij}|y_{i,j+1}) = N(m'_j + a'_j y_{i,j+1}, s'^2_j)$$

with parameters m', a' and s'_j which are easily derived from m, a and s_j.

In the following discussion we will use $\theta = (\theta_1, \delta_1, \dots \theta_K, \delta_K, \sigma, m_1, a_1, s_1, \dots, m_M, a_M, s_M)$ to denote the vector of all unknown model parameters.

We use the Copenhagen Stroke Database (Jorgensen et al. 1994) to derive informative (conjugate) priors for m', a' and s'_j, allowing us to obtain conditional posterior distributions in closed form. All data, including that from patients who have not yet completed the full 13-week study period, are used in updating the longitudinal model parameters. To simplify computations, in the current implementation we start updating the longitudinal data model only when some minimum fraction of the patients have completed final observations (currently 35% of the data set).

8 Investigator/System Interface

8.1 *Introduction*

The purpose of the present section is to describe how the program described in the previous sections can be used in an actual clinical trial. The investigator/system interface incorporates the information received from the investigators, orchestrates the running of the statistics programs, transmits to the investigators the details of what dose to administer the patient, and provides a means of, monitoring the progress of the study.

8.2 *Using the System: The Investigator*

Upon admitting a patient suitable for the study, the investigator:

1. Assesses their weight and their baseline score on the Scandinavian Stroke Scale,

2. Enters this information and the patient's initials and date of birth on a single page pre-printed form,

3. Faxes the form to the data system,
4. Receives a confirmation of her fax, with the number for the patient within the study, and the details as read from the investigator's form by the system so they can be checked, and
5. A minute or two later, receives a fax with instructions on how to make up the dose to give the patient.

System Overview

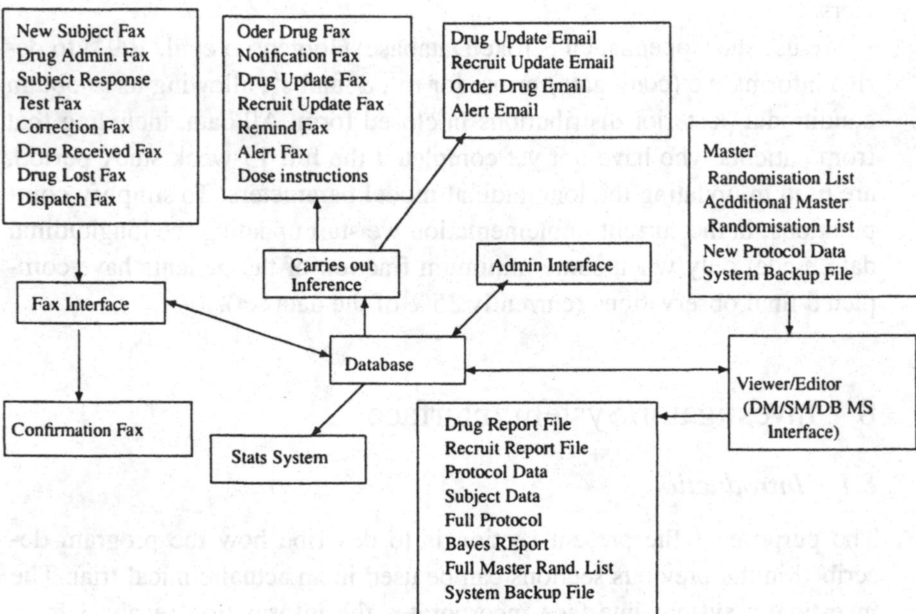

The dose assigned to a patient is specified to the investigator, or the investigator's pharmacy, by identifying two vials and instructing how much is to be taken from each and combined to make the dose. A unique number identifies each vial. Vials contain one of three concentrations of the drug or placebo.

This allows a wide range of doses to be used and a larger than usual number of different dose levels within the range is available for the dose allocation algorithm. The system is also able to scale the dose according to the patient's weight.

As well as the fax forms with their minimum of information, the investigators complete more detailed study forms and pass them to the CRAs in the usual way.

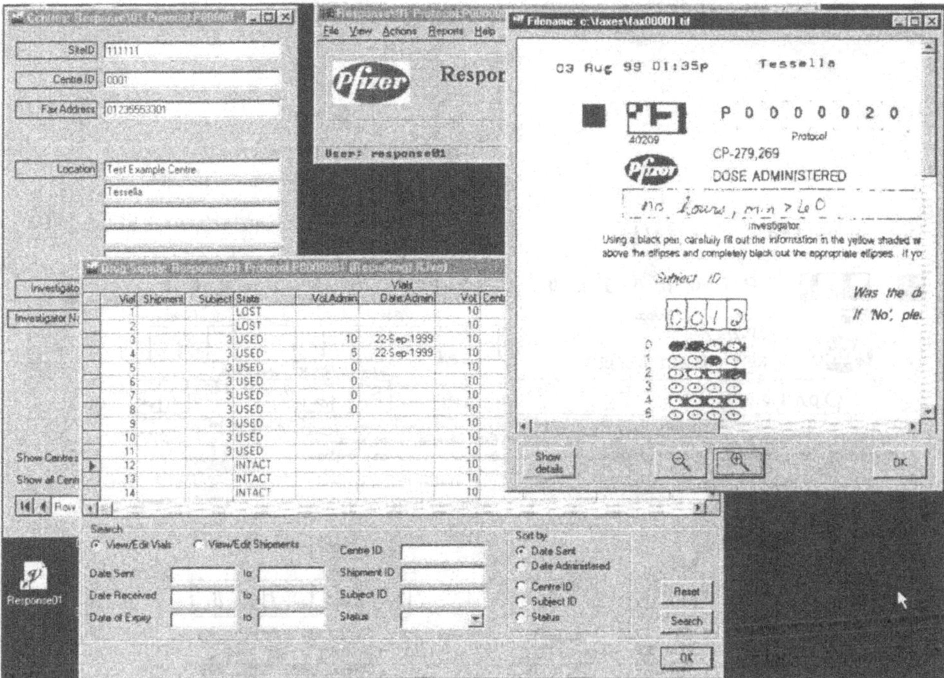

FIGURE 2.8. Screen shot of program

To enable updating the current distribution of the relevant parameters, the investigators complete a form indicating the results of follow-up visits for all patients in the study. They also fax these to the system. Should a patient develop complications the investigator can request being unblinding as regards that patient, again by completing a form and faxing it to the system. The system replies with details of the dose assigned to the patient.

Faxed forms are being used rather than an internet and Web browser-based interface because of the lack of availability of such facilities in some participating centers, and the fear that if the study sponsor provided the necessary facilities they would be abused or stolen. Faxed forms are slower and less reliable than an internet-based solution, and implementing an interface would be easier if internet access could be effected.

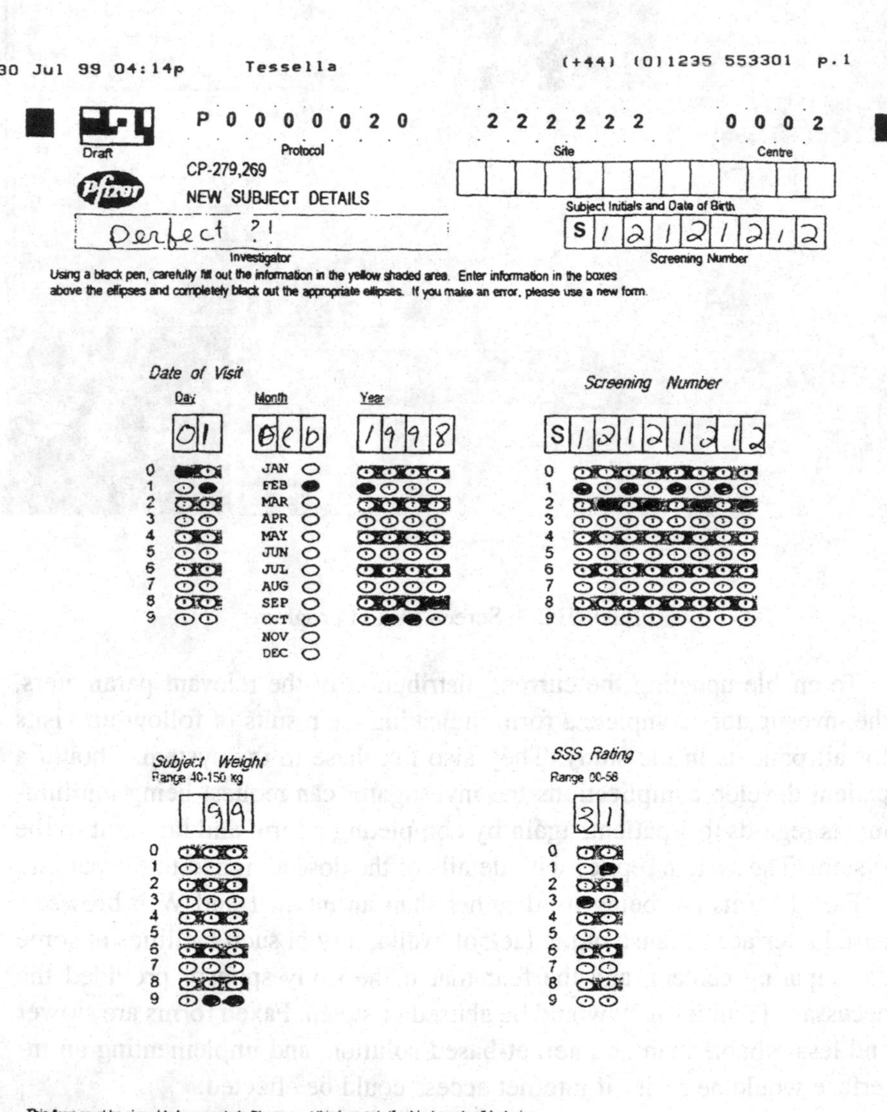
30 Jul 99 04:14p Tessella (+44) (0)1235 553301 p.1

Draft

P 0 0 0 0 0 0 2 0
Protocol

2 2 2 2 2 2
Site

0 0 0 2
Centre

CP-279,269

NEW SUBJECT DETAILS

Perfect ?!

Investigator

Subject Initials and Date of Birth

S 1 2 1 2 1 2 1 2

Screening Number

Using a black pen, carefully fill out the information in the yellow shaded area. Enter information in the boxes above the ellipses and completely black out the appropriate ellipses. If you make an error, please use a new form.

Date of Visit

Day: 01 Month: Feb Year: 1998

JAN FEB MAR APR MAY JUN JUL AUG SEP OCT NOV DEC

Screening Number

S 1 2 1 2 1 2 1 2

Subject Weight
Range 40-150 kg

99

SSS Rating
Range 00-56

31

This form must be signed to be accepted. Please send this form only if subject meets all inclusion and exclusion criteria. After you have completed the form, please send it to (+44) (0) 1235 53 84 25 and wait for a confirmation fax. If you do not receive confirmation within 5 minutes, please phone (+44) (0) 1304 64 81 67 to report the problem.

SIGNATURE

0 0 0 2
Centre

P 0 0 0 0 0 0 2 0
Protocol

2 2 2 2 2 2
Site

FIGURE 2.9. An example fax

8.3 *Using the System: The Study Administrators*

The system will hold a large amount of relevant and up-to-date information of interest to the Study Administrators (Statisticians, Data Managers, Study Manager and the Data Safety Monitoring Board). This is made available to them through a conventional PC-based program that allows the data to be navigated and viewed.

As well as the data, all the fax images can be viewed. This allows the Data Manager to check the data on the form against that read in by the system and correct it where necessary. A number of events, e.g. invalid faxes, receipt of correction forms, and requests to unblind patients, are flagged to the Data Manager to check and make corrections if necessary.

The data are available more quickly and more accessibly than for conventional trials. An additional advantage of having such a system is that it can proactively send reminders to investigators. For example, it can remind them when follow-up checks are due on patients. And it can let them know when their pharmacy is running low so they can request further packs of vials. The information on doses, effective doses and the status of the study as determined by the terminator algorithm are blinded to all but the Data Safety Monitoring Board.

8.4 *Using the system: The Pfizer pharmacy*

The pharmacy at Pfizer is responsible for providing a Master Randomization List of all the vials that may be used in the study, their number, the concentration they contain and their expiration date. Whenever the pharmacy dispatches a shipment of vials to a center they send a fax to the system indicating the vials sent. Thus, the system has the information it requires to specify doses to investigators in terms of the blinded vial numbers.

The system is informed that the vials are available at a center by the investigator faxing in a form acknowledging receipt of a shipment. The investigators also have a form they can fax should a dose administered be other than that directed and a form indicating that vials have been lost or broken.

The system sends a fax or e-mail to the Pfizer pharmacy when a center's supply of any particular concentration of drug is getting low. It also provides regular reports to the Pfizer pharmacy indicating what vials are at which center and which vials are in transit.

8.5 *How the system is constructed*

The system comprises two servers and a number of connected client PCs. There is one Fax server and one Statistics server. The Fax server runs the Data System and the Statistics server runs the Statistics Programs. The Data System comprises:

- A database built using Oracle 8,
- A fax reception package Zetafax,
- The form recognition package Teleforms with the forms designed specifically for this system and the associated Basic scripts to validate the data and enter into the database,
- The Study Administrators client program, written in Visual Basic,
- The Fax-Statistics interface programs written in C and UNIX shell script.

8.6 *The development process*

The system has been developed in two phases.

1. The first phase concentrated on the core areas of:
   (a) Receiving patient data on faxes,
   (b) Storing the data in a database,
   (c) Passing the information to the statistics programs, running the programs and storing the results,
   (d) Providing facilities to browse and edit the data.
2. The second phase extends the first phase system by:
   (a) Sending dose information to the investigator,
   (b) Managing the vial data,
   (c) Extending the Study Administration facilities,
   (d) Providing proactive reports, warnings and reminders.

For each phase:

- Detailed Use Cases and User Requirements were drawn up that detailed what the system had to achieve.

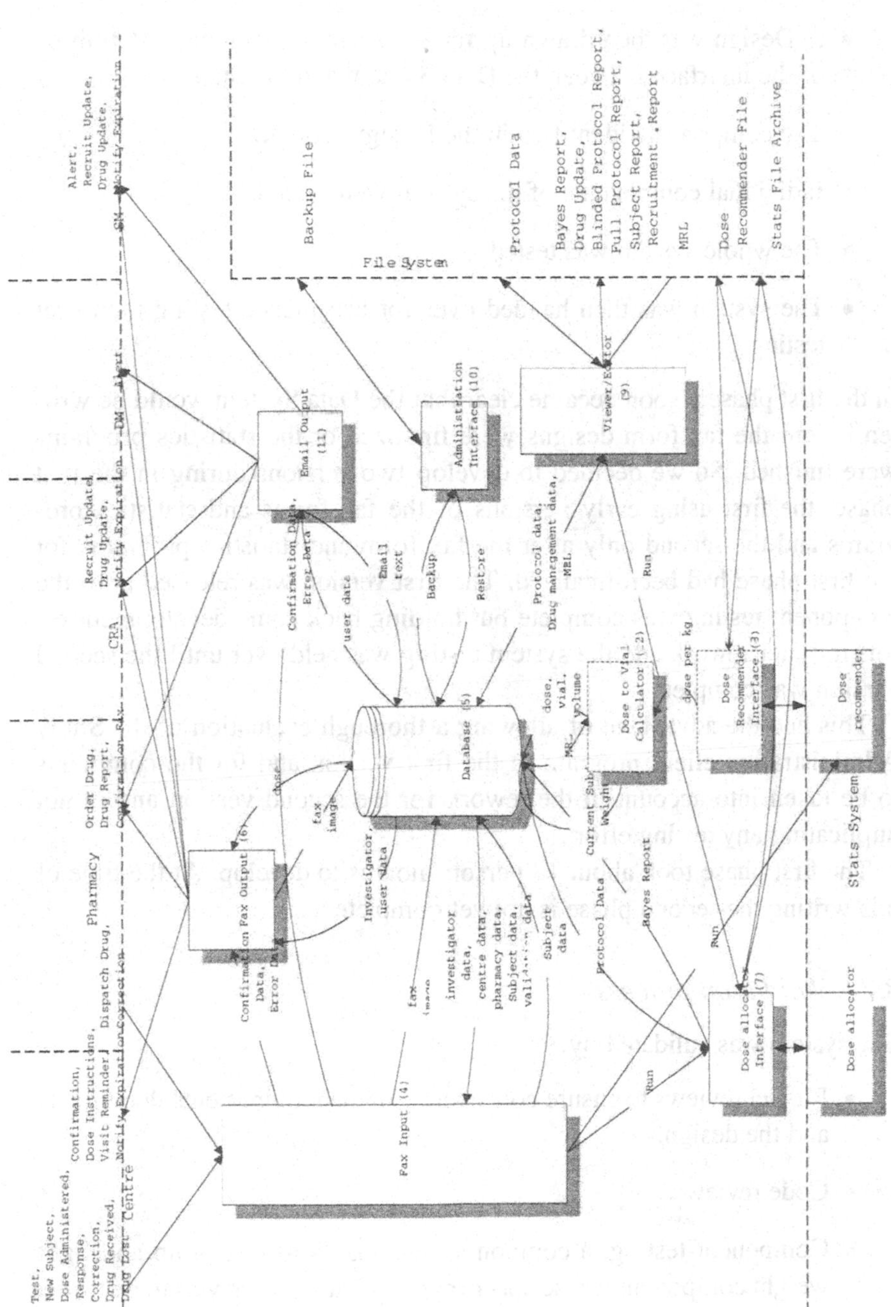

FIGURE 2.10. The System dataflow model

- These were refined into System Requirements that detailed the functionality and properties of a System.
- A Design was then drawn up for the system, including a definition of the interface between the Data System and the Statistics
- The components identified in the Design written.
- Individual components of the system were tested.
- The whole system was tested.
- The system was then handed over for acceptance testing then user testing.

In the first phase it soon became clear that the Data System would be written before the fax form designs were finalized or the statistics programs were finished. So we decided to develop two versions during in the first phase, the first using early versions of the fax forms and statistics programs and the second only after the fax form and statistics programs for the first phase had been finalized. The first version was released after the component testing was complete but holding back some development effort for later rework and the system testing was held over until the second version was complete.

This had the advantage of allowing a thorough evaluation of the Study Administration client program in the first version and for the comments to be taken into account in the rework for the second version and in not duplicating any testing effort.

The first phase took about 18 person-months to develop. At the time of this writing the second phase is not yet complete.

8.7 Validation process

The system was validated by:

- Project reviews to ensure consistency of the requirements documents and the design.
- Code reviews.
- Component testing, a component test plan was drawn up and light weight component test scripts drawn up stating what was to be tested but not how it was to be tested. The tests were then performed by the developers themselves.

- System testing, a system test plan was drawn up, enumerating the test scripts to be written and cross referring the test scripts to the system requirements. The system test scripts contained more detailed instructions on what steps to take to perform he tests and on the expected outcomes. Independent testers ran the system tests. The system testing also included some unscripted testing of the user interfaces by an independent tester, this proved particularly effective at finding faults.

9 Simulation Results

9.1 Dose range, data properties and underlying curves

The number of doses used is arbitrary, with a greater number being better from the perspective of learning about the dose-response curve. But logistical considerations limit the number in practice. We used sixteen possible doses in all simulations: $0, 0.1, 0.2, \ldots, 1.5$. These doses are realistic for the actual trial; but the set of doses to be used is not yet final. Equal spacing is natural for the NDLM but unequally spaced doses can be accommodated.

We investigate the properties of our design by assuming a variety of dose-response curves. These were chosen to reflect different scenarios that might be expected in a dose-response study (Figure 2.11).

- Curve 1: Null or flat dose-response, corresponding to no drug effect.

The shape of the next three curves is logistic with the lower 4 doses having no improvement over placebo, and the highest 5 doses forming a plateau at the maximal drug effect.

- Curve 2: Minimal clinically relevant benefit over placebo (2 points on SSS) at the highest dose.
- Curve 3: Modest benefit over placebo (4 points) at the highest dose.
- Curve 4: Very large benefit over placebo (8 points) at the highest dose.

The next two curves reach their maxima earlier (Curve 5) and later (Curve 6) than those above.

- Curve 5: Modest benefit over placebo (3.3 points) but none until the last 6 doses.

- Curve 6: Modest benefit over placebo (4 points) where the increasing is over the range of the first 6 dosessimilar to Curve 3 but with increase occurring 0.5 sooner in terms of dose.

The last curve we consider is the only one that is not monotonic.

- Curve 7: Bell-shape with increasing effect up to the middle doses and then the effect drops back to placebo level.

In generating weekly patient data we use the Copenhagen Stroke Database (Jorgensen et al. 1994). The SSS data for patients assigned to placebo are as in the CSD. A patient assigned to a positive dose responds as in the CSD, but with the improvement depending on the response at that dose for the curve assumed. For example, if Curve 3 or Curve 6 is assumed and the patient is assigned to dose 1.5 then the patient's change from baseline SSS is 4 points greater than actual.

9.2 *Process and goals*

We simulate a number of trials from a given curve. The statistical system incorporates software to enable us to generate patient data from the underlying dose-response curve, simulate the adaptive allocation of patients to doses, estimate the dose-response curve given the current patient data and make a decision regarding stopping the dose-finding phase of the trial.

We examine general properties of the algorithm, as follows.

- Doses allocated, depending on the underlying dose-response curve.
- Accuracy of estimate of dose-response curve.
- Sample size in comparison with standard designs.

We also wish to learn about the effect of different settings of the initialization files on the time taken to run the allocator and terminator functions. To use the system in an actual trial it must allocate a dose to the next patient in a matter of a few minutes.

There are a great deal of variable parameters in this approach and it is the aim of simulation to find out the effect of these parameters and identify optimal settings which allow best inferences to be made about the majority of the underlying curves.

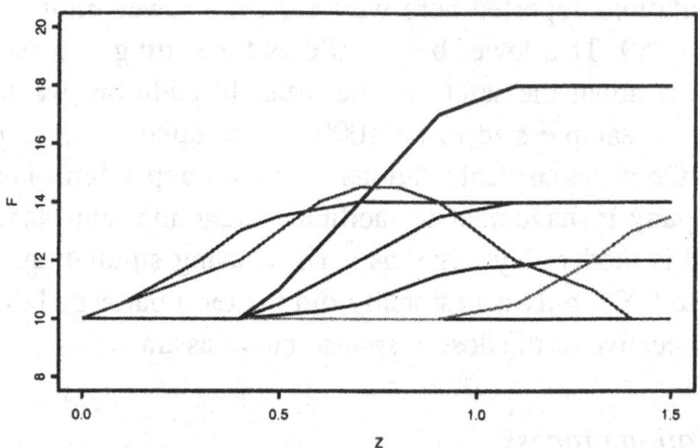

FIGURE 2.11. Nominal dose-response curves used in simulations

9.3 *Initialization file settings*

An initialization file controls the critical aspects of running the algorithm. For example, the default values for the prior estimates and longitudinal model analysis are from the Copenhagen Stroke Database. Default settings for Markov chain Monte Carlo (MCMC) iterations have been chosen to provide reasonable estimates of the parameters of interest but also to keep the calculation time within reasonable limits. Many options that can be set in the initialization file allow fallback positions should the software not function correctly. Default allocations include allocation at given posterior quantiles and random allocation at prespecified doses.

Other settings are invoked in generating data for simulations. In running a "trial" a patient database is created. If the settings indicate that the database should include deaths then this is incorporated. The patient data file is then read in sequentially (one week's worth of responses at a time) and the doses are adaptively allocated to the patients in trial. The adaptive nature of the design means the simulation must be run with patients being allocated to doses and patient responses being generated sequentially.

The simulation process allows for setting the dose range to be used, the inclusion and exclusion criteria for baseline scores, the patient horizon (maximum number of patients), the maximum number of observations per patient and properties of the distribution and scoring method for deaths in the generated patient database. The initialization file also controls the specification of the prior knowledge about the form of the dose-response curve and controls some aspects of the eventual design of the study – for example, the proportion of patients allocated to placebo.

In all simulations reported here we have set a lower limit of the sample size to be 250. This lower bound allows for getting at least a modest amount of data about the safety of the drug. In addition, we set the upper limit on the sample size to be 1000. Some upper limit is helpful in carrying out the necessary calculations. Also, an upper limit is appealing to drug company management to facilitate budgeting and planning. The limit of 1000 is moderately large, as verified using simulation. When we set the limit to 5000, only occasionally did the terminator go beyond 1000 patients, irrespective of the dose-response curve assumed.

9.4 Simulation process

In most of the simulations reported here, 100 trials were generated for each curve. When initialization settings were changed, these were made one item at a time in order to investigate the effects and to keep the number of sources of change to a minimum at any one step. For example for any one nominal curve we simulated 100 trials with no deaths and with one setting of minimum and maximum sample sizes, one setting of placebo proportion and one setting of the different NDLM parameters. Then we introduce deaths with a particular method of scoring death, holding all other factors constant. In this way we have attempted to build up a picture of the influence of different factors on our estimation of the underlying curve and on the inferences derived when we stop the trial.

To evaluate the overall and long-term performance of the algorithms and for comparison with standard designs we performed 1000 additional trial simulations on selected curves and for selected settings of the initialization file. These larger simulations are used to calculate Type I error probabilities (significant dose-response when the underlying dose-response curve is flat) and power of the procedure (proportion of trials with statistical significance for a nominal dose-response curve with the hypothesized clinically relevant difference).

The statistical package S-Plus was used to display the data from each simulation run – we examined the overall picture across the 100 trials and also observed the results for the individual trials.

Although we can evaluate the Type I error and power of our procedure there are very few metrics which will allow us to explicitly define its performance. As a result most of the evaluation from simulations has involved graphical presentation of the simulation results and comparing the dose allocation, estimation of the curve and stopping decision against what we would expect given knowledge of the underlying curve. We also compare the results of simulations against each other in order to identify any "oddities."

9.5 *Conventional dose-finding study*

We compare our results against conventional dose-finding study, one with three doses plus placebo. For a clinically relevant difference we wish to detect, for an estimate of variability, and for appropriate power and significance levels, we calculate sample sizes for each group. There are many assumptions underlying this calculation. The variability in the actual data may be greater than that assumed (taken from the Copenhagen Stroke Database). The effect of treatment may not be as great as assumed, or it may be greater. The location of the best effect may be outside the dose range assumed. And characteristics of the patient population may be different from those in other studies.

Any conventional study is saddled with such assumptions. If any are wrong then it may not be possible to draw a firm conclusion or the sample size may in retrospect be too large. Conventional studies may be able to recalculate variability and adjust sample sizes accordingly, or there may be other ways to adjust for some of the effects above while the study is ongoing, but in principle many factors are fixed at the outset of the study.

Table 2.1 shows sample sizes for a conventional study powered at 90% to detect a difference of 2 points or 4 points improvement over placebo. This assumes a 5% significance and testing using Dunett's correction for three active doses and placebo. Estimates of variance in change from baseline have been drawn from our analysis of the Copenhagen Stroke Database and reflect variability seen in an untreated patient population.

Improvement over placebo to detect	Assumed variance; change from baseline	Total sample size (in 4 groups)
2 pts.	150	3216
4 pts.	150	804

TABLE 2.1. Sample size estimates for a conventional dose-response study design.

We polled three clinicians about their choice of doses for the conventional dose-response study described above in the absence of any prior information about our underlying dose-response curves. Given those doses, the sample sizes above, an estimate of variability in the responses and the known responses at each of the doses we were able to construct true power calculations of Dunnett's test for the study. These are shown in Table 2.2.

It is quite clear from these results that the choice of doses has a large impact on the true power of the test. Clinician 1 adopted a conservative approach in his choice of doses and as a result ends up being underpowered for most of the underlying dose-response curves which we looked at. Clinicians 2 and 3 chose very similar doses and so on the whole their true powers are similar across curves. The overall picture changes for different curves. For curve 6, which changes early in the dose range, clinician 1's choice of doses picks up the critical part of the dose-response curve and so his true power is approximately 90%. In this curve clinicians 2 and 3 are overpowered. In curve 5, which changes late in the dose range, none of the clinicians have adequate power.

Improvement over placebo to detect	Assumed variance	Clinician 1 0 0.06 0.2 0.5	Clinician 2 0 0.5 0.75 1.0	Clincian 3 0 0.33 0.66 1.0
2	150	0.13	>0.99	>0.99
2	50	0.27	>0.99	>0.99
2	250	0.09	>0.99	>0.99
4	150	0.07	0.88	0.86
4	50	0.11	>0.99	>0.99
4	250	0.07	0.66	0.63

TABLE 2.2. Power calculations for the conventional dose-response study design. Curve 3 – Maximum improvement over placebo = 4pts.

For curve 3 it can be seen quite easily that if we had powered up to detect a difference of 2 points improvement over placebo then we would have been grossly overpowered under the choices of doses by clinicians 2 and 3. Interim analyses may have been performed in order to stop the study early in this trial. However, it is still true that even when the variability is substantially larger than we originally anticipated we have a total sample size which is too big given the observed underlying dose-response curve. This seems wasteful of resources and time.

In the case where we sample sized to detect a difference of 4 points improvement over placebo we would hope that our estimate of the true power is around the 90% used in the sample size calculation. Unfortunately clinician 1 misses most of the ascending part of the dose-response curve and so is underpowered. Clinicians 2 and 3 almost achieve the 90% power. In both cases though their choice of doses would mean that if there were any safety concerns about the top dose (at which we have a significant difference from placebo) then neither of their lower doses would achieve significance. This may lead to the study being repeated. We can also see that if the observed variability is different from our prior estimate then their true power is affected.

This illustrates that even in the "objective" world of frequentist sample sizing our judgements about likely effect size, estimate of variability and choice of doses can have a substantial impact on the inferences we make at the end of the study.

9.6 *Results of simulations*

Consider Curve 3, where the maximal improvement over placebo is 4 points. The ED95 for this curve is approximately 1.0. Placebo is allocated with a fixed minimum proportion, in this case 0.15. One patient accrues per day. Figure 2.12 shows the frequency distribution of doses assigned in a single simulation (left panel) and the actual doses allocated by time over the course of the trial (right panel). The trial lasted 36 weeks and 250 patients were accrued. Therefore, at the time of stopping, complete information concerning the 13 weeks of the study was available on about 2/3 of the 250 patients.

Early in the trial the algorithm seeks to identify the minimum and maximum effects for these are critical references against which the ED95 must be evaluated. In the latter phases of the study the algorithm homes in on the ED95 value, while also getting information about the minimum and maximum. The histogram shows that a small fraction of doses are allocated in the dose range from 0.1 to 0.7. The nominal dose-response curve in this range is nearly flat and these doses are away from the area of principal interest, the region containing the ED95.

Figure 2.13 shows the distribution of doses across 100 simulations. Superimposed on this is a summary of the fitted dose-response curves across these simulations. The diamonds represent the true (assumed) dose-response curve. The middle line shows the median of the fitted values with the quartiles of the distribution of the fitted values at each dose. This figure shows

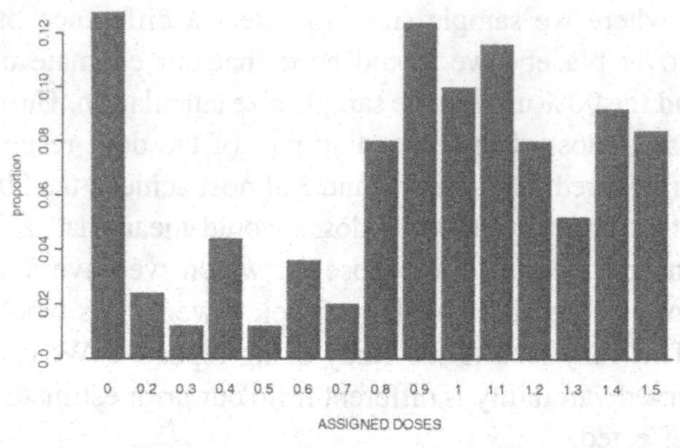

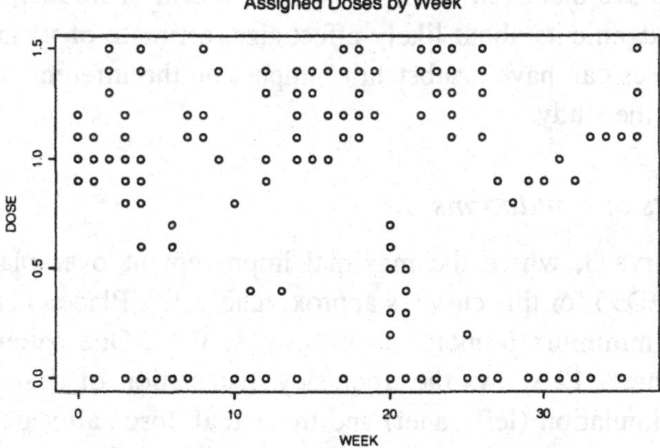

FIGURE 2.12. Results of a single simulation assuming Curve 3. Distribution of doses (top panel) and doses allocated over time (bottom panel).

that the low doses have a low probability of being allocated. It also shows that pre-specified minimal proportion of placebo doses of 15% is achieved. The fitted curves accurately estimate the underlying dose-response curve.

Figure 2.14 (left panel) shows the distribution of observed effect sizes (m) at the ED95 – measured as the number of points improvement over placebo on the Scandinavian Stroke Scale – against the standard error (s) of this estimate. The triangles represent those simulations in which the decision was made to go into a confirmatory phase and the circles correspond to simulations in which the decision is made to not go confirmatory. The sizes of the symbols are proportional to the trial sample sizes. Recall that the terminator function is called only after 250 patients have accrued and that the maximum study size is 1000. The figure shows that when the trial is stopped early it is because the effect is probably large. Naturally, for smaller sample sizes there is a bias in the observed response at the ED95.

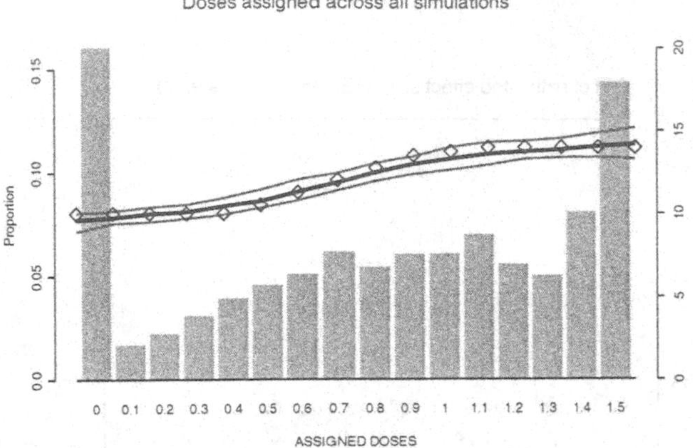

FIGURE 2.13. Allocation of doses across 100 simulations, assuming Curve 3 for the dose-response relationship.

For Curve 3, with its maximum of 4 points improvement over placebo the algorithm makes very few decisions to terminate the study because of lack of an observed benefit. In Figure 2.14 we see a "funnel" shape. This may be because in order to identify a small effect we will need to continue the trial to a larger number of patients and so we will end up with better precision in this estimate (smaller "s").

Figure 2.14 (bottom panel) shows distributions of the sample size taken to reach a decision using the stopping rule, the observed effect size, the standard error of the effect size and the posterior estimate of sigma. The bias in observed effect size at the ED95 can be seen from the distribution here, as can the fact that in the majority of cases the decision to stop has been taken comparatively early. The fact that some simulation runs have continued on to 1000 patients may be influencing the standard error of the estimated effect size causing the bimodality we see in this distribution – the longer we continue the more certain we can be about the estimate of the effect size.

Figure 2.15 shows similar results for 200 simulations assuming the null or flat dose-response curve (Curve 1). The proportion of simulations in which the decision is to go into a confirmatory study is made is very low (2.2%). In the terminator function if the observed effect is less than 2 points improvement over placebo then we conclude that the drug has insufficient clinical effect to warrant further development. The bottom panel of Figure 2.15 (upper left histogram) shows that the decision to stop was made after relatively few patients (median of 258 where the minimum is 250). As

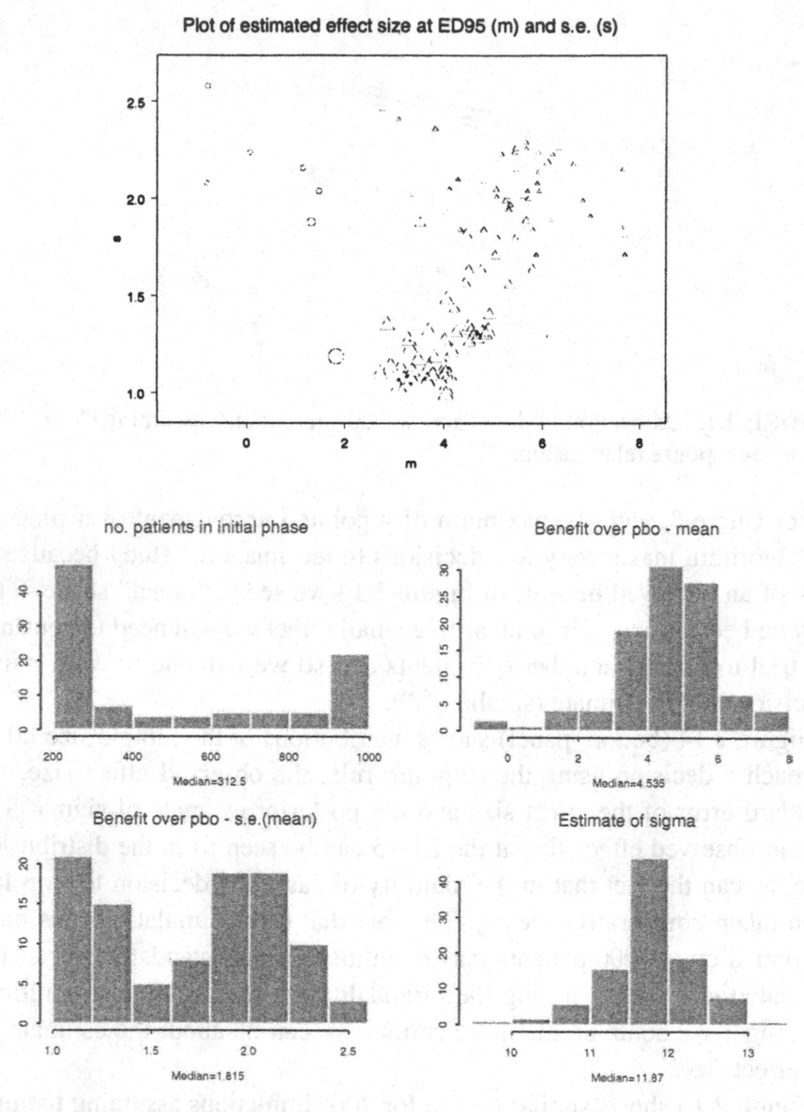

FIGURE 2.14. Distribution of observed effect sizes at the ED95 with corresponding precision. Dose-response is that of Curve 3. In the top panel symbol size is proportional to trial sample size at stopping. The bottom four panels show frequency distributions of sample size, posterior mean benefit over placebo at the ED95, its standard error, and the error standard deviation σ.

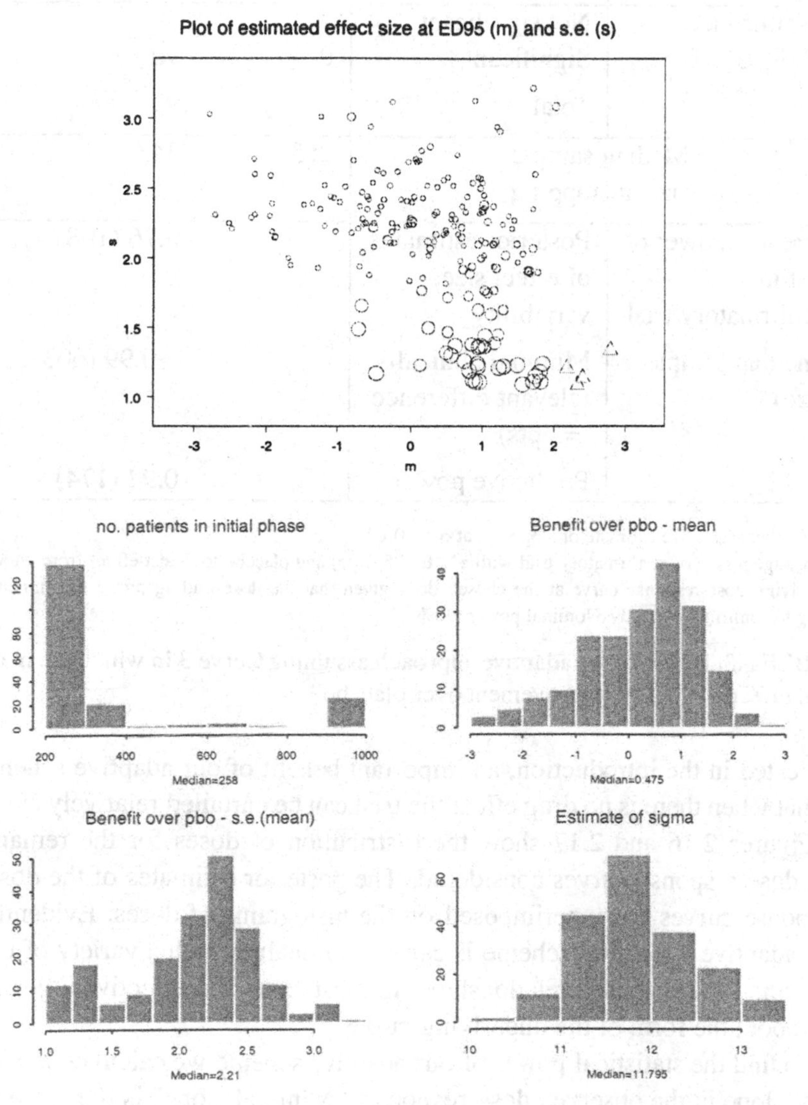

FIGURE 2.15. Distribution of observed effect sizes at ED95 with corresponding precision. The underlying dose-response is null, that of Curve 1. The symbol size in the top panel is proportional to sample size in the dose-finding phase.

		Terminator decision	
		Stop (no effect)	Continue to confirmatory trial
Significance of slope*	Not significant	7	3
	Significant	0	90
	Total	7	93
Median sample size at stopping		255	343
Average power of test in confirmatory trial (median sample size)$^{+}$	Posterior estimates of effect size, variability		0.76 (108)
	Minimal clinically relevant difference (=2 pts)		>0.99 (603)
	Predictive power		0.92 (174)

* Test of conservative estimate of slope = 0 at $\alpha = 0.05$.

+ Average power of confirmatory trial with z^* (ED95 dose) and placebo to detect effect from known underlying dose-response curve at the chosen dose given that the dose-finding phase recommends going to confirmatory study. Nominal power=90%.

TABLE 2.3. Power of the adaptive approach assuming Curve 3 in which the maximal effect is 4 points improvement over placebo.

indicated in the introduction, an important benefit of our adaptive scheme is that when there is no drug effect the trial can be curtailed relatively early.

Figures 2.16 and 2.17 show the distribution of doses for the remaining dose-response curves considered. The posterior estimates of the dose-response curves are superimposed on the histograms of doses. Evidently, our adaptive allocation scheme is capable of dealing with a variety of underlying dose-response relationships. In most cases it is effective in learning about the form of the underlying curve.

To find the statistical power of our adaptive scheme we calculate a minimal slope in the observed dose-response. "Minimal slope" is the slope of a line spanning the whole dose range from placebo to the size of effect at the dose that will be selected to take into the confirmatory trial. This is minimal in the sense that if the maximal effect occurs anywhere other than at the maximum dose then the slope of the line will be greater. The slope is calculated only for those simulations that call for going forward into a confirmatory study. We can test this minimal slope for significance.

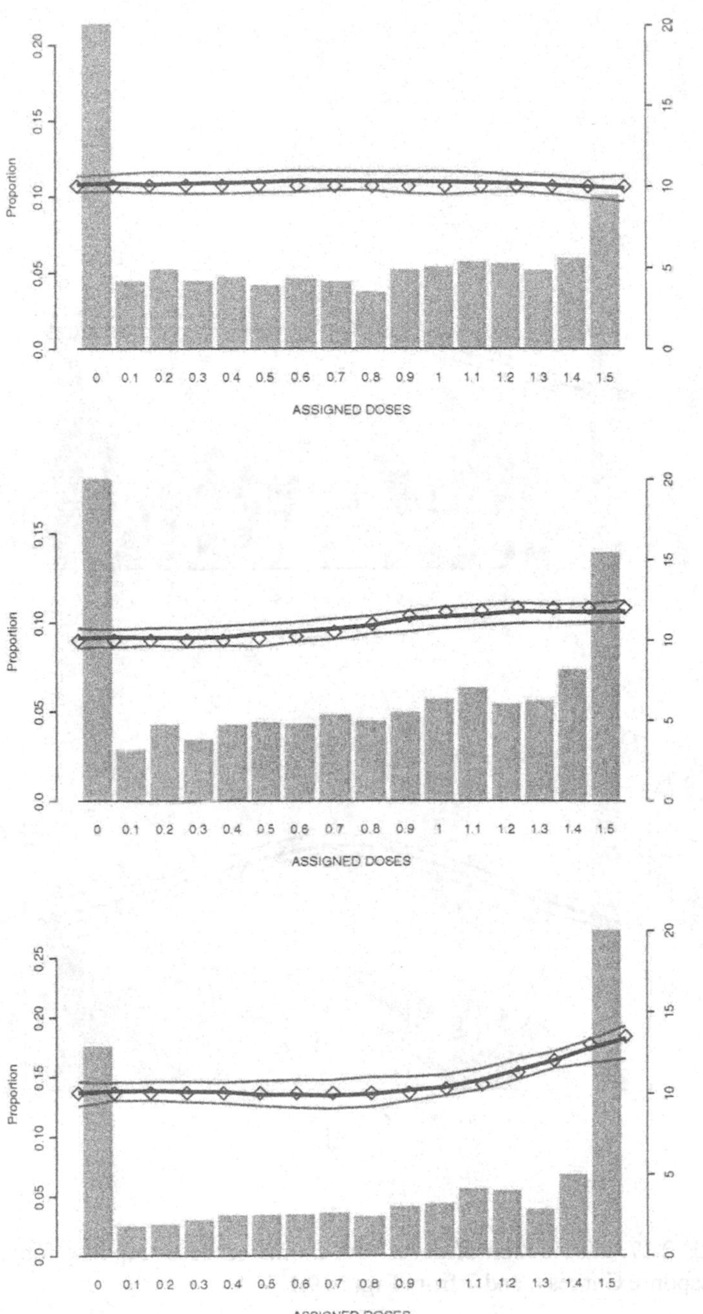

FIGURE 2.16. Distribution of doses and estimated dose-response functions for dose-response Curves 1, 2, and 5 from Figure 9.1.

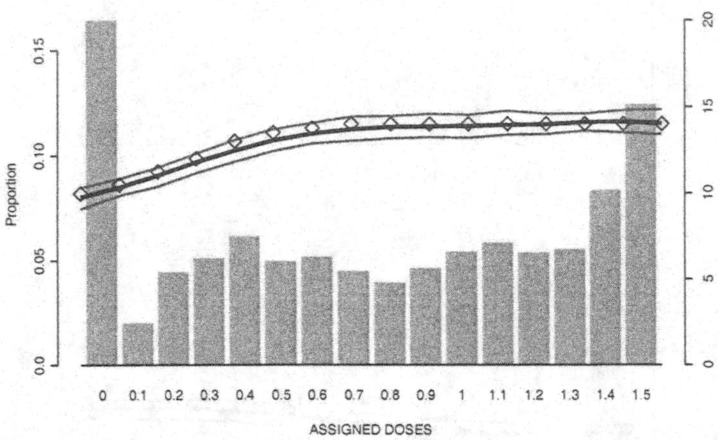

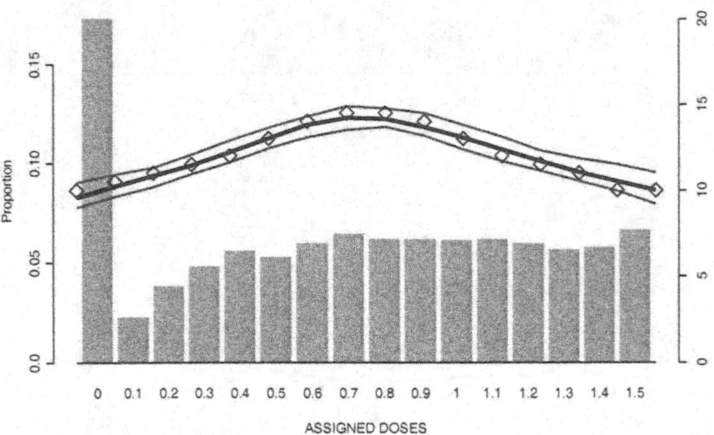

FIGURE 2.17. Distribution of doses and estimated dose-response functions for dose-response Curves 6 and 7 from Figure 9.1.

Table 2.3 gives a summary of the results from 100 simulations assuming Curve 3. In only 7 of 100 trials was the decision against going to confirmatory phase. As regards statistical significance of the minimal slope, only 3 out of the 93 simulated trials that go forward into the confirmatory phase did not show a significant slope.

The dose chosen from the adaptive scheme will be taken forward into a confirmatory trial. Since we know the true response at that dose we can easily calculate the power of the confirmatory study to show a significant difference from placebo. We can select a sample size for this confirmatory trial in a number of ways: using the effect size and estimate of variability from the dose-finding study, using a minimal clinically relevant difference along with the estimate of variability, and using predictive power to account for uncertainty in the estimate of effect from the dose-finding study. Estimates of the average power of the test for the confirmatory study using these three methods of sample sizing is also shown in Table 2.3.

The power to show a difference from placebo in the confirmatory trial shows that if we use the minimal clinically relevant difference of 2 points improvement over placebo then we will overpower the study since the true effect is approximately 4 points improvement. Using the posterior estimate of the effect size at the ED95 leads to less than the nominal 90% power since we are not allowing for uncertainty in our estimate of effect size in the calculation of sample size. This is incorporated using the standard error around the estimate of effect size (as shown previously) and using predictive power methods. We can see that using this approach our average power is much closer to the nominal 90% power. Of course in practice all of these sample size estimates may be overridden by the need to have a substantial amount of safety data for regulatory submission.

		Terminator decision	
		Stop (no effect)	Continue to confirmatory trial
Significance of slope*	Not significant	196 (98%)	0
	Significant	0	4 (2%)
	Total	196 (98%)	4 (2%)
Median sample size at stopping		257	1000

* Test of conservative estimate of slope = 0 at $\alpha = 0.05$.

TABLE 2.4. Type I error of the adaptive approach compared to the conventional design. Curve 1–no effect.

Table 2.4 shows the same summary for the null curve (Curve 1) in which there is no drug effect. This analysis provides the Type I error of this approach: any significant difference is a false-positive. In only 2% of the 200 simulated trials do we make the (incorrect) decision to go forward into a confirmatory study.

		Terminator decision	
		Stop (no effect)	Continue to confirmatory trial
Significance of slope*	Not significant	8	92
	Significant	0	0
	Total	8	92
Median n		1000	257
Power of test in confirmatory trial$^+$ (median n)+	Posterior estimates; effect size, variance		0.84 (42)
	Min. clin. relevant difference (=2 pts)		>0.99 (213)
	Predictive power		0.93 (60)
Significance of slope *	Not significant	14	3
	Significant	0	83
	Total	14	86
Median n		277	413
Power of test in confirmatory trial$^+$ (median n)+	Posterior estimates; effect size, variance		0.67 (150)
	Min. clin. relevant difference (=2 pts)		>0.99 (847)
	Predictive power		0.84 (238)

* Test of conservative estimate of slope = 0 at $\alpha = 0.05$

\+ Average power of confirmatory trial with z^* (ED95 dose) and placebo to detect effect from known underlying dose-response curve at the chosen dose given that the dose-finding phase recommends going to confirmatory study. Nominal power=90%.

TABLE 2.5. Power of the adaptive approach vs. conventional design. Curve 3 – maximal effect = 4 points improvement over placebo, different observed variances.

Table 2.5 shows a summary similar to that shown above for Curve 3 but for different underlying variance. A major benefit of the adaptive design is that it adapts to the variability in the data. As a result, the confirmatory study will be powered appropriately based on the best estimate of variability from the data observed in the dose-finding phase. If variability in the data is substantially less than we would expect a priori then our inferences can also be made much sooner and with fewer patients.

However, Table 2.5 shows that when the variance is smaller, statistical power is not greatly increased. Rather, upon learning that the variance is small the algorithm stops the trial more quickly. The median sample size required to reach the decision to go forward into the confirmatory trial is smaller, and in fact is close to the minimum possible sample size of 250.

These power calculations show that again the adaptive design achieves close to the nominal 90% power for the confirmatory trial for either assumed variance. The predictive power approach takes into consideration the uncertainty in the estimate of the treatment difference and so achieves better power overall, albeit with larger average sample size. In the adaptive design the impact on our study of different observed variability is in the sample size taken to make a decision and not on the power of the test – which is shown in the calculations of true power shown for the conventional study.

From these simulations we conclude that our sample sizing and final inference across both trials are appropriate. The algorithm does what it is supposed to do: adapt to the underlying dose-response curve and to the variability in accumulating data. The dose-finding trial allows for stopping the drug's development early – either because it has no effect, or because the evidence is sufficiently clear to conclude that the treatment is effective. The simulations also show that when using the results of the dose-ranging phase to sample size a confirmatory trial, the latter will very likely confirm that the dose effect seen in the earlier trial is significant.

The simulations summarized above are (loosely) based on information from the Copenhagen Stroke Database but ignoring deaths. There are a number of issues we face in incorporating deaths into the generated data set, and we must deal with these as well when handling deaths in our analysis of the primary endpoint, change in SSS. Death must be included in some way.

To include mortality in the generated data we specify the distribution of deaths for different baseline scores – patients with mild initial stroke are less likely to die. We must also address the relationship between dose and death. The former we may evaluate from our analysis of the Copenhagen Stroke Database assuming that dose levels do not interact with baseline severity. The latter we must hypothesize at present because we have no information about whether and how the drug may affect the incidence of death. It may be that the effect of dose on death is not monotone, especially when averaging over severity of stroke. Perhaps low doses have no effect on mortality, medium doses offer protection against death and higher doses are linked to a mechanism that hastens death. This would give us a U-shaped incidence of deaths across the dose range.

We want to incorporate death into the SSS endpoint, down-weighting any benefit of the treatment at doses should there be an increase in mortality rate. This will allow for addressing the role of mortality in the dose-response relationship. An alternative would be to evaluate the dose-response among survivors only and analyze deaths separately. But the two analyses – SSS improvement by dose and death by dose – must be combined eventually in some fashion. We feel it best to incorporate mortality at the earliest opportunity, but of course this does not prohibit considering the two measures separately after the trial is complete.

It is not clear how to best handle deaths as part of the principal endpoint and we have not yet settled on a final strategy. We consider two approaches here. In terms of neurological function one obvious solution would be to set SSS = 0 should the patient die. This would give a negative change from baseline, which seems appropriate. A patient who enters the trial with an SSS of 20 and dies has a change from baseline of -20. However, a patient who arrives with SSS = 40 and at the end of the study has a score of 10 has a worse change from baseline (-30) and therefore a greater negative impact on the effect at the dose assigned. Perhaps this is appropriate and perhaps it is not.

An alternative approach is to regard the score at death as the minimum SSS over the period of the trial, including baseline. A patient who improves but then dies nets a 0. A patient who declines by 10 points and then dies scores -10. But it might make more sense to count them the same. Indeed, the first patient may have passed through a change of -10 while dying but happened not to be tested at that time. Again, survivors may have lower scores than those who do not survive.

The plots of the data (Figure 2.18, top two panels) show the distribution of deaths across doses. The top panel shows that scoring death as zero separates the distribution of scores for patients who die from that those who survive. The second panel (center) shows slightly more overlap by setting the score for death to the minimum SSS over time.

The assumed dose-response curve in this example is the flat dose-response curve (null Curve 1) and so the effect of including deaths in the data can easily be seen. For this reason we have also investigated dose-response

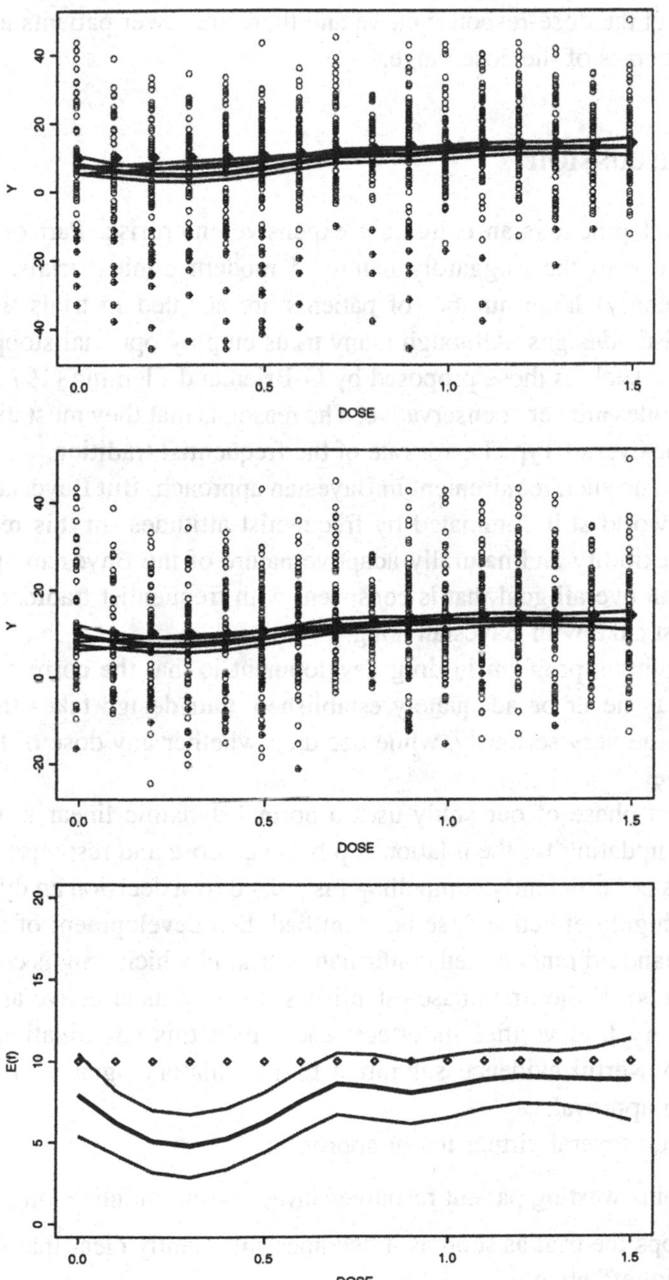

FIGURE 2.18. Example of generated data including deaths. Upper panel: death scored as zero. Center panel: death scored as minimum score observed in the trial. Deaths are indicated as circled pluses. Bottom panel: estimated dose-response curve for Curve 1 (flat dose-response curve) but including deaths.

curves that are not monotonic (Curve 7). Figure 2.16 shows results of our simulations (without deaths) in this curve. Doses assigned have adapted to the form of the dose-response curve and there are fewer patients allocated in the extremes of the dose range.

10 Discussion

Drug development is an extremely expensive enterprise. Part of the expense is due to the sluggardly nature of modern clinical trials. A fixed (and generally) large number of patients are accrued in trials that have deterministic designs. Although many trials employ optional stopping via boundaries such as those proposed by O'Brien and Fleming (1979), these stopping rules are very conservative. The reason is that they must diligently protect the overall Type I error rate of the frequentist tradition.

There is no such requirement in Bayesian approach. But Bayesians must live in a world still dominated by frequentist attitudes. In this report we use the flexibility and naturally adaptive nature of the Bayesian approach but with an overall goal that is consistent with frequentist tradition. It is a frequentist cake with Bayesian icing.

A ubiquitous problem in drug development is that the correct dose of a drug may never be adequately established. Our design takes the dose-finding issue very seriously, while deciding whether any dose of the drug is effective.

The first phase of our study uses a normal dynamic linear model and Bayesian updating for the relationship between dose and response. Should the results be sufficiently compelling (as judged by a decision analysis) and should a highly effective dose be identified then development of the drug enters a standard randomized confirmatory trial in which drug is compared with placebo. If the first phase establishes the drug as effective and if the confirmatory trial verifies its effectiveness then this combination would be very powerful evidence submitted to a regulatory agency in seeking marketing approval.

There are several virtues to our approach:

(i) It avoids wasting patient resources investigating uninteresting doses.

(ii) It stops the trial as soon as it becomes sufficiently clear that the drug is not effective.

(iii) It moves into confirmatory phase when the evidence suggests that such is appropriate.

(iv) It adapts to the actual variability in the accumulating data, perhaps stopping earlier than would be predicted (small variance) and perhaps continuing longer than would be predicted (large variance).

(v) It uses a longitudinal model for the course of stroke in predicting final results for those patients having incomplete data, and learns about this model on the basis of the patients accrued to the trial.

The only disadvantage in our approach as opposed to conventional designs is that it requires a reliable and rapid system for communicating with participating centers. This system must be able to update the statistical software based on recent information and it must be able to relay the doses assigned in very short order. Such a system is expensive, but its cost is tiny in comparison with the potential savings that accrue from using an efficient design.

References

Berger, J.O. (1985). *Statistical Decision Theory and Bayesian Analysis*, New York: Springer-Verlag.

Berry, D.A. (1993). A case for Bayesianism in clinical trials (with discussion), *Statistics in Medicine*, 12, 1377-1404.

Berry, D.A. and Stangl, D.K. (1996). Bayesian methods in health-related research. In *Bayesian Biostatistics*, (D.A. Berry and D.K. Stangl, eds.), New-York: Dekker, 3-66.

Carter, C. K., and Kohn, R. (1994). On Gibbs sampling for state space models, *Biometrika*, 81, 541-553.

Chaloner, K. and Verdinelli, I. (1995). Bayesian experimental design: a review, *Statistical Science*, 10, 273-304.

DeGroot, M. (1970). *Optimal Statistical Decisions*. New York: McGraw Hill.

DeHaan, R., Horn, J. et al.(1993). A comparison of five stroke scales with measures of disability, handicap, and quality of life. *Stroke*, 24, 1178-1181.

Dirnagl, U., Iadecola, C., and Moskowitz, M.A. (1999). Pathobiology of ischaemic stroke: an integrated view. *Trends Neurosci.*, 22, 391-397

Frühwirt-Schnatter, S. (1994). Data augmentation and dynamic linear models, *Journal of Time Series Analysis*, 15, 183-202.

Gilks, W.R., Richardson, S., and Spiegelhalter, D.J. (eds.) (1996). *Markov Chain Monte Carlo in Practice*, Chapman and Hall.

Hill, M.D. (1998). Stroke treatment: Time is brain, *Lancet*, 352(SIII), 10-14.

Jorgensen, H.S., Nakayama, H., Raaschou, H.O., Vive-Larsen, J., Stoier, M. and Olsen, T.S. (1994). Outcome and time course of recovery in stroke. Part I: Outcome. The Copenhagen Stroke Study, *Archives Physical and Medical Rehabilitation*, 76, 399-405.

Lees, K.R. (1998). Does neuroprotection improve stroke outcome? *Lancet*, 351, 1447-1448

Lindenstrom, E., Boysen, G, Christiansen, L.W., Rogvi Hansen, B. and Nielsen, B.W. (1991). Reliability of Scandinavian Stroke Scale. *Cerebrovascular Disease*, 1, 103-107.

O'Brien, P.C., Fleming, T.R. (1979). A multiple testing procedure for clinical trials, *Biometrics*, 35, 549-556.

Scandinavian Stroke Study Group (1985). Multicenter trial of hemodilution in ischemic stroke. Background and study protocol. *Stroke*, 16, 885-890.

Thall, P.F., Simon, R.M., and Estey, E.H. (1995). Bayesian sequential monitoring designs for single-arm clinical trials with multiple outcomes, *Statistics in Medicine*, 14, 357-379.

Thall, P.F., and Russell, K.E. (1999). A strategy for dose-finding and safety monitoring based on efficacy and adverse outcomes in phase I/II clinical trials. to appear in *Biometrics*.

Tierney, L., (1994). Markov chains for exploring posterior distributions, *Annals of Statistics*, 22, 1701-1762.

Verdinelli, I. (1992). Advances in Bayesian experimental design. In *Bayesian Statistics IV*, (J. M. Bernardo, J. O. Berger, A. P. Dawid and A. F. M. Smith, eds), pp. 467-48, Oxford: Oxford University Press.

Warlow, C.P., Dennis M.S. et al. (1996). *Stroke: A practical guide to management*. Oxford, Blackwell Scientific.

West, M. and Harrison, P.J. (1997). *Bayesian Forecasting and Dynamic Models*. Second Edition. New York: Springer-Verlag.

Whitehead, J. and Brunier, H. (1995). Bayesian decision procedures for dose determining experiments, *Statistics in Medicine*, 14, 885-893.

Whitehead, J., and Williamson, D. (1998). Bayesian decision procedures based on logistic regression models for dose-finding studies, *Journal of Biopharmaceutical Statistics*, 8, 445-467.

Discussion

Kathryn Chaloner, University of Minnesota

1 Introduction

The trial is a combination of a trial to seek the best dose (a dose seeking phase) and a second trial comparing a selected dose to placebo (a confirmatory phase). The dose response relationship is assumed to be a quantitative, normally distributed, response with independent errors between patients. Of particular interest is the dose at which the expected response is 95% of the maximum possible response.

It is impossible to evaluate a trial design without knowing something of the condition being studied, and I will therefore summarize a very brief (and superficial) literature review of recent stroke trials. There are many novel aspects of this design, and many issues that could be discussed, and I will choose just a few.

2 Stroke Trials

There are two common approaches to treating stroke with drugs: first thrombolytic agents and, second, more recently, neuroprotective agents (like the one in this trial). There is also a role for brain imaging in determining the type of stroke and, in some cases, surgery is indicated.

Most of the trial literature is on thrombolytic agents and there are two very recent and very large trials of thrombolytic treatment for stroke. The International Stroke Trial (IST, IST Collaborative Group, 1997) enrolled 19,435 patients with stroke and was a factorial design of "aspirin" against "avoid aspirin" and "heparin" against "avoid heparin." The Chinese Acute Stroke Trial (CAST, CAST Collaborative Group, 1997) enrolled 21,106 patients and the design was aspirin against placebo. In addition there are several, recently published, large trials of other thrombolytic therapy. Some of these are: the European Cooperative Acute Stroke Study (ECASS, 620 patients, Hacke et al, 1995) and ECASS II (800 patients, Hacke et al, 1998), the National Institute of Neurological Disorders and Stroke Study Group Trial (NINDS, 624 patients, NINDS rt-PA Stroke Study Group, 1995), the Trial of Low-molecular Weight Heparin (312 patients, Kay et al 1995), the Trial of ORG 10172 in Acute Stroke Treatment (TOAST, 1281 patients, Publications Committee for TOAST Investigators, 1998),

Overall the results indicate that the benefit of the drugs studied, if any, is small. Results of the two largest trials (IST and CAST), as a meta-analysis, indicate that there is a very small survival advantage to using aspirin and also a small advantage in functional outcome. Beneficial results for other therapies have been observed in a few small subgroups, depending on the type and severity of stroke and also on the timing of the intervention. Thrombolytic therapy is, in general, associated with serious hemorrhaging as a side effect and, at best, the benefit on survival and on functional outcome is small. Several trials of streptokinase as a treatment for stroke were terminated early due to serious toxicity.

Trials of neuroprotective agents are reviewed in Hill and Hachinski (1998) and Lees (1998) and a number of trials are currently underway. Development of some neuroprotective agents has been abandoned because of toxicity and increased mortality on active drug compared to placebo. No neuroprotective agents are documented as clearly efficacious although some appear promising.

After my review of the literature on controlled stroke trials I tentatively concluded that in most trials the safest arm to be assigned was placebo, and that most trials have not demonstrated that the active arm is superior. I am therefore a little cautious, and a little skeptical, of this proposed trial for a new neuroprotective agent. The proposal assigns only 15% of patients to placebo and the statistical models used to motivate the method (in Figure 2.1) and the models used to simulate from (in Figure 2.11) do not allow for placebo being better than active. I find both these aspects questionable and disturbing.

3 Safety and Toxicity

3.1 *Placebo May be Better than Active*

The models used for motivation in Figure 2.1 and the models used for simulation in Figure 2.11, assume that the active drug is no worse than the placebo: that is it is assumed that $df(z,\theta) = f(z,\theta) - f(0,\theta)$, with $f(z,\theta)$ as in (1), is always non-negative. Given the history of many drugs being abandoned in development because they lead to increased mortality over placebo, I find this disconcerting. Perhaps this new drug is safe, I hope that it is, but I am a little skeptical. In my experience it is very natural for people working on developing a drug for a serious condition to be optimistic about both safety and efficacy: but other possibilities must be considered. My prior opinion, based on my literature review, is that the drug might be unsafe and I would like to see some data to convince me otherwise.

How does the algorithm work if the best response is on placebo? If placebo is the ED100 then the ED95 is worse than placebo.

3.2 *Safety*

There must be an underlying safety issue with thc drug in the proposed trial or else there would be no reason to choose the ED95 as the dose of interest. If safety werc not a concern why not choose the ED99? Or the ED100 and maximize the response? Presumably the answer is in safety and toxicity: too much drug is toxic. The authors state in Section 9.6 that "we have no information about whether and how the drug may affect the incidence of death".

The paper does refer to safety, for example in the review by the Data Safety and Monitoring Board (DSMB), and in choosing the dose for the confirmatory phase. None of the methodology in the paper, however, for dose allocation or for monitoring, explicitly considers toxicity. This raises the question: why not?

3.3 *Univariate or Multivariate?*

The univariate endpoint in this trial measures efficacy alone. All decisions (the next dose, whether to continue) are framed in terms of the single univariate measure of change in stroke scale from baseline. Scientifically, however, this is not the whole picture. In addition to learning about efficacy we also want to learn about safety and toxicity. The drug is no good if the patient is cured of all stroke symptoms but subsequently dies of toxicity or suffers some other permanent impairment. It is unclear in this particular example how much safety is a concern but previous trials with neuroprotective agents have raised serious problems (for example sedation and hypertension, see Lees, 1998). Thrombolytic agents also have safety concerns: in the TOAST trial for example serious intracranial bleeding occured in 14 patients given active drug and 4 patients given placebo (p-value 0.05). The serious consequences of toxicity warrant consideration. Although such toxicities are not necessarily expected with this new drug, it is impossible to be sure, based on the limited testing in drug development, exactly what the safety concerns will be.

Safety issues are best examined in a placebo controlled trial and an advantage of having a sample size fixed ahead of experimentation is that the power for detecting infrequent, but very serious, complications can be easily calculated. Consider mortality as an example. The percentage of patients who will die on placebo will depend on the eligibility requirements which are, as yet, unspecified. Suppose for illustration, however, that 2.5% of patients will die on placebo (assume this trial has broad eligibility and includes patients with mild strokes). Suppose further that the probability is doubled on active to 5%, and consider the probability of detecting this with a test of level 5% in a fixed sample size trial. Suppose also that patients on any dose of active drug can be pooled, and note that the proposed design allocates only 15% of patients to placebo. If the proposed lower limit of a sample of size 250 is used as a fixed sample size (38 on placebo and 212 on active) the probability of detecting a mortality difference of 2.5% on placebo and 5% on active is approximately 0.12. Even with the maximum sample size of 1000 (150 on placebo and 850 on active) the probability is only about 0.33. Is this enough?

The paper discusses the issue, as yet unresolved, of how to incorporate death into the stroke scale. Figure 2.18 shows that incorporating mortality into the stroke scale can make placebo better than some doses in some of the models used for simulation. In any decision, however, about using the drug, there is potentially a trade off between efficacy on the stroke scale and safety. In thrombolytic therapy this is clearly recognized: for example Hill and Hachinski (1998) quote Riggs (1996) who recommends asking patients and their families: "Would you risk a 12% chance of improved neurological outcome at 3 months following ischaemic stroke against a 3% chance of early death from rt-PA-induced intracerebral haemorrhage?" Both outcomes, change in stroke scale and mortality, are important separately: the outcome is bivariate.

Most phase II trials look at both safety and efficacy, usually as separate objectives. The methodology is still new, but there has recently been some interesting work in design for phase II trials incorporating both efficacy and toxicity into the model as a bivariate response. See for example Conaway and Petroni (1995, 1996), Bryant and Day (1995) and Heise and Myers (1996). For a bivariate Bernoulli response of efficacy (yes/no) and toxicity (yes/no) it may be possible to specify a region where the trade off is acceptable between the probability of response and the probability of toxicity (see for example Figure 4 of Conaway and Petroni, 1996). The larger the probability of efficacy the larger probability of toxicity may be tolerated. Thall and Russell (1999) take a different approach and use a model with a trinomial dose-response: no reaction/cure/adverse reaction. See also Fan (1999).

Perhaps some of these ideas could be useful here? Especially for the dose seeking phase? Although methodologically challenging this would seem to be preferable to the proposed approach where only efficacy is used for dose allocation and design decisions. In the proposed approach toxicity is not quantified to enter into any decision.

Alternatively, a more cautious sequential design could be used where only small increases or decreases in dose are permitted ("up and down methods"). The patient is protected from a large increase in dose which might be toxic.

3.4 *Summary*

To reiterate: if not much is known about efficacy then it is likely that not much is known about safety. For this specific trial it is a little disconcerting that although toxicity is an issue (or why else choose the ED95 rather than the ED100?) the design does not take toxicity considerations into account for either the sequential allocation of doses or for monitoring. In addition the models used for motivation in Figure 2.1 and for simulation in Figure 2.12 do not consider for the possibility that the active drug gives a worse outcome than the placebo.

4 Adaptive Allocation

The proposed trial uses adaptive allocation and so the probability of being randomized to a particular dose changes over time: in particular patients randomized towards the end of the trial will be more likely to be randomized to the ED95 dose. One problem with this procedure is that the population randomized might change over time during the trial, a phenomenon observed in other trials (see Byar et al, 1976, for example). For illustration suppose that as the trial progresses clinicians become more willing to randomize victims of more severe strokes. If this happens then the patients at different doses will not be directly comparable. Although parametric modeling can partially account for some of these imbalances, additional assumptions are required that may not hold and bias might result.

This seems especially concerning when the effect that is being looked for is small: a small bias might be important. In other treatments for stroke only small improvements over placebo have been found. Small differences in adverse experiences may also be important.

With non-adaptive allocation to doses, randomization ensures that the groups are likely to be directly comparable.

5 Transition to Confirmatory Phase

The authors originally proposed a "seamless transition" between the dose seeking and confirmatory phase. At the end of the dose seeking phase a single dose is selected and, after that, patients are randomized to either that single dose or placebo. It was proposed that neither the clinicians or the patients be aware that the transition had occurred. Combining the phases of drug development is a novel idea that is becoming more popular in clinical trials with the pressure to make drugs available earlier in the development process.

I believe there are, however, several complex issues in such a transition that need to be addressed. Statistically it is very appealing to combine the two phases and use the data from the first stage for the second stage but there are some logistical, scientific and ethical issues that arise.

Here are some of the issues:

1. How toxic is toxic enough to cancel the confirmatory phase? Many toxicities are unexpected and it is hard for the protocol team to prespecify all possible outcomes. It is asking a lot of a DSMB, possibly asking too much, to make a such a determination without carefully specified guidelines: especially if the safety issues of the drug are not yet documented.

   Data is usually confidential until the trial is over (as the authors proposed for this trial). In this case patients and investigators would be unaware when the transition to a confirmatory phase happens. I believe, however, that both review by the investigators, and also independent peer review of dose seeking information and safety information is very important in drug development and in the design of a confirmatory trial. Peer review is clearly impossible without publication of the dose seeking data. The decision of whether or not to proceed to a confirmatory phase might be best made with thorough, and public, review of the safety and dose seeking data.

2. A closely related concern is ethical: when a patient consents to enter a trial they must be informed of the objective of the trial. But the two phases have different objectives: the early phase has an objective of trying to determine the best dose and the confirmatory phase has an objective of demonstrating efficacy. One interpretation of ethical requirements would require different informed consent information for the two parts of the trial. This requirement seems very reasonable as a dose seeking trial may have different risks than a confirmatory trial: in a dose seeking trial, where little is known about safety, it is more likely that a patient will be given a toxic dose.

Other issues are more complex and may not all apply to this particular trial. In early phases of drug development, when little is known about the drug, it may be wise to make more intensive monitoring or laboratory measurements. In addition it may be advisable to perform the trial at sites at selected research institutions where access to specialized care for complications is available (for example access to emergency neurosurgery). In the confirmatory phase, when more is known about the drug, monitoring of laboratory values might be less intensive and sites might be included which are more representative of general clinical practice.

The dose-seeking phase, therefore, might potentially measure a slightly different intervention (drug plus intensive laboratory monitoring versus placebo plus intensive laboratory monitoring at selected research institutions) than the confirmatory phase which might examine the implications of the therapy if used in wide clinical practice. The first phase might be more of an "explanatory trial" (as defined by Schwartz and Lelouch, 1967) looking at a narrow scientific question whereas the second phase might be a "pragmatic trial" looking at how the drug might work in broad clinical practice.

The distinction between an "explanatory trial" and a "pragmatic trial" is useful: although any one trial may not completely fall into one category. An explanatory trial is more basic science looking at the activity of the drug and characterizing its safety profile. Eligibility requirements may be quite narrow to reduce variability. A pragmatic trial, however, usually answers a the public health question of whether or not an intervention will have an impact, if widely used in a diverse population in a real world setting. Typically the magnitude of the effect of the drug is greater in the more controlled setting of an explanatory trial than in a pragmatic trial.

Potentially, therefore, the early phase and the later phase of the trial are served best by different designs.

6 Endpoints

The proposed primary endpoint is score at 13 weeks on the Scandinavian Stroke Scale. As the authors acknowledge, different trials of treatment for stroke have used different endpoints. Some, like the trial proposed by Berry and colleagues, use a stroke scale. Others such as the very large IST use a binary outcome for "death or dependency" as a primary outcome. The TOAST trial used a binary outcome "favorable" or "not favorable" as a primary outcome and also a "very favorable" or "not very favorable" outcome. These binary outcomes were defined using two stroke scales. Different endpoints have different advantages: a stroke scale is typically very sensitive to small effects, whereas binary, or categorical, outcomes are typically more clinically meaningful and appropriately incorporate mortality. In the proposed trial described in this paper it would be best to use the stroke scale as the endpoint for the dose seeking phase: small differences in outcome are important. A binary, or some other categorical outcome, that takes account of survival, is best for the confirmatory phase.

7 System Interface

Before concluding I would like to congratulate the authors on providing the implementation details of their system interface. These kind of practical details are often missing in abstract discussions of sequential randomization and it refreshing to see these details addressed. Timeliness in passing the dose to the clinical site is clearly critical in a sequential design such as this.

8 Conclusions and Summary

To conclude I will summarize my questions for the authors. Most have to do with why they chose to take this approach for this particular problem, and may well be due to my lack of understanding of the research question.

1. Are safety issues adequately addressed? Should safety and toxicity be incorporated into the design?

2. Why has the possibility of placebo being better than drug not been given more consideration? (Especially given the history of drug development for stroke).

3. The adaptive allocation is motivated as minimizing the total number of patients required for the trial. The disadvantage is that groups of patients at different doses may not be directly comparable. Stroke is very common, however, and many stroke trials have been successful in recruiting large numbers of patients very quickly. Would non-adaptive allocation be better in this problem?

4. Is the seamless transition really best?

5. Is the endpoint appropriate for both phases?

There are so many new ideas in this paper I am overwhelmed with questions. This is a very complicated design and there are many fascinating aspects of this trial. I congratulate the authors on their creative hard work and I look forward to answers to my questions above. I also look forward with interest to seeing the trial implemented.

Acknowledgements

I thank Robert Woolson, University of Iowa, Jim Neaton and Lisa Fosdick, University of Minnesota, Isabella Verdinelli, Carnegie Mellon University and Samuel Wieand and Stephanie Land, University of Pittsburgh, for some helpful input.

References

Bryant, J. and Day, R. (1995). Incorporating toxicity considerations into the design of two-stage phase II clinical trials. *Biometrics*, 51, 1372–

1383.

Byar, D.P., Simon, R.M., Friedewald, W.T. et al. Randomized clinical trials: perspectives on some recent ideas. *New England Journal of Medicine*, 295, 74–80 (1976).

Chinese Stroke Trial collaborative group (1997). CAST: a randomised placebo controlled trial of early aspirin use in 21,106 patients with acute ischaemic stroke. *Lancet*, 349, 1641–1649.

Conaway, M. and Petroni, G. (1995). Bivariate sequential designs for phase II trials. *Biometrics*, 51, 1372–1383.

Conaway, M. and Petroni, G. (1996). Designs for phase II trials allowing for a trade-off between response and toxicity. *Biometrics*, 52, 1375–1386.

Fan, S. (1999). *Multivariate Optimal Designs*. PhD Thesis, School of Statistics, University of Minnesota.

Hacke, W, et al (1995). Intravenous thrombolysis with recombinant tissue plasminogen activator for acute Hemispheric Stroke. *Jour. of Amer. Med. Assoc.*, 274, 1017–1025.

Hacke, W., et al (1998). Randomised double-blind placebo-controlled trial of thrombolytic therapy with intravenous alteplase in acute ischaemic stroke. *Lancet*, 352, 1245–1251.

Heise, M.A. and Myers, R.M. (1996). Optimal designs for bivariate logistic regression. *Biometrics*, 52, 613-624.

Hill, M.D. and Hachinski, V. (1998). Stroke treatment: time is brain. *Stroke*, 352, 10–14.

International Stroke Trial Collaborative Group (1997). The international stroke trial (IST): a randomised trial of aspirin, subcutaneous heparin, both, or neither among 19,435 patients with acute ischaemic stroke. *Lancet* 349, 1569–1581.

Kay, R. et al (1995). Low-molecular-weight heparin for the treatment of acute ischemic stroke. *New England Jour. of Med.*, 333, 1588–1593.

Lees, K.R. (1998). Does neuroprotection improve stroke outcome? *The Lancet*, 351, 1447–1448.

The National Institute of Neurological Disorders and Stroke rt-PA Stroke Study Group (1995). Tissue plasminogen activator for acute ischemic stroke. *New England Jour. of Med.*, 333, 1581–87.

The Publications Committee for the Trial of ORG 10172 in Acute Stroke Treatment (TOAST) Investigators (1998), *Jour. of Amer. Med. Assoc.*, 279, 1265–1272.

Riggs, J.E. (1996). Tissue-type plasminogen activator for acute ischemic stroke. *Arch. Neurol.*, 53, 1581–1587.

Schwartz, D. and Lellouch, J. (1967) *J. Chron. Dis.*, 20, 637–648.

Thall, F.P. and Russell, E.K. 1998. A Strategy for Dose-Finding and Safety Monitoring Based on Efficacy and Adverse Outcomes in Phase I/II Clinical Trials. *Biometrics*, 54, 251-264.

Discussion

Stephanie R. Land, NSABB, Pittsburgh Cancer Institute and the University of Pittsburgh
H. Samuel Wieand, NSABBP and the University of Pittsburgh

Introduction

We first want to thank the authors for a very thought-provoking paper. We liked the fact that they were comfortable using frequentist methods in the confirmatory trial, albeit for regulatory reasons, while demonstrating the potential of Bayesian analyses to facilitate the dose-seeking phase and sample size computations. This approach may be unsatisfying to a "pure" Bayesian or a "pure" frequentist, but is a practical way to reach statisticians who normally do not use Bayesian methods. We believe that the reason many biostatisticians (including us) seldom use Bayesian analyses is not philosophical, but rather represents a discomfort with utility functions and priors that are often overly simplistic (to facilitate computations) and a reliance on models which are hard to validate. A second factor for some statisticians (including the second author of this discussion) is a lack of familiarity with the recent advances made in the development of algorithms which have allowed for the use of more general priors and utility functions than was once the case. We believe that most statisticians will be comfortable with (and interested in understanding) the use of Bayesian methods in the dose-seeking phase of the trial (where frequentist methods have had moderate success at best), since they still have the comfort of the randomized Phase III trial at the end.

A major strength of the authors' work is that it elucidates many of the difficult practical issues that arise in initiating and completing dose-seeking and confirmatory trials and proposes solutions for nearly all of these issues. Any statistician involved in dose-seeking trials would benefit from carefully reading this manuscript and incorporating some of these ideas into his or her own work. Potentially useful features include the continual reanalysis of the dose-seeking data, the use of a utility function in determining whether to continue or stop the dose-seeking phase, the incorporation of imputation using partial data, and the use of predictive probability methods in determining sample size. Of course, the devil is in the details, but the authors use powerful machinery to make the problem tractable. They provide some details regarding implementation, but gaps remain, particularly with respect to some of the statistical algorithms.

This discussion will first focus on the efficacy measure and the range of doses. We can think of few situations in which the authors' choice is applicable. In the subsequent section, we present a simulation comparing the operating characteristics of a frequentist method and the authors' method. In the final section we summarize our thoughts regarding the strengths and weaknesses of the authors' work.

Range and Efficacy Measure

As the authors state in the Introduction, "There have been a number of failures in recent trials investigating neuroprotective agents in acute stroke and these may be due to an imperfect understanding of the dose-response relationship and a consequent incorrect selection of dose." This is a valid issue to address, but the authors' approach seems more likely to lead to under-dosing than an approach that continues increasing the dose until a dose-limiting toxicity is reached. Since the authors' efficacy measure does not incorporate toxicity, one anticipates an increasing dose-response curve, much like the figures shown in Berry et al. Figure 2.1. Then the range of assigned doses (particularly the maximum) may be more important than the shape of the curve in the range, a fact not discussed in the manuscript. Consider the curves in Berry et al. Figure 9.1, but assume the selected range of doses was 0.0 to 0.75. In that range, curves 1, 4, and 7 would all have a maximum benefit of approximately 4 units and the algorithm would converge to an estimated ED95 dose with a benefit of approximately 3.8 units. However, the benefit at the correct ED95 dose for curve 4 is 7.6 units. The authors' approach would converge to the ED47.5 rather than the ED95. Of course, a similar argument applies to Curve 5, where the authors' method would lead to a solution that would be closer to the ED50 dose than the ED95 dose. The sophisticated algorithms will lead to incorrect solutions if the range is chosen incorrectly. For most data sets, the shape of the estimated dose-response curve would be likely to reveal that the ED95 is outside the selected range, but this important topic is not addressed.

This raises an equally important issue. For many drugs, toxicities (including, but not limited to, death) become unacceptably high and the benefit/risk (efficacy/toxicity) ratio reaches a maximum at doses lower than the ED95. Perhaps a better use of Bayesian techniques would be to elicit risk and benefit utilities prior to starting the study and to use the benefit/risk curve rather than the efficacy curve in the algorithm. In this case, one might look for the dose that gives maximum benefit/risk rather than the ED95. Curve 7 of Berry et al. Figure 2.12 would be very relevant for this approach and deserves further study.

We do not mean to imply that the authors' method is never appropriate. If one were considering a drug that had minimal acute side effects at low doses, e.g., aspirin to treat a headache, and the dose-response curve began to level off, stopping at the ED95 dose might minimize cost and avoid possible long-term toxicities.

Comparison with a Conventional Design

It is difficult to define a "conventional competitor" for the authors' design. However, we have taken the conventional designs used by the National Surgical Adjuvant Breast and Bowel Project (NSABP) and tried to modify them as we would if we were conducting a dose-seeking trial under the assumptions given by the authors. These assumptions include the following: 1.) There is a defined range of "safe" doses (doses not exceeding a maximum tolerated dose) which include the optimal (ED100) dose. We will use the range 0 to 1.5 to be consistent with Berry et al. Figure 2.12. 2.) The mean effect of the drug at a specified dose (d) is μ_d and the distribution of response is Normal (μ_d, σ^2) where σ^2 may be unknown, but is independent of the dose. Again, we will match the authors' manuscript and assume $\sigma^2 = 150$. 3.) The dose is given in one bolus so that the amount specified will in fact be the dose given. 4.) There is one response for each patient (since we do not have adequate information to perform imputation). We also ignore the fact that there is a time lag (13 weeks) between patient entry and attainment of outcome.

Specifications of the conventional design:

1. We will randomize equal numbers of patients to doses of 0.0, 0.5, 1.0 and 1.5.

2. We will design our trial to have specified power (.9) against a clinically relevant (one-sided) alternative.

3. After 250 patients have been randomized, the variance will be estimated using the average of the s^2 across the four treatments. The estimated variance will be used to estimate the total sample size required for the correct power and alpha-level.

4. The initial 250 patients will still be used in the final efficacy analysis as the variance analysis has essentially no effect on the type I error.

5. Our upper bounds (measuring efficacy) will be standard O'Brien-Fleming bounds (Fleming, et al., 1984), modified for the fact that there are three non-placebo doses (all being compared to placebo). If the O'Brien Fleming bound is crossed for any of the treatments, accrual will be terminated. The dose with the largest response (among the doses for which the O'Brien-Fleming bound was crossed) will be considered to be the optimal dose. If the trial is not stopped early, the response for all doses that show a statistically significant benefit at completion of the trial will be compared. The dose with the largest response will be considered to be the optimal dose. To allow stopping for lack of efficacy, we will use a method (Green, S. et al, 1997) employed both by the Southwest Oncology Group (SWOG) and the NSABP. This approach is to do a one-sided test of the null hypothesis: $\delta = \delta_A$, where δ_A is the clinical benefit of interest, versus the alternative that $\delta < \delta_A$. The test is conducted at a small significance level, in our case alpha=.005 and computed at each interim analysis. If at any point this test is rejected, accrual to the experimental arm corresponding to that dose is terminated.

6. We will perform three interim analyses, one when 250 patients have been analyzed and the other two at equally spaced points in accrual between the 250 patients and the estimated final possible sample size.

This clinical trial will be simulated for four scenarios.

<u>Scenario 1</u>: The clinically meaningful benefit is assumed to be 2 units so that our sample size is based on having power .90 to detect that difference, and the true dose-response curve is Curve 1 (no dose-response

	curve 1		curve 3	
	N	Type I Error	N	Power
conventional (csd=2)	2132	0.051	1252	1
conventional (csd=4)	514	0.052	472	0.95
Berry et. al	257+	negligible	691-	0.86

TABLE 2.1. Comparison of conventional trial (for two values of the clinically significant difference) with Berry et al. design (median total sample size[1] and probability of concluding the drug is effective).

relationship). The results of 10000 simulations of the design indicate that the overall probability of erroneously concluding that there is a dose that is better than placebo is 0.052. At the first stage, there is a 0.087 chance that one experimental treatment will be dropped; there is a 0.021 chance that two treatments will be dropped; and there is a 0.005 chance that the trial will be terminated with the conclusion that there is no treatment effect. The situation becomes a little more complicated at subsequent looks, but the average number of patients required for the trial, based on 10000 simulations of the entire process, is 2159. The observed median sample size is 2132. These results are provided in Table 1 of our discussion.

If the alternative of interest is larger, a smaller sample will be required.

Scenario 2: The clinically meaningful benefit is assumed to be 4 units so that our sample size is based on having power .90 to detect that difference, and the true dose-response curve is Curve 1 (no dose-response relationship). The median sample size is 514, and the Type I error is 0.051 (in 1000 simulations).

The authors' approach led to a median sample size of 257 for Curve 1 and appeared to have a Type I error near 0. Their method correctly identifies a lack of dose response in the range of interest much more efficiently than our conventional method.

Scenario 3: The clinically meaningful benefit is assumed to be 2 units so that our sample size is based on having power .90 to detect that difference, and the true curve is Curve 3 (i.e., is 0.3 units when the dose is 0.5, 3.94 units when the dose is 1.0 and 4.0 units when the dose is 1.5). The average computed maximum total sample size for the conventional design is 3352. However, the O'Brien-Fleming bound for efficacy was frequently crossed in our 10000 simulations and our average total sample size is 1081 (median 1252). Benefit for therapy was shown in all 10000 simulations. The dose

[1]Median sample sizes for Berry et al. method are not exact because the paper provides medians separately for trials that did and did not go to confirmatory trials.

1.5 was selected as optimal 5159 times; the dose 1.0 was selected 4836 times; and the dose 0.5 was selected 5 times. The latter choice should be considered to be a failure as the dose was not actually associated with a clinically meaningful benefit. Unfortunately, the current version of the authors' manuscript does not indicate what doses they selected as the ED95 doses.

Scenario 4: The clinically meaningful benefit is assumed to be 4 units so that our sample size is based on having power .90 to detect that difference, and the true curve is Curve 3. Using the conventional method, the average maximum total sample size is 836 and the average number of patients entered is 486 (median 472).

The median sample size for Curve 3 simulations using the authors' method was 691, so it is not clear which method is more efficient.

Our simulations indicate that the authors' approach offers a clear reduction in sample size when there is no dose-response effect and matches up well with the frequentist methods when there is a treatment effect. The authors provide a clean two-armed confirmatory trial that avoids the multiple comparison issues. It also allows the possibility of using the dose-seeking data as a secondary trial, although it is not yet clear whether that will be acceptable to a regulatory agency.

The authors' method is less likely to identify a clinically meaningful effect than the frequentist approach. The frequentist method has at least a 90% chance of identifying the pre-specified clinical difference and a greater chance of identifying a larger difference. The predictive power method does not offer this assurance. Table 2.3 of Berry et al. indicates that for a drug with a maximal effect of 4 units, when 2 would be clinically meaningful, there is a 93% chance that the drug will be carried to the confirmatory trial and a 92% chance that the confirmatory trial will be significant. Thus, this important effect would be identified with probability 0.86 (0.93 × 0.92). Using the conventional method, this effective dose would have been identified with a probability exceeding 0.99. The results in Table 2.5 of Berry et al. (where only the variance was different) indicate that the authors would have missed this important effect 23% of the time.

Advantages, Limitations, and Work to be Done:

The numerous potential benefits of the authors' work include:

1. The method could potentially be applied to identify any dose meeting a criterion for clinical benefit, e.g., the ED95 or an optimal benefit-risk ratio.

2. The combined dose-seeking and confirmatory trials generally use fewer patients than traditional (primarily frequentist) methods.

3. The sample size is further reduced by imputation using intermediate results.

4. The transition from the dose-seeking to the confirmatory trial requires no interruption in accrual.

Some potential limitations of the methods (as presented) are listed below. Several of these are minor issues that can be addressed rather easily. Others will require more extensive work.

1. The maximum safe dose must be known in advance. In particular, the current method may result in under-dosing of patients if the initial range does not include the ED100 (or at least the ED99+) dose.

2. Little detail is provided describing how to incorporate non-lethal adverse events.

3. The procedure makes nonstandard requirements of pharmacies and statistical centers. The nonstandard pharmaceutical requirements could increase the possibility of incorrect dosing.

4. Several model assumptions are required, and priors must be determined. The NDLM model includes parameters that must be either specified by the statistician or elicited from the medical investigators. These parameters include n_0, S_0, w_j, m_0, C_0, r and V. It would seem difficult to explain the authors' method clearly enough to obtain good information about these parameters reflective of medical expertise. The longitudinal model is based on experience with untreated patients (in the Copenhagen Stroke Database) and does not incorporate the dose. It seems plausible that there could be a short-term response to a drug given at high dose, in which case the time profile of response would be different for a patient given a high dose than for a patient given placebo. Such an interaction could lead to biased imputations that would affect dose selection and application of the "Terminator." It will be important to know how sensitive the authors' methods are to these assumptions. No post-hoc validation was described in the manuscript.

5. An issue related to the preceding is that the authors' method requires relevant historical data, such as the Copenhagen Stroke Database.

6. An extremely effective dose (providing twice the clinically meaningful effect) was not carried into the confirmatory setting 7-8% of the time (as shown in Tables 2.3 and 2.5 of the manuscript). Note: no example was given where both the clinically meaningful effect and the actual effect of the optimal dose were the same, e.g., 2. We did not understand the algorithm well enough to know whether this would result in a higher number of failures (in the sense of terminating without continuing to confirmatory trial) than in the examples given, but suspect this will be the case. If so, the utility criteria should be reexamined.

7. The estimators of mean effect size and variance are biased. Perhaps the "Recommender's" modification of the dose assignment is designed to overcome this bias, but this has not been demonstrated.

8. In the current implementation of the authors' approach, the comparison of interest must be with a placebo. Modifications of the method would be required when the new drug is a candidate to replace a standard treatment, and the disease is sufficiently serious that it would be unethical to give only placebo. Changes would be required in the automatic assignment of 15% of patients to placebo and in the dose-ranging. (An ethical lower bound would need to be specified.) In addition, the "Terminator" is based on the estimated increase in effect of dose z^* over the placebo, where the effects of both are determined by the NDLM model. The effect of the standard drug is not likely to be well-modeled by $f(0, \theta)$, where f is fit using the data from the new drug.

Summary: We have pointed out potential weaknesses of the authors' work and believe that some of them are serious enough that more work is required before this approach can be implemented widely. In fact, we believe some of the concerns identified by us (and Dr. Chaloner) should be addressed before the procedure is used in the authors' specific setting. However, this is bold, pioneering work in its early stages. It offers considerable promise as a rational and efficient way to address the problem of identifying an "optimal" dose and confirming its utility. We hope that the authors and others continue to pursue and improve on this approach.

References

Green, S., Benedetti, J. and Crowley, J. (1997). Clinical Trials in Oncology, Chapman & Hall, London.

Fleming, T.R., Harrington, D.P. and O'Brien, P.C. (1984). Designs for group sequential tests, *Controlled Clinical Trials*, **5**, 348-361.

Rejoinder

We thank the discussants for their kind remarks and for their diligent efforts: Drs. Land and Wieand in evaluating our procedure and Dr. Chaloner in working to understand stroke and its treatment. We first address the issues that the two reviews have in common.

Common Issues

Both discussants are concerned that we focus on efficacy while ignoring safety. Obviously, safety is important. It cannot be ignored and we do not ignore it. We have both general and specific answers. First, the point of presenting our method in a public forum was to advertise its versatility and wide applicability. The method is ideally suited for handling a general endpoint that is a combination of efficacy and safety, a point we tried to make at the Workshop. Simply view "response" in "dose-response" as a combination of safety and efficacy and representing the overall utility of a dose to a patient. Our procedure allows for non-monotonic dose-response curves for precisely this reason. Drs. Land and Wieand and also many of Dr. Chaloner's references are in the context of cancer. The effectiveness of a typical cytotoxic agent increases with dose, but so does its toxicity. At some point the increased toxicity more than balances any increase in effectiveness, and the utility of the overall response decreases. Our methods apply to anti-cancer agents with a simple reinterpretation of the endpoint.

Another way of handling safety is being implemented in a trial that was designed by the first author and is now underway at M.D. Anderson Cancer Center. The disease is leukemia and the issue is dose. Dose assignment is adaptive, but the procedure is different from the one considered in the present paper. In the Anderson trial, toxicity is critical. We move gingerly up the dosage scale, adding the next higher dose to the set of admissible doses if the toxicity profile is satisfactory. Dose assignment is adaptive within the admissible dose range. Such a modification would be trivial to effect in the algorithm described in the present paper.

Finally, as regards the trial at hand, stroke is not cancer and the compound in our trial is not cytotoxic. There is extensive information regarding the compound from preclinical and phase I and II studies. For confidentiality reasons it has not been possible to give details of this information. However, we can say that prior to taking the compound into the present trial, a large phase II dose-escalation safety trial was conducted in stroke patients. A data safety monitoring committee found the compound to be well tolerated in the population and doses studied (K Lees, 4th World Stroke Congress, Melbourne Nov 2000). Were there a reasonable chance of serious side effects then we would modify the endpoint to include them, as described in the first paragraph. This modification could include mortality and any other such measure, whether related to stroke or not.

We chose not to directly incorporate toxicity or tolerability into the dose-allocation algorithm, given the complexity of weighing efficacy and a priori unknown safety issues in the indication studied, and also weighing individual utilities versus population utilities. However, toxicity and tolerability issues are incorporated indirectly: a) We assume that safety and tolerability have been extensively studied prior to conducting a dose-ranging study, say by a dose-escalation study, providing a rationale for fixing an upper limit to the dose-range. b) We assume that all safety and efficacy data will be reviewed on an ongoing basis by an independent data monitoring committee (IDMC) throughout the course of the study. The IDMC could incorporate safety issues e.g. by withdrawing unsafe doses, by recommending changes to the entry criteria etc. We felt that the potential complexity of efficacy/toxicity issues and the weighing of individual utility versus utility for the overall study population would be best addressed by human expert assessment.

Land and Wieand

Drs. Land and Wieand worry that "sophisticated algorithms will lead to incorrect solutions" if the true ED95 is not within the range of doses considered. This is obviously true. It is also true for simple algorithms. But we love the comment because it gives us the opportunity to extol a great virtue of our approach. Namely, we could use an open dose range and add higher doses during the trial. We have not done so only for practical reasons. Actually, in the trial at hand, sentiment among some scientists involved in designing the trial was that the therapeutic range is a good deal smaller than 1.5. We argued that there is little loss for increasing the range of doses to well beyond that thought to be therapeutic. Should it turn out that the response curve plateaus at doses smaller than 1.5, as these scientists believe, we will learn that, and the cost of this information will be modest. Anything is possible, of course, and no trial protocol can accommodate every eventuality. Suppose that to everyone's great surprise, an interim analysis suggests that dose-response is flat over most of the range but it increases dramatically at the upper endpoint of the dose range (see curve 5). The data safety monitoring board will know this and if the effect is sufficiently pronounced then they can choose to inform the trial's steering committee. Among the possibilities for this or any other unexpected eventuality is stopping the trial.

Tolerability, bioavailability and cost-of-goods issues can play an important role in determining the upper dose limit. On a practical note we define the ED100 with respect to the preset dose range, and think of it as the dose that yields the best possible effect within the dose range that can actually be explored. We do not regard it to be the theoretical dose that would yield the best effect, if doses greater than the existing dose limit could be studied.

Drs. Land and Wieand compare our approach with a "conventional design." It is more than a little interesting that they had to change the ground rules in an important way to make this comparison. In the example in our paper we considered 16 doses, including placebo. Any conventional design adapted to 16 doses would lose and lose big in comparison with our design—especially if it adjusts for multiple comparisons. They considered only 4 doses, including placebo. One cannot hope to learn as much about a dose-response curve with only 4 doses as with 16 doses. With doses restricted to 0, 0.5, 1.0 and 1.5, both designs will fail miserably should the true dose-response be curve 7, or some other curve in which the peak occurs between doses considered. As a hockey team asked to play basketball, we welcome the challenge. And we are happy with the result.

Drs. Land and Wieand make a false comparison between their overpowered design with a power > 0.99 and what they calculate as $0.86 = 0.92 \times 0.93$ for our design. Firstly, the power of > 0.99 is conditional on their having chosen the right doses. If they are to criticize us for assuming that the ED95 is within the range of doses that we choose, then such a criticism is even more relevant for their design. Secondly, what they fail to appreciate is that we have selected a dose and confirmed it while they have only selected a dose. In their approach, the regulatory authorities will require confirmation that the dose is effective in at least one further trial.

There is still another way that Drs. Land and Wieand ask us to play by different ground rules. The discussants are tied to conventional notions of sizes of errors of Types I and II. They indicate that our procedure has lower power than does the conventional designs they use for comparison. We can adjust the parameters of our design to obtain any preset error sizes. But that is not our objective. We address these errors—especially that of Type I—because they are important in a regulatory environment. Our design is Bayesian. We stop the dose-ranging phase based on decision analysis (although our computer program allows the option of stopping based on posterior probabilities alone). Utility is monetary. (In the view of the first author, a pharmaceutical company maximizing expected profit is consistent with the overall societal aim of maximizing public health.) Power is an incidental derivative. It may be greater than or less than what is "conventional," depending on trial costs and potential profit. They point to one particular setting of our algorithm and say that we have less than some arbitrarily selected power level; we point to their conventional design and say that for the setting in question their design is greatly overpowered. Neither of us is right in any absolute sense. But their attitude is tied to a power setting that is completely arbitrary while ours is based on explicit considerations of real-world quantities. We continue the dose-ranging phase when and only when it is economically appropriate to do so. (Whether utility is monetary or measured in terms of public health, choosing the same power level in both rare and common diseases makes no sense: In the former case what is learned from the trial has less utility—in both the English and technical senses of the word.)

In the same vein, the discussants suggest that we should reexamine our utility criteria because the results do not correspond with their notions about false positive and false negative error rates. We are always open to reexamination and have done so. But we ask that the discussants reciprocate: they and other frequentists should reexamine their notions of what these error rates should be. There may be settings in which the usual (0.05, 0.10) and (0.05, 0.20) are appropriate. But the same values cannot and should not apply in every setting. We dare say that (0.20, 0.30), (0.001, 0.001) and even (0.40, 0.40) are sometimes appropriate. We are not concerned that the conclusions are different for our design as compared with the conventional design. Actually, we are happy that there is a difference because it provides an example of the folly of consistency in clinical trial design.

In the following we comment on the specific "limitations" indicated by Drs. Land and Wieand that we have not previously addressed.

3. We agree that our nonstandard requirements increase the possibility of incorrect dosing. Recognizing this, we have put an enormous amount of effort into insuring that this does not happen.

4. Yes, it is true that we do not consider different longitudinal models depending on dose. We agree that is a limitation. We have discussed incorporating into our model the possibility of an interaction of the type indicated by the discussants. This remains work for the future. However, we are satisfied that moderate forms of interaction have little effect on the performance of our procedure. Dramatic forms of interaction—such as when patients on positive dose get to their asymptote within one week while those on placebo require ten weeks—are very unlikely. Should they arise then the DSMB may take some action, such as recommending stopping the trial.

5. Actually, we do not require "relevant historical data." In fact, the discussants used our method without considering any aspect of the Copenhagen Stroke Database. It is a virtue of the Bayesian approach that it can incorporate historical information. We used early endpoints to help in predicting the final endpoint. We exploited historical information about the extent of correlation between the endpoints. But we could also model this relationship and learn about it in an on-line fashion.

7. This comment seems to refer to the simulations with Curve 3 summarized in Figure 9.4. and Curve 1 shown in Figure 9.5. We noticed the bias in both simulations. However, the sample sizes and noise in the data mean that we would not expect to see large-sample consistency of posterior distributions (cf. Figure 9.8 which shows some simulated data sets). Also, some bias results from our use of informative prior distributions, in terms of both level and correlations across doses. The additional randomization in the 'recommender' was not introduced to address bias but to mitigate concerns about the lack of randomization, which is extraneous to the formal decision problem. A feature of our approach to dose allocation is that we not only find the optimal dose, we also learn about the expected utility function. This allows us to include additional randomization in the 'recommender' without any significant loss in expected utility at the dose that is eventually assigned.

8. Yes, modifications of our method would be required in settings where using a placebo is not possible. This modification would not be straightforward, and indeed establishing a dose-response relationship in such settings

is always difficult.

Finally, Drs. Land and Wieand suggest that their concerns should be addressed before we use our method in an actual trial. We are sorry that our original paper was not sufficiently clear as regards their concerns. We hope and believe that we have now successfully addressed them. In addition, the algorithm to be deployed in our dose-ranging trial has been improved in several ways since the one presented in our paper. Moreover, the entire system has undergone a thorough validation.

Chaloner

Dr. Chaloner is disturbed because none of the dose-response curves from which we simulate have negative slopes. The flat curve (curve 1) that we do consider is the most difficult case among those relationships showing no or negative benefit. If the true curve is actually decreasing then the trial stops sooner than it does for this case.

Dr. Chaloner would like to see some data to convince her that the drug is safe. She need not be concerned: Before the trial could start, Institutional Review Boards and regulation agencies had to see data addressing this issue. However, as we have indicated, this data is confidential.

The principal reason we focus on the ED95 and not on the ED100 is that the slope of the curve may be positive indefinitely to the right of the ED95. Moreover, because of the variability in similarly treated stroke patients, distinguishing between the ED95 and the ED100 would be essentially impossible.

We agree that the types of patients in the trial may change over time. This is of course true in conventional trials as well, and this phenomenon also complicates the analysis of those trials. We address the concern in the same way that conventional trials address the concern: By randomizing patients to placebo in a uniform fashion. Regarding the 15% figure, this is a minimum with the possibility of higher proportions depending on the accumulating data. This proportion is not small in a dose-finding study. For example, similar proportions would be used in balanced randomized designs with five or six positive doses.

Regarding a seamless shift into confirmatory phase, our study is in fact not using this feature. However, informed consent should not a roadblock in such a design. A patient or the patient's representative can be informed about receiving either placebo or some positive dose if the final dose has not yet been identified or placebo and some positive dose if the latter has been identified. It is not clear whether they will understand the distinction between the two and it is not clear that they need to understand the distinction. But the ethical issues are no more problematic than for a traditional study.

Dr. Chaloner distinguishes between "explanatory" and "pragmatic" trials, suggesting one for early phase and the other for late phase. Our design has the benefits of both. She goes on to suggest that a binary outcome "is best for the confirmatory phase." We disagree. Moreover, we regard pronouncements of and establishing traditions of "best" as responsible for making drug development inefficient and counter to delivering good medicine. Innovation is impossible when everyone accepts what is "best." In the case at hand, suppose the binary endpoint requires a clinical trial with 10,000 patients. The trial would never be done. One view is that such a drug would not be worthwhile and that it should fall by the wayside. But not a small number of stroke experts regard a 2-point increase in Scandinavian Stroke Scale as clinically important. And a stroke patient who is not given the opportunity of walking a little better might disagree with Dr. Chaloner.

Our reaction to Dr. Chaloner's third conclusion is the same as that in the previous paragraph. She suggests that non-adaptive allocation may be better in our trial because stroke is very common and it is easy to recruit large numbers of patients. We could not disagree more strongly. First, as a group, patients in an adaptive trial receive better treatment. Second, in reducing trial size we move effective therapies into the clinical setting more rapidly. Third, saving patients means that pharmaceutical companies can move more quickly through the development process and consider a greater number of compounds, again improving public health. Finally, clinical trials are extremely expensive and are the primary reason for the high cost of prescription drugs. Slowing the drug process increases the overall cost of health care.

Again, we thank the discussants for their erudite comments and for their diligent and thorough efforts.

Modeling the Impact of Traffic-Related Air Pollution on Childhood Respiratory Illness

Nicola G. Best
Katja Ickstadt
Robert L. Wolpert
Samantha Cockings
Paul Elliott
James Bennett
Alex Bottle
Sean Reed

ABSTRACT Epidemiological studies of the health effects of outdoor air pollution have suffered from a number of methodological difficulties. These include major problems of estimating exposure; confounding due to socioeconomic deprivation and other factors; failure to account for possible spatial dependence induced by unmeasured covariates with smooth spatial variation; and the potential for bias arising from approximation and aggregation in the data. We present a flexible modeling framework using recently developed Bayesian spatial regression methods to address some of these issues. We apply this approach to a study of the relation of London traffic pollution to the incidence of respiratory ailments in infants using the recently available Hospital Episodes Statistics (HES) dataset.
Key words: Ecological bias, Environmental exposure, Hospital admissions, Marked point process, Poisson regression, Spatial correlation.

1 Introduction

The possible harmful health effects of outdoor air pollution from industrial sources and road traffic are the focus of great public concern and scientific attention in the United Kingdom. Indeed, the goal of reducing the harmful health impact associated with outdoor air quality is a principal issue raised in the UK government's "National Environmental Health Action Plan," drawn up under the auspices of the World Health Organization. Key scientific questions affecting UK policy decisions are whether and how recent apparent increases in the rates of childhood respiratory illness are related to atmospheric pollution arising from motor vehicle emissions.

The evidence for such a link remains equivocal, due in large part to a host of methodological difficulties that have beset scientific investigations of the relationship. These include major problems of estimating exposure; confounding due to possibly unknown or unmeasured risk factors; and the potential for bias arising from approximation and aggregation in the data. These methodological obstacles must be addressed to enable quantitative assessment of the effects of traffic pollution on health to inform public health policy, transport planning and environmental legislation regulating vehicle emissions.

In this Case Study we present an epidemiological investigation of hospital admission rates for respiratory illnesses in children under one year old living in Greater London, and relate these rates to measures of exposure to traffic-related air pollution. We address the scientific question of whether childhood respiratory illness rates are associated with motor vehicle emissions by studying four specific issues:

1. Do infants living in areas of high traffic density have higher rates of respiratory illness?

2. Which pollutants or vehicle types are associated with the greatest risk?

3. Do factors such as socioeconomic deprivation of the area of residence, or individual characteristics such as gender increase or modify this risk?

4. Can we account for residual spatial confounding due to unknown or unmeasured risk factors?

To answer these questions we construct a hierarchical Bayesian semiparametric spatial point-process regression model supporting inference about the relationship of children's respiratory illness rates to traffic pollution and other risk factors. We address the exposure estimation problem by employing sophisticated geographical information system (GIS) methods to build exposure extrapolation models; we approach the confounding problem by including in our model a latent spatially varying risk factor; and we reduce or avoid the problem of ecological bias by using each source of data at its natural level of spatial resolution, without any unnecessary approximation or aggregation.

In the next section (2) we provide a detailed discussion of the epidemiological background to this Case Study. The available data, including our use of GIS methods to develop suitable measures of exposure based on the raw data (traffic flows and vehicle emission factors) are described in Sections 3 and 4, and an exploratory analysis of the data is presented in Section 5. In Section 6 we review briefly existing methods for the epidemiological analysis of exposure-response relationships and discuss some of the associated methodological problems. This will motivate the modeling approach described in this section and in Section 7. Section 8 contains a final discussion.

2 Epidemiological Background

There is evidence of marked geographical variation in the occurrence of many diseases, including cancers and respiratory illnesses (Gardner et al., 1984; Parkin et al., 1997). In particular, urban-rural differentials have been noted, with higher rates of disease in urban areas (Greenberg, 1983; Howe et al., 1993), and a strong association is typically observed with measures of socioeconomic deprivation (Jolley et al., 1992). Onc possible explanation is differences in smoking patterns. Another is the possible adverse health effects of urban air pollution derived from road traffic emissions. Indeed, the search for possible environmental causes of ill health, including a putative effect of traffic related air pollution, is currently a major international focus of epidemiological research.

Several factors have contributed to these developments. One is the apparent increase in prevalence of respiratory illnesses, especially in children, which has characterized almost all countries in the western world, and many developing countries, over the last two to three decades (Anderson et al., 1994; Burney, 1988). Another is the inexorable rise in traffic volumes and associated emissions of air pollutants, which has led to serious declines in air quality in many cities (HMSO, 1994). A third is the increasing scientific evidence for a link between exposure to traffic pollution and a range of health effects, including cardiorespiratory mortality and morbidity and reduction of lung function (Committee of the Environmental and Occupational Health Assembly of the American Thoracic Society, 1996).

Much of the growing literature on the health effects of ambient air pollution has focused on the acute effects of exposure; recent reviews include those by Schwartz (1994) andBrunekreef et al. (1995); Anderson et al.

(1998) provide a summary of time series studies of air pollution and hospitalization for asthma. These findings provide relatively clear evidence of an association between respiratory illness and a range of pollutants (Committee of Medical Effects of Air Pollution, 1995a,b). Specific attention has focused on the effects of fine particulates(Schwartz, 1994; Dockery et al., 1993; Pope et al., 1995).

While, on balance, it is considered that these acute associations with air pollution represent real effects, the underlying mechanisms remain unclear; they may reflect exacerbations of pre-existing disease rather than the occurrence of new events (Committee of Medical Effects of Air Pollution, 1995a,b). To what extent, if any, the findings of the acute studies can be extrapolated to the possible effects of chronic exposure to air pollution is unknown. From a public health perspective, these potential chronic effects are vitally important because of the large numbers of people potentially exposed, *i.e.*, there may be large population attributable risk due to low levels of traffic-related outdoor air pollutants, even if the excess relative risk of such exposures is small. The limited evidence from chronic lung function studies suggests a small but significant negative association with particulate air pollution (Schwartz, 1989; Chestnut et al., 1989). There is also some evidence of a link between PM_{10} exposure and prevalence of chronic respiratory symptoms, especially in children (Schwartz, 1993; Dockery et al., 1993; Roemer et al., 1993). However, there are contradictory findings from a number of recent studies examining relationships between exposure to markers of traffic-related pollution and chronic respiratory health. For example, while several have reported positive associations between, variously, respiratory symptoms, hospitalization or lung function of children living close to roads compared to those living further away, or living in streets with high levels of traffic, self-reported truck-traffic or high predicted NO_2 levels (Murakami et al., 1990; Brunekreef et al., 1997; Osterlee et al., 1996; Wjst et al., 1993; Weiland et al., 1994; Duhme et al., 1996; Edwards et al., 1994; Pershagen et al., 1995), others have reported no association (Livingstone et al., 1996; Waldron et al., 1995; Briggs et al., 1997).

Unfortunately, most population studies of the health effects of chronic exposure to traffic-related pollution suffer from major problems of exposure assessment. Of the studies cited above, many used non-specific indicators such as proximity to main roads. Other studies which have focused on specific pollutants, such as particulates, nitrogen dioxide or ozone, have generally used levels of pollutants recorded at local monitoring stations to classify subjects into exposure groups. The key assumption underlying these approaches is that cumulative personal exposure correlates with a static measure of average ambient pollution concentrations or source activity (road traffic) in the vicinity of the individual's home location. The validity of this assumption clearly depends on the spatial resolution of the exposure measurements and the time-activity pattern of the subjects concerned. A number of studies have found poor correlation between measurements of ambient air quality from outdoor monitoring stations and indoor air pollution and/or personal exposure (for example, see Urell and Samet (1993) and Ozkaynak and Spengler (1996) and references therein). However, there is some evidence to suggest that residential indoor air quality and personal exposure of children both show a reasonable correlation with measures of ambient air pollution (nitrogen dioxide concentration) and road traffic density close to the home (Shima and Adachi, 1998; Hoek et al., 1989). Compared with working adults, children are likely to spend a significantly greater proportion of their time in the microenvironment where they live. It is therefore reasonable to assume that exposure estimates based on home locations are more likely to reflect personal exposure to air pollution in children than adults. There is also evidence that it is possible to reliably estimate either community average or individual multi-day (chronic) personal exposures using available outdoor air quality measurements, even when the cross-sectional correlation between acute personal exposure and ambient air quality is poor (Mage and Buckley, 1995; Wallace, 1996).

Other recent approaches to air pollution exposure assessment include time-activity studies (Noy et al., 1990), which typically lead to more reliable estimates of personal exposure. However, the expense and time demands of such studies makes the design infeasible for a population-based analysis. Vostal (1994) discusses the use of probabilistic models to estimate the chance of individual exposure to high pollution concentrations. This approach requires data on specific activity patterns for population sub-groups and is more relevant for estimating acute rather than chronic exposure to high levels of air pollution.

Other difficulties faced by epidemiologists studying the effects of chronic exposure to traffic-related pollution on respiratory health include the potential for confounding by risk factors that are unrelated to pollution but exhibit similar spatial patterns (*i.e.*, are correlated with pollution exposure). Such risk factors may include smoking, home environment (indoor air quality, presence of allergens etc.), and socioeconomic status. Seasonal effects, weather variables and infectious disease epidemics may also represent important risk factors for respiratory illness. However, to confound studies of the *chronic* effects of exposure to traffic pollution, the average or typical values for these variables would have to show spatial variation in a similar way to the pollution measures, which is unlikely. By contrast, weather and seasonal variables are major confounders for acute studies since they will tend to be correlated with short-term temporal variations in air quality.

2.1 Work leading to the present study

Given the above discussion, it is clear that great care is required when designing, analyzing and interpreting studies of the health effects of chronic exposure to traffic-related air pollution. Young children represent a particularly suitable group to study in this context for a number of reasons. Their respiratory and immune systems are immature, making them particularly vulnerable to the potentially harmful effects of air pollution; their activity patterns tend to be localized so that the bulk of their exposure to traffic-related pollution may be reasonably assumed to occur within the neighborhood of their home; and by definition, their cumulative lifetime exposure takes place over a relatively short period, thus eliminating concerns about long latency times and the need for historical exposure data.

This Case Study therefore aims to investigate the effects of exposure to vehicle emissions on chronic, non-malignant respiratory health in young children. Our motivation arises from a previous analysis of hospital admissions for childhood respiratory illnesses in relation to traffic density in London, UK, which was carried out by some of the present authors (Best et al., 1998a). Hospital admissions represent a valuable and as yet underexploited source of routine health data with which to investigate population morbidity in the UK. These data have only recently been made available to us (see Section 3 for further details), and our original analysis considered small area variations in admission rates by electoral wards in the Thames region (which includes London). Four health outcomes were considered: admissions for all respiratory illnesses in children aged less than one year and 1–4 years, bronchiolitis admissions in children under one year old and asthma admissions in 1–4 year old children. Information on exposure to traffic pollution was obtained from measured and modeled estimates of vehicle flows on all major and minor roads in Greater London (see Section 4). These were averaged over electoral wards to provide a measure of average annual traffic density in each small area. These data were analyzed using ecological regression models to relate ward-level counts of hospital admissions to the ward level summaries of traffic density and other area-level covariates such as socioeconomic deprivation. A Bayesian hierarchical modeling approach was adopted, using log-linear Poisson regression with Gaussian conditional autoregressive (CAR) smoothing priors (Besag et al., 1991) to capture the effects of unmeasured confounders and allow for spatial dependence in residual disease risk between neighboring areas. Our findings provided some evidence that, for infants under one year old, risks of respiratory illness, and specifically bronchiolitis, were related to high density of medium and heavy goods vehicles in the ward of residence. This is in accord with previous findings of reduced lung function in children related to diesel traffic (*e.g.*, Brunekreef et al. (1997)).

2.2 *The need for further analysis*

The Bayesian ecological regression approach used above is becoming increasingly popular as a method for investigating the relationship between geographical variations in disease and suspected environmental and other risk factors (Richardson et al., 1995; Bernardinelli et al., 1997; Knorr-Held and Besag, 1998). This approach offers the advantages of requiring only environmental exposure averages over geographical areas (which are often available even when individual level exposure data are not) and of exploiting spatial correlation in the data. However, inference may be sensitive to the choice of geographical units selected for analysis, and aggregation may distort or mask the true exposure-response relationship for individuals - a phenomenon known as ecological bias (see Section 6.1). By averaging traffic density over electoral wards (which range in area from 0.3 km^2 to 23.8 km^2 in London), it is possible that important local variations in exposure may be obscured in such a way as to distort the analysis. Since we are in the fortunate (and unusual) position of having exposure data available at a high level of spatial resolution, one of our aims in the present Case Study is to develop methods which are better able to exploit the raw data at their natural level of aggregation.

One advantage of using traffic density rather than measures of individual pollutants as the exposure estimate in our original study is that the former may be viewed as a proxy for the synergistic effect of the complex mix of pollutants emitted by vehicle exhausts. Nonetheless, investigation of the effect of specific substances such as particulates and acid gases like nitrogen dioxide is important if we are to better understand the mechanisms underlying the putative relationship between traffic-related pollution and respiratory health. Emissions data for a range of compounds associated with vehicle fuel combustion have recently been linked to our London road traffic database and these additional variables will also be included in the present Case Study.

The models we develop in this Case Study therefore aim to address some of the methodological limitations outlined above and extend our original analysis to provide a more accurate assessment of the effects of traffic pollution on children's health. For practical reasons, we choose to focus on just one of the respiratory outcomes investigated in our original analysis, that is, hospital admission for bronchiolitis in infants under one year old. A brief description of this illness is given below, together with a discussion of the possible factors influencing the likelihood of hospitalization for a child with bronchiolitis.

2.3 Hospital admissions for bronchiolitis

Bronchiolitis is a common disease of the lower respiratory tract of infants, resulting from inflammatory obstruction of the small airways. It occurs during the first 2 years of life, with a peak incidence at approximately 6 months of age. In many areas, it is the most frequent cause of hospitalization of infants: in our original study, 14% of all emergency admissions, and 37% of respiratory-related emergency admissions for children under one year old in London were diagnosed as bronchiolitis. The illness is viral and occurs both sporadically and epidemically; incidence is highest during winter and early spring.

In order to model the relationship between exposure to traffic pollution and incidence of bronchiolitis, we need to consider the sequence of events that must occur before a child is admitted to hospital with such a diagnosis. Figure 3.1 shows a schematic representation of this pathway. The first stage represents potential risk factors that may provoke an episode of bronchiolitis in an exposed child. An infant with bronchiolitis will first exhibit symptoms of mild upper respiratory tract infection such as runny nose and sneezing, which may be accompanied by fever. After several days, wheezy cough and dyspnea may develop. In mild cases, symptoms usually disappear in 1 to 3 days, but in more severe cases the child will be in severe respiratory distress and the parents may decide to seek medical help. This decision is represented by stage 2 in Figure 3.1. The parents may take the child directly to the Accident and Emergency (A&E) department of a local hospital, or may first seek advice from their family doctor (called a General Practitioner or GP). This choice is represented by stage 3 in Figure 3.1, and may be influenced by factors unrelated to the health status of the child. These include the relative accessibility of hospital and GP services, the parents' perception of the quality of care offered in each case and the parents' relationship with their GP. If the child is first taken to the GP, there are two subsequent pathways: either the GP will decide to treat the infant, in which case no record of the episode will appear in our hospital admissions database. Alternatively, the GP will refer the child to hospital (stage 4). Again, the factors governing the GP's decision may be unrelated to the health status of the child, but typically reflect the degree of diagnostic uncertainty, the facilities and treatments available to the GP, and the GP's training and experience and relations with hospital consultants. However, infants with respiratory distress should usually be hospitalized, hence most children with bronchiolitis who are first taken to their GP will end up in hospital. Given that the child presents at hospital (by whichever route), a decision will then be made as to whether or not to admit him/her for treatment (stage 5). At this stage, a formal diagnosis should be made and recorded in the patient's notes. In most cases, the diagnosis will be bronchiolitis, although the illness may be confused with bronchial asthma or bronchopneumonias. It is also possible that infants presenting with other respiratory symptoms will be mistakenly diagnosed as having bronchiolitis. The final stage (6) in our pathway from exposure to recorded diagnosis of bronchiolitis in our hospital database is the data entry process, whereby the diagnosis entered in the patient's notes is transcribed onto the computerized database. The potential for errors at this stage is discussed in more detail in Section 3.2.

The complexity of factors influencing the pathway shown in Figure 3.1 offers clear potential to distort our investigation of the relationship between exposure to traffic-related pollution and risk of hospitalization for bronchiolitis in young children. While our routine data sources do not provide sufficient information to learn about every stage shown in the figure, we attempt to account for as many aspects as possible in the statistical models described in Sections 6 and 7. The possible mechanisms underlying this pathway should also be bourne in mind when attempting to interpret the results of our analyses.

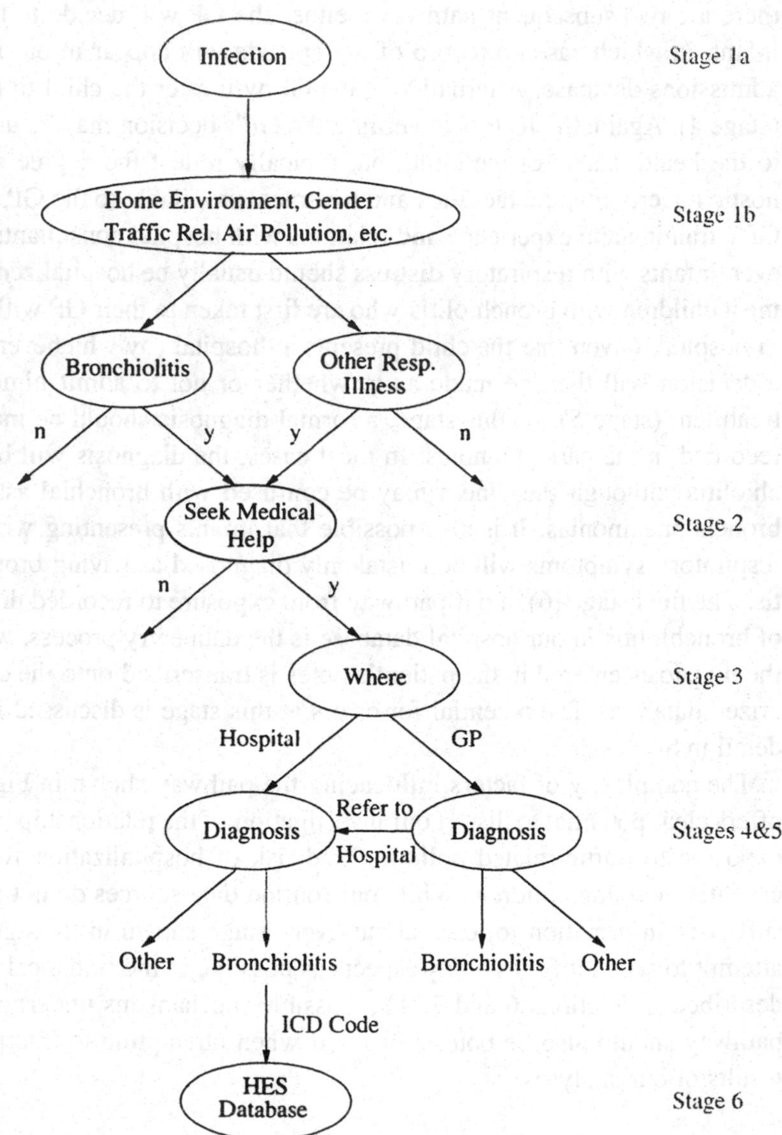

FIGURE 3.1. Schematic representation of the stages involved between a child developing bronchiolitis, being admitted to hospital with such a diagnosis, and the diagnosis being recorded in the Hospital Episode Statistics (HES) database used in our study.

3 Health, Population and Sociodemographic Data

The data used in this study were obtained from several different sources and were recorded at a variety of different geographic scales. In this section we define the study region and describe the health event data, population data and sociodemographic data and method of linkage; the traffic-related exposure data is described later in Section 4.

3.1 Study region geography

The study region chosen for analysis comprises the part of North West Thames Regional Health Authority bounded by the River Thames on the south and by the M25 London orbital motorway on the north and west. This subset of the Thames region studied earlier (Best et al., 1998a) was chosen to maintain a geographically diverse study area so that spatial features might emerge, but reduce the number of health events to a manageable enough size to permit a more detailed spatial analysis. Additionally, traffic exposure data were only available to us on roads within the M25.

Each form of spatial data we use has its own level of spatial aggregation. At the finest resolution we have the postcode of residence for each child. Postcodes are used for the delivery of mail within the UK (Raper et al., 1992). Each building or residence in the UK is assigned to some postcode. Approximately 1.4 million of the approximately 1.6 million UK postcodes are comprised entirely of residential addresses, with a median of twelve residences per postcode. Areas covered by a single postcode have no formal boundaries and can vary widely in size: in urban areas they are typically a few hundred meters across, although rural postcodes often cover a much larger area. The easting and northing (x and y) coordinates of the south-west corner of the 100m National Grid square containing the "first" residence (this is not systematically defined) of each postcode are available from the Central Postcode Directory (`http://census.ac.uk/cdu/Datasets/Lookup_tables/Postal/Central_Postcode_Directory.htm`); we convert these to a nominal centroid by adding 50m to both coordinates, as suggested in Gatrell et al. (1991). Figure 3.2 shows an example of how these postcode coordinates and centroid are defined for one postcode.

Census and many other socioeconomic data in the UK are available only at aggregated levels of geography. The smallest geographical unit for which Census data are available is termed the *Enumeration District*, or ED, of which there are 151,719 in the UK. The geography of EDs does not reflect the underlying socioeconomic characteristics of the population;

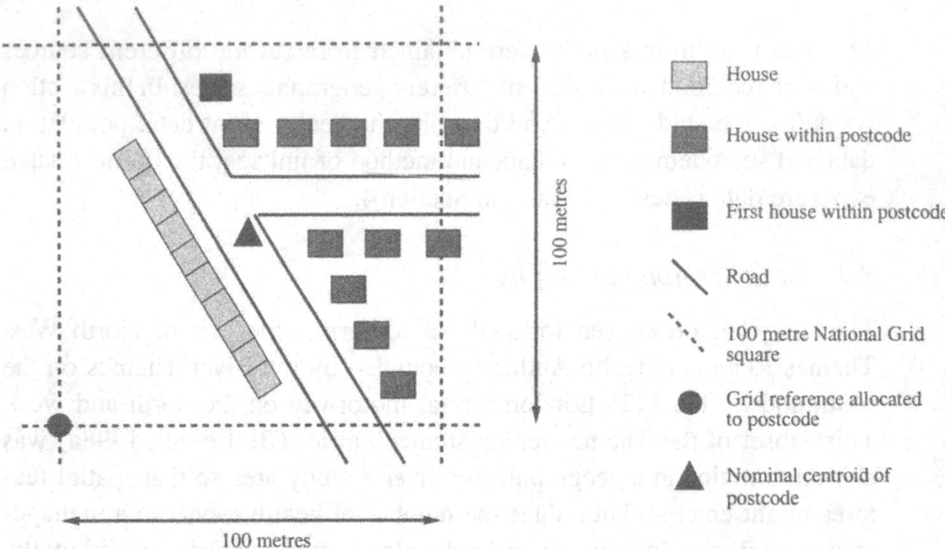

FIGURE 3.2. Example of how the postcode coordinates and nominal centroid are defined for one postcode.

rather, they are administrative boundaries arbitrarily imposed for ease of collection of the Census data. The geographical extent of EDs therefore varies considerably, with rural EDs often much larger than urban ones. On average an ED contains around 10–15 postcodes. A postcode can be linked uniquely to the ED within which its centroid is located through the use of annually updated Postcode to Enumeration District lookup tables (`http://census.ac.uk/cdu/Datasets/Lookup_tables/Postal/Postcode_Enumeration_District_Directory.htm`), available from the data archive at the Manchester Information and Associated Services (MIMAS) (`http://mimas.ac.uk`).

The next larger unit of geographical aggregation in the Census hierarchy are termed electoral wards, of which there are 10,933 in the UK. EDs nest exactly within wards and it is therefore straightforward to identify which ward an ED lies within, and also to link postcodes to wards. Within the study area defined for this Case Study, there are 269 wards, 5,060 EDs and 74,622 postcodes.

3.2 *Health events*

Hospital activity data have been collected in the UK since 1949 from all National Health Service (NHS) hospitals, originally as paper records and more recently in digital format. The current computerized reporting system, known as the Hospital Episode Statistics (HES), came into operation in 1987. The HES database is maintained by the UK Department of Health with a copy held by the Small Area Health Statistics Unit (SAHSU) in the Department of Epidemiology and Public Health at Imperial College, UK. We use the HES database to obtain information on hospital admissions for infants living in the study region.

It is generally recognized that the HES transition period of 1987–91 was disruptive, compromising data coverage, completeness and accuracy (Williams et al., 1994). The improvement in accuracy in recent years probably reflects the fact that, since 1991, HES data have been used for contracting in the internal NHS market. In the UK, hospitals act as 'providers' who supply medical services to 'purchasers' such as other hospitals and general practitioners, using information recorded in the HES database to bill the purchasers.

Background to HES data

The basic unit of the HES database is the *finished consultant episode*, or FCE, covering the period during which a patient is under the care of one consultant. Every NHS hospital in England has to submit electronically the forty or so HES data items for each FCE in each patient's hospital stay. The data items are entered to the hospital's own digital Patient Administration System by trained clinical coders working with either patient notes from the ward trolleys or with discharge summaries. The required data items include some demographic data, such as date of birth, sex and postcode of residence, but to ensure confidentiality they do not include the patient's name, NHS number, or any other unique identifier. Clinical data such as primary and secondary diagnoses and dates and details of any operations performed within the patient's stay are also recorded on the HES system. Diagnoses are coded using the International Classification of Diseases (ICD) codes, often with the help of computer software. Clinical coding at this crucial stage is somewhat subjective and is not subject to any external audit, so clearly there is some scope for inter-coder variation (Walshe et al., 1993; Dixon et al., 1998). The HES database grows by some ten million records per year.

HES data quality

There are three principal aspects of data quality:

- Coverage: the proportion of total hospital inpatient activity recorded in the database;
- Completeness: the proportion of records that have an entry in any specified field;
- Accuracy: how well the database entries reflect the true value of a particular field.

Coverage: In addition to electronic entry of inpatient episode data for HES, hospitals are obliged to submit paper counts (called KP70s) of patient episodes, collected independently of the HES counts. The Department of Health use KP70 totals as the 'gold standard' against which HES is compared. Discrepancies are investigated by the Department. Williams (1997) found that HES was nationally within 1.2% of the KP70 figure for the period 1989–95, and Sheldon et al. (1994) generally found HES and KP70 totals to be in close agreement.

The HES database includes only NHS (*i.e.*, publicly supported) admissions, systematically excluding all private hospitals and private patients treated in NHS hospitals. Unlike in the USA however, the majority of patients in the UK still receive NHS treatment and so will be captured on the HES database if admitted to hospital. The proportion of emergency admissions in young children that are treated privately is also likely to be small.

Completeness: HES data are subject to a large number of automated validation checks during data cleaning (Department of Health, 1998). A recent descriptive analysis of the HES data held by SAHSU found close to 100% completeness for the sex, date of birth, postcode and admission date fields, and over 98% completeness for the primary diagnosis field. However, some hospital Patient Administration Systems insert the vaguest ICD code ('799', in version 9 of the ICD) by default in the absence of any specified condition, thereby having fewer missing primary diagnoses but commensurately more completely ambiguous codes (McKee and Petticrew, 1993).

Accuracy: Most studies that have investigated the accuracy of clinical coding relate to data collected before the current HES system was introduced. One recent study focused on the reproducibility of clinical codes at two large acute hospitals within the former North West Thames region between 1991 and 1993, comparing local, external (with at least four years' experience) and the most senior coding manager (Dixon et al., 1998). Exact (four-digit ICD code) agreements on the main diagnosis were 41% at one hospital and 59% at another hospital; the figures for approximate (three-digit ICD code) agreement were 52% and 69% respectively.

Diagnostic ICD codes are not the only fields subject to error. In particular, the HES database demands that the patient's postcode field not be left blank. This leads to the use of "dump" codes, such as the postcode of the patient's general practitioner (GP) or the hospital's own postcode, when the patient's postcode is unknown. Such a practice may lead to false local 'clusters' of disease if the postcode is used to locate cases in a spatial analysis of HES data.

In summary, the HES database provides a relatively complete record of all NHS hospital admissions in England. In particular it includes essentially all emergency admissions for 0–1 year old children in our study region. The fraction of missing fields within each record is also negligible. However, the (in)accuracy of the data values entered on the HES database is of much greater concern, especially for the primary diagnosis code. Our modeling approach will therefore attempt to address the problem of inter-hospital variation in diagnostic coding and the use of vague ICD codes.

Data used for the Case Study

HES data are collected by fiscal year, which runs from April to March in the UK. Due to concerns over data quality for early years of HES data, we restrict our analysis to the three year period from April 1992 to March 1995. This period also coincides with the time frame over which the road traffic exposure data were collected (see Section 4). Data for more recent years are still being checked and cleaned by the data providers and are not yet available to us.

The subset of data considered for this study therefore comprises all emergency hospital admissions in the period April 1st 1992 to March 31st 1995 for children aged under one year on the day of admission and whose postcode of residence lay within the northwestern part of London that defines our study region. We defined an admission to be the first episode of each hospital stay which ended within the study period. Thus a child admitted on two separate occasions (stays) and having two separate first episode entries would constitute two admissions in our analysis. Conversely, a child who was admitted just once during the study period but was referred to a number of different consultants would acquire more than one episode but would only constitute one admission. Ideally, we would have preferred to carry out a person-based analysis, but in the absence of a unique patient identifier in the HES database it is not possible to link multiple admissions for the same patient. We considered trying to link such admissions by matching records (either exactly or probabilistically) according to the sex, date of birth and postcode fields. However, potential problems with coding errors and changes of address (postcode) were deemed to make this approach unreliable. Our data are thus admission-based, and so are not strictly independent (the same child may be admitted multiple times during our study period). Nonetheless, we argue that it is a reasonable working hypothesis to assume independence for modeling purposes, since a rough estimate based on exact matching of sex, date of birth and postcode suggested that repeat admissions comprised less than 20% of our data.

Data items extracted from the HES database for our study subset included postcode, sex, age, primary and secondary diagnosis codes and hospital. Before classifying the subset into groups for analysis, the following exclusion criteria were applied:

- All admissions for accidents and poisonings —this criterion was imposed to minimize possible confounding between the putative effect of exposure to traffic-related air pollution associated with living close to busy roads, and the increased risk of accidents on or near such roads;
- All birth-related admissions —this criterion was imposed since birth-related admissions were deemed to be qualitatively different from other types of infant emergency admissions.

The remaining admissions were then classified into three groups according to the primary diagnosis field:

Bronchiolitis cases (B):	ICD-9 code 4661
Non-respiratory controls (N):	Any ICD-9 code except codes 460–519 (respiratory diseases) and those covered by the exclusion criteria above.
Unspecified diagnosis (U):	ICD-9 code 799

This gave a total of 2,107 cases of bronchiolitis, 9,904 non-respiratory controls and 1,220 admissions with an unspecified primary diagnosis. As a crude check on the use of "dump" postcodes in these data we calculated the number of admissions with a postcode corresponding to that of the hospital to which the child was admitted: only 5 such admissions were identified, suggesting that there is no systematic tendency for hospital postcodes to be used when the patient's postcode is unknown.

3.3 Population denominator data

In order to make inference about population disease rates we require an estimate of the spatial distribution of the population at risk, *i.e.*, the distribution of 0–1 year old children living in our study region. There are several problems with the obvious approach of using the most recent UK census counts. In particular, the population counts vary from year to year, and census data are only available for 1991 while our study period covers 1992–1995; furthermore, the smallest aggregation unit by which census counts are available is the ED, whereas our health event data are geo-referenced to the much finer postcode level. We therefore elected to use the national register of births as our source of population data. Registration of births is mandatory in the UK, and all records, including the sex, date of birth and home postcode of the baby, are held on a database by the UK Office for National Statistics. A copy of this database is also held by SAHSU. These data are thus available at the same spatial resolution and cover the same time period as the health events; they are also likely to be more complete than the census counts.

For the purposes of this Case Study, we therefore employ as a population estimate the number of babies born in each postcode between October 1st 1991 and September 30th 1994. This gave a total of 159,706 births. If no child moved house during this period, this would represent the cumulative number of children aged 0–1 year living in each postcode on the mid-point of each financial year (September 30th) of our study period. Clearly some migration of families with newborn babies will occur, but we expect this to represent only a small fraction of the total. Often the family will only move house within the same residential area and so interpolation or aggregation of the birth totals over localized areas (as is done in Section 5) should reduce errors in our population estimates due to migration.

3.4 Sociodemographic and other covariate data

In addition to the putative effect of traffic-related air pollution on respiratory health in young children, there are a number of other variables which may represent important risk factors for respiratory illness and which may confound any relationship with exposure to traffic pollution. In particular, the indoor environment of the child's home may represent an important source of exposure to air pollution: combustion of fossil fuels from any unvented source (*e.g.*, gas stove or water heater) and parental cigarette smoking may raise concentrations of pollutants such as nitrogen dioxide (NO_2) and fine particulates (Dockery et al., 1981; Quackenboss et al., 1986, 1991). Mould/mildew associated with damp housing conditions may also exacerbate risk of respiratory illness (Strachan, 1991). Boys have generally been found to be more susceptible to respiratory illnesses than girls (Strachan et al., 1994), and illness rates may also vary by race. Seasonal variations in temperature and pollen counts may also affect the risk of respiratory illness (Anderson et al., 1998); however, the present study is concerned with the effects of *chronic* exposure to traffic-related air pollution, and so short-term temporal variations in risk are less relevant.

Unfortunately, the routine datasets used to obtain the health events and population data for this study contain little individual-level covariate information. For the present purposes, the only relevant individual attribute recorded with each birth or hospital admission was gender. For the health event data, the name and postcode of the hospital to which the child was admitted were also recorded as individual attributes for possible study of hospital effects.

In the absence of individual-level covariates, we must attempt to use aggregate measures of the relevant risk factors to reflect the 'typical' value of that covariate among individuals in the group. For this purpose, we group individuals according to their area of residence. Census data, available at the ED level, may then be used to provide information on the ethnic mix and socioeconomic characteristics of each area (ED). Specifically, we calculate the proportion of 0–4 year old children (the finest age strata recorded on the Census) in each ED from each of the following ethnic groups: White, Black (including Black African and Caribbean), Asian, Indian (including Pakistani and Bangladeshi) and Other. Specific information on smoking rates and housing conditions are not available from the Census. However, the Carstairs measure of socioeconomic deprivation (Carstairs and Morris, 1991) has been shown to correlate strongly with cigarette consumption (Kleinschmidt et al., 1995) and may also be viewed as a proxy for poor housing conditions and lifestyle factors which are often correlated with ill health (Jolley et al., 1992).

Accessibility to hospital and to primary care (GP) facilities is likely to influence a parent's decision on choice of care for a sick child. The University of York carried out a small area analysis of hospital utilization rates (Carr-Hill et al., 1994) and derived measures of hospital and GP accessibility for electoral wards in England. These scores are distance-weighted counts of the numbers per capita of hospital beds and GPs, respectively. Maps showing the geographical distribution of these covariates are given in Section 5.

4 Traffic Exposure Data

In this section we describe the raw data and methods used to derive appropriate measures of exposure to traffic-related pollution at two levels of geographical resolution (lines and grids).

Geographical Information Systems (GIS) are tools that enable the integration, manipulation, analysis and display of datasets that are geographically referenced. GIS techniques are increasingly being used within health-related disciplines and applications due to the inherently spatial nature of many health issues ((Gatrell and Senior, 1999); (Smith and Jarvis, 1998); (Cummins and Rathwell, 1991)).

A GIS database containing the spatial datasets for this Case Study was developed using the GIS software packages ARC/INFO and ArcView (`http://www.esri.com/`). The raw datasets were obtained at varying levels of resolution, generalization and accuracy and thus various GIS techniques were required to assemble, integrate, overlay and process them. The road traffic GIS database involved three main datasets: a *stick network* of modeled flows and emissions on (mostly) major roads; a set of $1km \times 1km$ *squares* containing modeled flows and emissions on minor roads; and a *vector dataset* of roads in London.

4.1 Traffic flow data

Major roads (stick network)

The London Research Centre (LRC) produces modeled estimates of traffic flows from road traffic in major urban areas, including London. Major roads are loosely defined as roads for which traffic flow data are available from the London Transportation Studies (LTS) model (Buckingham et al., 1997). In practice, this includes most roads with a traffic flow in excess of 400 vehicles per hour (during a morning weekday peak period). In central London this gives intense coverage of roads, while in the outer areas it is mostly only main roads (*i.e.*, A roads and motorways) which are included. The data are modeled for a series of *links* or 'sticks' *i.e.*, straight lines connecting specified start and end nodes.

The LRC uses output from the LTS transport model as their basic activity data. The LTS model is a well-established tool, widely used for transport planning in London, and maintained on behalf of the Government Office for London (GOL) by the MVA Consultancy. The LTS model is the only source of comprehensive traffic flow data available on a London-wide scale and on an internally-consistent basis. The model gives coordinates for the start and end points of the links in the stick network; figures for the flow of cars, light vehicles and goods vehicles per hour per period; and link length and transit time. Validation of such models is non-trivial. However the LTS model has been validated against traffic surveys and appears to perform reasonably well (MVA Limited, 1999).

The LRC processes this basic activity data in a number of ways to build up a detailed picture of traffic flow on major roads in London. Initially the network is clipped to include only those links within and including the M25 orbital motorway. Average link speed is then calculated using the distance/transit time data. The basic vehicle-type figures from the LTS model (*e.g.*, motorcycle, car, heavy goods vehicle) are then refined to reflect the likely distribution of more specific vehicle types (*e.g.*, petrol engined car of < 1.4 liters and without catalyst). This is done with reference to Department of Environment Transport and Regions (DETR) traffic counts averaged across parts of the study area. Vehicle fleet statistics are also used to determine the proportion of petrol/diesel vehicles and the proportion of cars with catalytic converters.

The above data on modeled traffic flows for major roads on the stick network were supplied to us by the LRC for use in this Case Study. Further processing was required to produce estimates of the total annual traffic flows on each line segment for use in our statistical models. The data were provided as typical hourly bi-directional flows for weekdays during morning-peak (7am to 10am), evening-peak (4pm to 7pm) and off-peak (10am to 4pm) periods. These were converted to annual flows for each major road using factors provided for nightly flows (weighted by whether the road was in Central, Inner or Outer London) and for Saturday and Sunday flows. Bi-directionality was removed simply by summing the total flow for each link.

These estimates gave the annual flows for 1991 but they may also be used for more recent years since most of the major roads in London (except perhaps the M25) were considered to have reached capacity by 1991 and could not carry any more traffic.

Minor roads (grid squares)

LRC also produces modeled traffic flows on minor roads for 1km $\times$ 1km squares (based on the National Grid) approximately covering the area within the M25 motorway. A minor road is simply defined as one for which link-based traffic flow data is not available *i.e.*, a road which is not part of the assigned highway network used in the LTS model.

Detailed traffic flow data are rarely available for minor roads, therefore traffic flows usually have to be estimated on an aggregated basis. This means that the minor roads are essentially treated as area sources. The LRC estimates the total vehicle-kilometers driven on minor roads by taking an estimate of the total vehicle-kilometers driven on the entire road network and subtracting the kilometers already accounted for by the major roads in the LTS model. The estimate for the entire network is obtained from the DETR Rotating Traffic Census. This gives a total traffic flow for all minor roads in the network which must then be apportioned across the grid squares. The LRC achieves this by estimating road network density using a combination of the Ordnance Survey's OSCAR database of road center-lines and manual estimates from the Geographer's A-Z . The total kilometers for each grid square is then further broken down into the various vehicle categories as per the major roads.

For this Case Study, modeled traffic flow data were obtained from the LRC for the following vehicle categories: motorcycles, petrol cars, diesel cars, buses, petrol light goods vehicles, diesel goods vehicles, and medium and heavy goods vehicles. The data provided were yearly totals for 1995.

4.2 *Emissions data*

Traffic flows are a useful proxy for exposure to traffic-related pollution. However they do not provide information about the loads of specific pollutants to which the population is exposed. Ideally, this sort of information would be obtained either by air quality monitoring of ambient pollution levels or by monitoring of the pollutant load being received by individuals. In practice, however, air quality monitoring is expensive and only gives point-based observations, whereas emissions from traffic are along lines (*i.e.*, roads). Individual monitoring is also costly and can therefore usually only be carried out on a small number of subjects. In addition, monitoring does not identify the sources of the pollutants. An alternative method to using monitored pollution concentration data is to employ atmospheric emissions inventories. An air pollution emissions inventory is a schedule of the sources of pollutants within a particular geographical area (Buckingham et al., 1997). A comprehensive emissions inventory will usually include not only pollutants from traffic but also emissions from other mobile sources (such as rail, rivers and air) and emissions from stationary sources such as industrial processes, and waste disposal.

In the UK, a National Atmospheric Emissions Inventory (NAEI) is maintained by the National Environmental Technology Centre (NETC) on behalf of DETR. The NAEI contains data on a wide range of pollutants and is published for 10km × 10km grid squares for the UK. The NAEI, however, is a *top down* inventory whereby national data is allocated to smaller areas on the basis of the resident population and other indicators of activity levels. By comparison, the LRC produces detailed atmospheric emissions inventories for specific urban areas. These are *bottom up* inventories which compile data at the local level and are generally available at a finer level of geographical resolution than the national inventory.

In producing local emissions inventories, measures of activity relating to the emissions are collected and emissions factors are then applied to the activity data in order to estimate the likely emissions:

$$\text{Activity rate} \times \text{Emission factor} = \text{Emission rate}\,.$$

The LRC has produced an atmospheric emissions inventory for London (the London Atmospheric Emissions Inventory, LAEI, (Buckingham et al., 1997) which is available at the 1km × 1km square level. The activity data used in the inventory are derived from a wide range of sources and different time periods, but the nominal year of the inventory is 1995.

For this Case Study we focus on traffic-related emissions which comprise the major source of air pollution in London. Emissions from traffic on major and minor roads and also cold starts and hot soaks (see subsection *Cold starts/Hot soaks*) were obtained from the LAEI from the LRC for the following pollutants: sulphur dioxide (SO_2), nitrogen oxides (NO_x), carbon monoxide (CO), carbon dioxide (CO_2), non-methane volatile organic compounds (NMVOC), benzene, 1,3-butadiene, black smoke, total suspended particulates (TSP), particulate matter of less than $10\mu m$ aerodynamic diameter (PM_{10}) and methane (CH_4).

Major roads

The LRC calculates modeled levels of emissions for each link in the stick network described in Section 4.1. Using the traffic flows described above, emissions factors from the UK Emissions Factors Database (`http://www.london-research.gov.uk`) are applied to each link depending on the vehicle flow, speed and road type of the link. This gives the total amount (in metric tonnes per annum) of each pollutant type emitted from road traffic for each individual link in the stick network.

Using GIS line-in-polygon techniques the LRC also re-apportions the emissions data from the stick network to the 1km × 1km squares, based on the length of each link contained within each grid square. For each pollutant, this gives the total emissions from traffic on major roads per grid square in tonnes per annum per km^2.

Minor roads

The LRC uses a different technique to calculate emission levels for the minor roads because data on average speeds are not available for these roads and so speed-related emissions factors cannot be used. Instead, typical (urban) driving-cycle based emissions factors (from the UK Emissions Factors Database) are employed to estimate pollutant specific emissions (in tonnes per annum per km^2) on minor roads in each 1km × 1km square.

Cold starts/Hot soaks

When vehicles are first started they emit higher levels of pollutants than when they are at their normal operating temperature. These emissions occur at the start of trips (called *cold starts*) and need to be estimated separately from the major and minor roads emission levels. This is done by calculating the number of trip-starts in each grid square using information about trip origins and destinations from the LTS model. The LTS trip-matrix gives trip-ends data for a series of polygons which the LRC then re-allocates to grid squares based on the proportion of each polygon's area contained within each grid square. The number of trip-starts are then multiplied by relevant emissions factors to give total pollutant loads arising from cold starts for each grid square. These calculations are based on the assumptions that all cold start emissions occur within 1km of the trip start and that average ambient temperatures, catalyst performances and times since vehicle was last operated apply.

Similar methods are used to estimate emissions from *hot soaks* which occur as evaporative emissions when the engine is switched off. Benzene and NMVOCs are the only pollutants relevant to hot soaks.

Total emissions from all road traffic by 1km × 1km square are obtained by adding the estimated cold start and hot soak pollutant loads to the estimates for minor and major roads.

4.3 Exposure measures used for the Case Study

Major roads

One option for the traffic-related exposure measures used in this Case Study is to take the traffic flow and pollutant loads on the stick network and 1km × 1km squares as supplied by LRC. However, we were concerned that, at the spatial resolution of our intended analyses, there may be important inaccuracies in these data. Specifically, the stick network from which the LRC data are derived is only a straight line approximation to the major roads network. The stick network represents a simplified model of the true location of roads and, as such, may contain inaccuracies which are non-

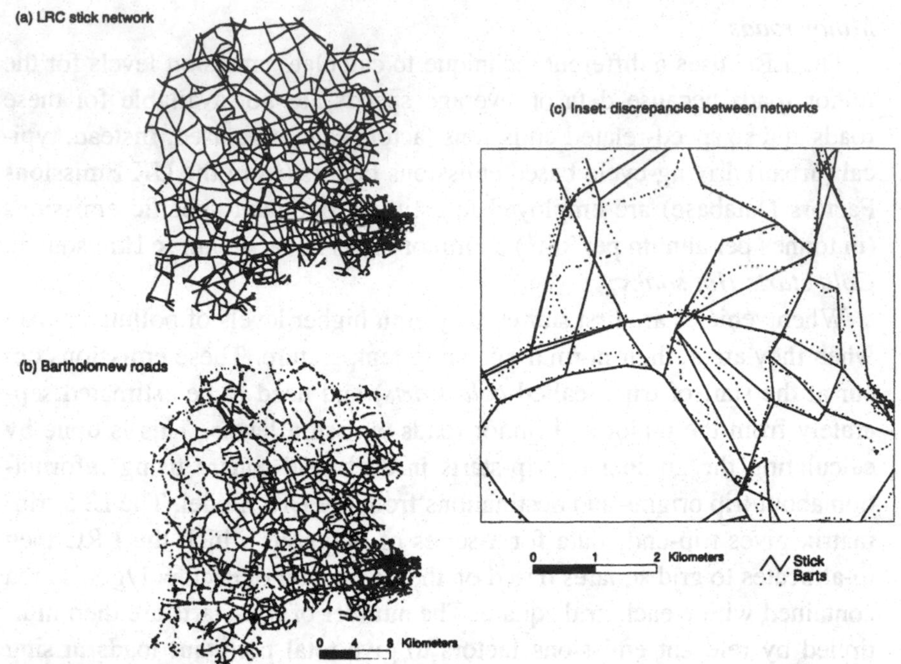

FIGURE 3.3. Road networks: (a) LRC stick network, (b) Bartholomew roads. Inset (c) demonstrates differences in networks in terms of generalization and accuracy of road locations.

negligible when the data are used for detailed small-area spatial analyses.

Simple examination of the degree of error suggests that some roads may be as much as 1km away from their true location; the average discrepancy is approximately 100–200m. We therefore obtained the Bartholomew digital road data (`http://www.bartholomewmaps.com`) for London. This is a very detailed dataset taken from original 1:5000 source data for London roads, and the degree of resolution and accuracy make it a much more realistic representation of the true road network than the LRC stick network.

Figure 3.3 shows the LRC stick network and the Bartholomew roads. The inset presents an example of the discrepancies between the two networks in terms of the generalization and accuracy of road locations.

Our goal was to link the LRC stick network with the Bartholomew road data and to then transfer the traffic flow and emissions data from the stick network onto the Bartholomew roads. The straight line links of the stick network did not match the Bartholomew roads sufficiently well to allow fully automated matching. A combination of automated and manual editing facilities within ARC/INFO was employed to manually link the sticks to the roads. This linkage of the stick and road networks was a large undertaking, but it was deemed to be worthwhile because of the potential usefulness of the dataset for the current study and future health-related research within the London area.

The matching work was shared between the Department of Epidemiology and Public Health at Imperial College (DEPH) and the London School of Hygiene and Tropical Medicine (LSHTM). The area was segmented into 5km grids (following the Bartholomew tiles) and a common strip down the middle was defined for both institutions to link. This allowed comparison of the links created by the two institutions, as a measure of quality control. Visual comparisons of the linkages for the whole area were also made at DEPH.

Using this link file, the attribute information on the stick network was transferred onto the Bartholomew road coverage, to produce annual traffic flow data for all major roads in London. For the subsequent analyses, the traffic flows were grouped into two vehicle categories: i) cars and light vehicles *(CLGV)*, and ii) medium and heavy goods vehicles *(MHGV)* (where buses were included in medium and heavy goods vehicles).

Using GIS *line-in-polygon* techniques, we also transferred the attribute data from the Bartholomew roads onto the set of 1km × 1km squares.

Minor roads

The minor roads data were supplied pre-processed by the LRC on the 1km × 1km grid squares. For the purposes of the subsequent analyses the traffic flows were aggregated into the same two vehicle categories as the major roads, namely: *cars and light vehicles* (comprising motorcycles, petrol cars, diesel cars, petrol light goods vehicles, diesel light goods vehicles) and *medium and heavy goods vehicles* (comprising buses and medium and heavy goods vehicles).

Emissions data

Eventually we aim to link the emissions data from the stick network to the Bartholomew roads in the same way as for the traffic flows. However, this work has yet to be completed and for the purposes of this Case Study, we use the line segment and grid square datasets derived from the stick

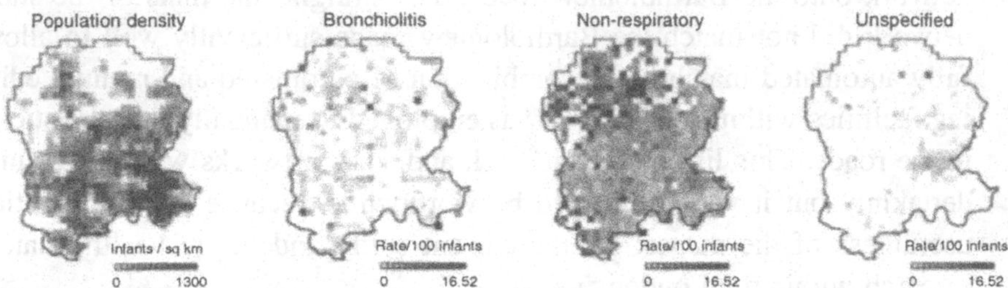

FIGURE 3.4. Population density (infants per square kilometer) and raw admission rates (per 100 population) in the training database for bronchiolitis, non-respiratory illness and unspecified causes.

network obtained directly from LRC.

5 Exploratory Analyses

For later model validation we divided our dataset of hospital admissions into separate training and prediction samples. Approximately 10% of the admissions in each diagnostic group (B, N and U) were flagged using uniform random numbers and set aside for later use as a validation sample. All exploratory analyses and model estimation reported in this and subsequent sections were carried out on the remaining 90% subset.

For a continuous estimate of population density across the study region we used the count of births within each kilometer grid cell, scaled by a factor of 0.90 for use with the training sample and by 0.10 for use with the validation subset.

Figure 3.4 presents maps showing the population density estimates and raw hospital admission rates per 100 population for the training sample on the kilometer grid.

Our data include admissions to the twenty-seven hospitals within the study region and to a small number outside the region: the latter were assigned a single hospital code (28) for the purposes of this analysis. Table 3.1 shows the number (and percentage) of admissions in each diagnostic group in our training database for each hospital. This indicates marked variations between hospitals, particularly in the proportion of unspecified ICD codes recorded by each hospital.

Table 3.2 shows the number of admissions stratified by method of inpatient admission: 'A&E' indicates referral by the hospital's Accident and Emergency department, 'GP' indicates referral by a general practioner, 'Consultant' indicates referral by another hospital consultant and 'Other' indicates other methods of admission (often a referral from the A&E department of *another* hospital). Hospital admission rates by each method seems consistent across diagnostic groups, although somewhat fewer GP-referred admissions end up with an unspecified diagnosis compared to admissions by the other three methods.

The remaining entries in Table 3.2 summarize the number of admissions in the training database by various covariate categories. For this purpose we have arbitrarily categorized the continuous covariates (deprivation, grouped by quintile, and access scores, group by tertile) and assigned to each area the ethnic group of the majority of children living there. This table highlights a number of differences between the covariate distribution for unspecified admissions compared to bronchiolitis or non-respiratory admissions: the former tend to live in more deprived areas with a high proportion of children from ethnic minorities and high access to hospital and GP services. Admissions for bronchiolitis and non-respiratory diagnoses show quite similar covariate profiles, although a slightly higher proportion of boys are admitted for bronchiolitis.

For the continuous models introduced in Section 7 the hospital identifier may be treated as an individual attribute for each subject, but for the discrete area-level analyses of Sections 5.1 and 6 we must aggregate all covariates to a common level of aggregation, for which we chose the kilometer grid. To achieve this, we first smoothed the hospital frequencies by averaging over a coarser 3km × 3km grid, calculating for each such 9km^2 cell the proportion of all admissions with a postcode centroid in that cell who attended each of the 28 hospital groups. This 28-dimensional vector was then assigned to each of the nine one-kilometer grid cells comprising a particular 3km × 3km cell. The geographical distribution of these *relative admission frequencies* for the 14 hospitals receiving most admissions are displayed graphically in Figure 3.5; not surprisingly, these maps reveal marked geographical variation.

A similar procedure was used to calculate area-level relative frequencies of hospital admission by each method ('A&E', 'GP' or 'Other'), after merging the rather small 'Consultant' and 'Other' groups. Maps showing the geographical distribution of these relative frequencies are presented in Figure 3.6. Again, a very strong spatial pattern emerges.

Hospital	Primary Diagnosis						Total
	Bronchiolitis		Non-Respiratory		Unspecified		
1	165	(11.7%)	1132	(79.9%)	119	(8.4%)	1416
2	250	(17.0%)	1078	(73.4%)	141	(9.6%)	1469
3	212	(16.6%)	1049	(82.2%)	15	(1.2%)	1276
4	213	(17.5%)	916	(75.2%)	89	(7.3%)	1218
5	147	(15.1%)	792	(91.5%)	33	(3.4%)	972
6	226	(21.9%)	780	(75.6%)	26	(2.5%)	1032
7	205	(21.4%)	714	(74.5%)	40	(4.2%)	959
8	112	(19.0%)	468	(79.2%)	11	(1.9%)	591
9	91	(15.5%)	412	(70.3%)	83	(14.2%)	586
10	74	(9.9%)	289	(38.8%)	381	(51.2%)	744
11	60	(18.0%)	265	(79.3%)	9	(2.7%)	334
12	43	(15.1%)	186	(65.5%)	55	(19.4%)	284
13	17	(8.3%)	189	(91.7%)	0	(0.0%)	206
14	6	(4.9%)	115	(93.5%)	6	(4.9%)	123
15	1	(1.1%)	59	(65.6%)	30	(33.3%)	90
16	15	(22.7%)	48	(72.7%)	3	(4.5%)	66
17	6	(11.3%)	44	(83.0%)	3	(5.7%)	53
18	8	(15.4%)	44	(84.6%)	0	(0.0%)	52
19	4	(8.7%)	29	(63.0%)	13	(28.3%)	46
20	2	(8.0%)	21	(84.0%)	2	(8.0%)	25
21	0	(0.0%)	19	(86.4%)	3	(13.6%)	22
22	2	(12.5%)	14	(87.5%)	4	(25.0%)	16
23	4	(18.2%)	14	(63.6%)	3	(13.6%)	22
24	0	(0.0%)	15	(83.3%)	3	(16.7%)	18
25	7	(36.8%)	11	(57.9%)	1	(5.3%)	19
26	1	(11.1%)	5	(55.6%)	3	(33.3%)	9
27	2	(25.0%)	4	(50.0%)	2	(25.0%)	8
28	33	(12.3%)	214	(79.6%)	22	(8.2%)	269
Total	1902	(15.9%)	8926	(74.9%)	1097	(9.2%)	11925

TABLE 3.1. Number of admissions (percentage of hospital total) for each diagnostic group in the training database, split by hospital.

	Primary Diagnosis						Total	
	B		*N*		*U*			
Admission Method								
A&E	1162	(61.1%)	4868	(54.5%)	706	(64.4%)	6736	(56.5%)
GP	577	(30.3%)	2417	(27.1%)	155	(14.1%)	3149	(26.4%)
Consultant	30	(1.6%)	282	(3.2%)	39	(3.6%)	351	(2.9%)
Other	133	(7.0%)	1359	(15.2%)	197	(18.0%)	1689	(14.2%)
Hospital Access Score								
Low	566	(29.8%)	2605	(29.2%)	134	(12.2%)	3305	(27.7%)
Medium	889	(46.7%)	3928	(44.0%)	535	(48.8%)	5352	(44.9%)
High	447	(23.5%)	2393	(26.8%)	428	(39.0%)	3268	(27.4%)
GP Access Score								
Low	575	(30.2%)	2654	(29.7%)	158	(14.4%)	3387	(28.4%)
Medium	880	(46.3%)	3890	(43.6%)	493	(44.9%)	5263	(44.1%)
High	447	(23.5%)	2382	(26.7%)	446	(40.7%)	3275	(27.5%)
Gender								
Girls	722	(38.0%)	3715	(41.6%)	455	(40.5%)	4882	(41.0%)
Boys	1180	(62.0%)	5199	(58.3%)	648	(59.1%)	7027	(58.9%)
Missing	0	(0.0%)	12	(0.1%)	4	(0.4%)	16	(0.1%)
Deprivation								
Affluent	224	(11.8%)	1098	(12.3%)	87	(7.9%)	1409	(11.8%)
Quintile 2	294	(15.5%)	1309	(14.7%)	113	(10.3%)	1716	(14.4%)
Quintile 3	440	(23.1%)	1986	(22.2%)	183	(16.7%)	2609	(21.9%)
Quintile 4	516	(27.1%)	2495	(28.0%)	316	(28.8%)	3327	(27.9%)
Deprived	428	(22.5%)	2038	(22.8%)	298	(36.2%)	2864	(24.0%)
Ethnic group								
White	1628	(85.5%)	7602	(85.3%)	800	(67.3%)	10030	(84.1%)
Black	61	(3.4%)	332	(3.7%)	141	(11.9%)	534	(4.5%)
Indian	200	(10.4%)	914	(10.2%)	139	(11.7%)	1253	(10.5%)
Asian	6	(0.3%)	32	(0.3%)	5	(3.6%)	43	(0.4%)
Other	6	(0.3%)	45	(0.5%)	12	(5.3%)	63	(0.5%)
Missing	1	(0.05%)	1	(0.01%)	0	(0.0%)	2	(0.02%)

TABLE 3.2. Number of admissions for bronchiolitis (B), non-respiratory illness (N) and unspecified diagnoses (U) in the training database by method of admission and other covariate categories. Numbers in parentheses refer to percentage of all admissions for a given diagnosis in each covariate category.

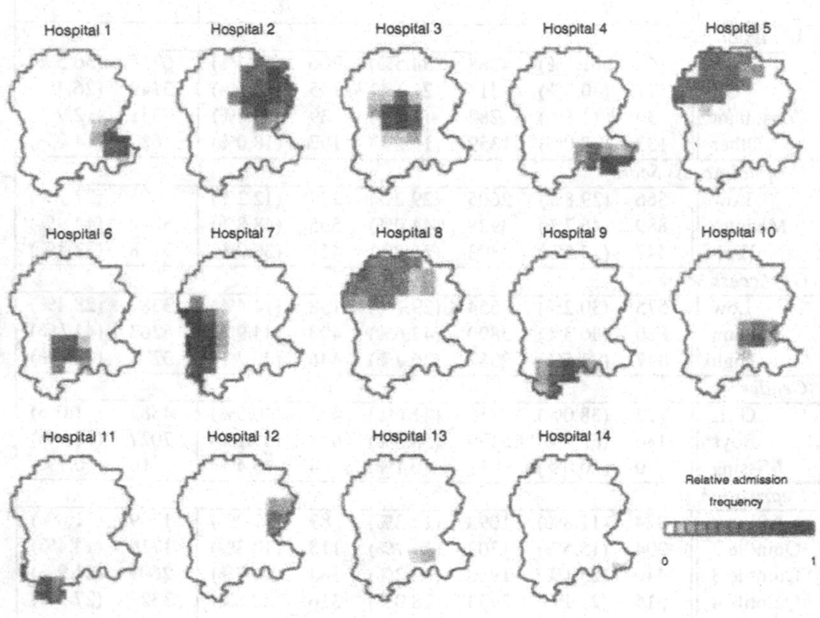

FIGURE 3.5. Relative admission frequency to each hospital according to place of residence.

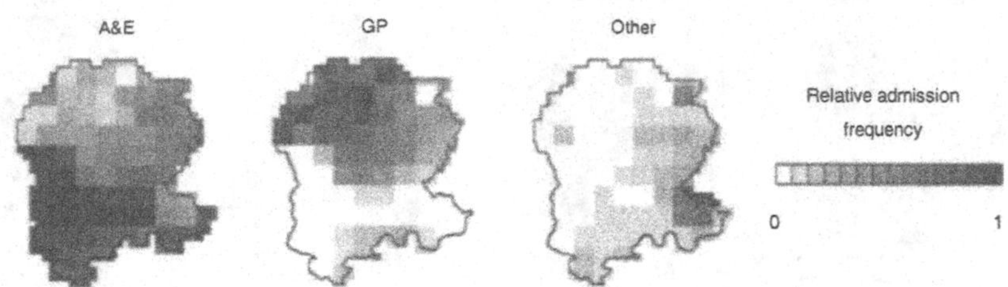

FIGURE 3.6. Relative frequency of hospital admission via referral from the A&E department, referral by GP or by other methods, according to place of residence.

Maps showing the geographical distribution of the other covariates listed in Table 3.2 are shown in Figure 3.7; covariates originally recorded at either ED or ward level have been reapportioned to the kilometer grid using area-weighted averages.

Figure 3.8 shows maps of the density of cars and light goods vehicles (CLGV), the density of medium and heavy goods vehicles (MHGV) and total vehicle emissions for various pollutants per km^2 across our study region. We focus on particulates (TSP, black smoke and PM_{10}) and acid gases (CO, NO_x and SO_2) since these are the pollutants most commonly associated with adverse respiratory health effects (Committee of the Environmental and Occupational Health Assembly of the American Thoracic Society, 1996), and not on the organic compounds (NMVOCs, benzene, 1,3-butadine, and methane) more often associated with longer-term adverse health effects such as cancers.

It is clear from these maps that the traffic variables are highly correlated; NO_x, SO_2, TSP, PM_{10} and black smoke are nearly co-linear, with pairwise correlations ranging from 0.97 to 0.99. We therefore decided to select just one of these five variables for use in our subsequent analyses: we chose black smoke, since this had the smallest correlation with the three other traffic exposure variables still considered *i.e.*, CLGV (correlation = 0.6), MHGV (correlation = 0.74), and CO (correlation = 0.72). Note that correlations between the traffic variables and the other potential explanatory variables shown in Figure 3.7 were much weaker (typically less than 0.4 in absolute value).

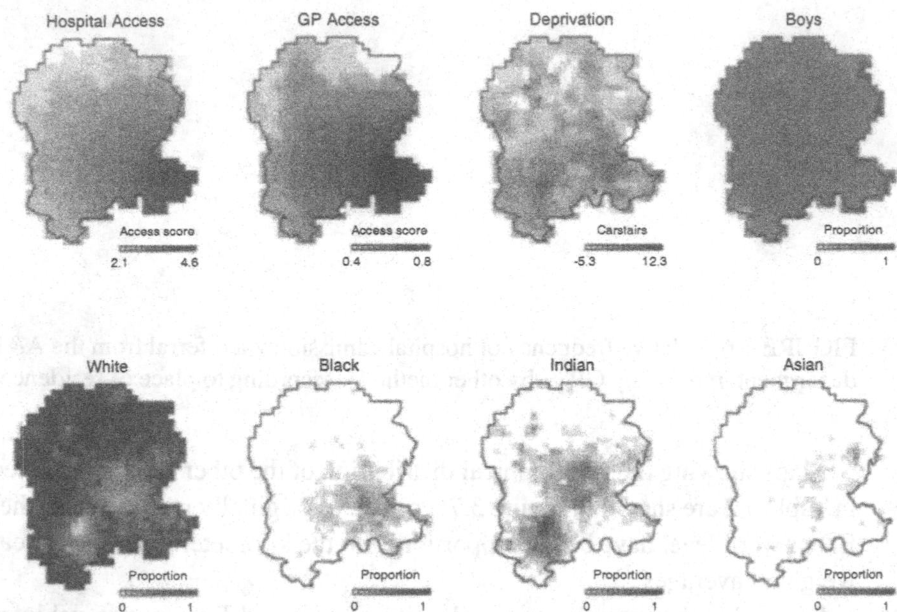

FIGURE 3.7. Geographical distribution of potential explanatory variables.

By assigning each child in our database the value of the traffic density and pollutant load for the kilometer grid cell containing their home postcode centroid, we then calculated the empirical distribution of each of the traffic exposure variables (CLGV, MHGV, CO and smoke) for the three diagnostic groups and for the population of 0–1 year olds as a whole. These are shown in Figure 3.9. There is considerable variation in exposure within each group, with some suggestion that children admitted to hospital with bronchiolitis typically live in *less* polluted areas than children admitted with other diagnoses.

Finally, we plot raw admission rates for bronchiolitis against selected covariate values on the kilometer grid to explore possible associations between the health outcomes and putative risk factors (Figure 3.10). To aid visual assessment of any trends in these relationships, we split each covariate into deciles and calculate the median and 5th and 95th percentiles of the empirical distribution of observed admissions rates for infants in each exposure decile; these are shown as the solid and lower and upper dashed lines on the plots respectively.

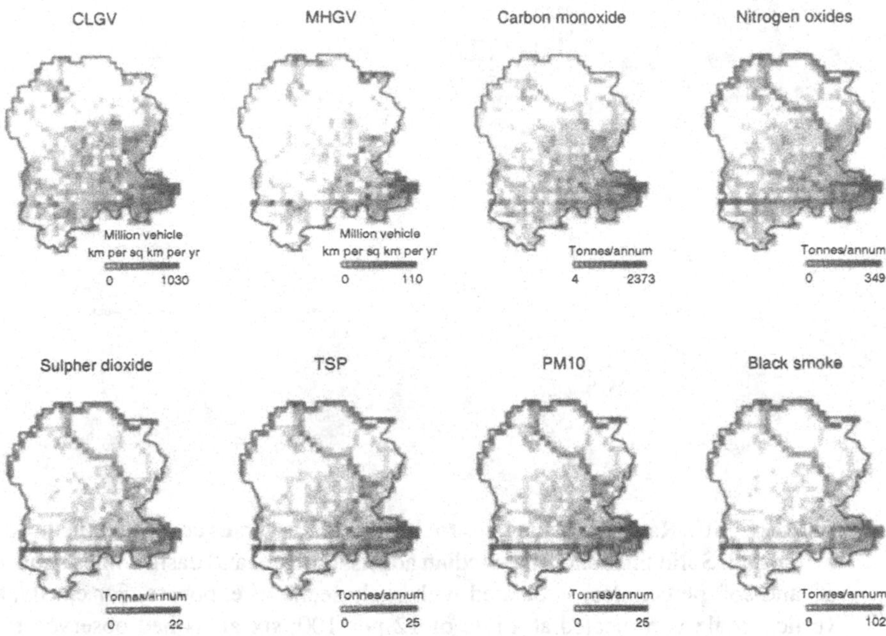

FIGURE 3.8. Traffic density and pollutant loads on the kilometer grid.

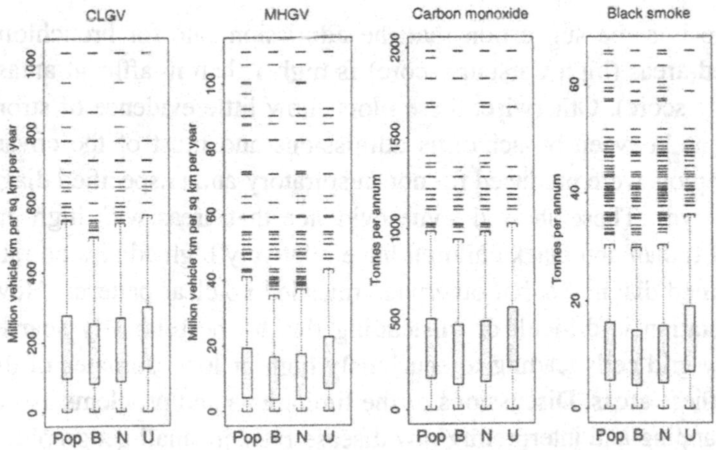

FIGURE 3.9. Empirical distribution of traffic density and pollutant loads for infants admitted to hospital with with bronchiolitis (*B*), non-respiratory illnesses (*N*) and unspecified diagnoses (*U*) and for the whole population of 0–1 year olds.

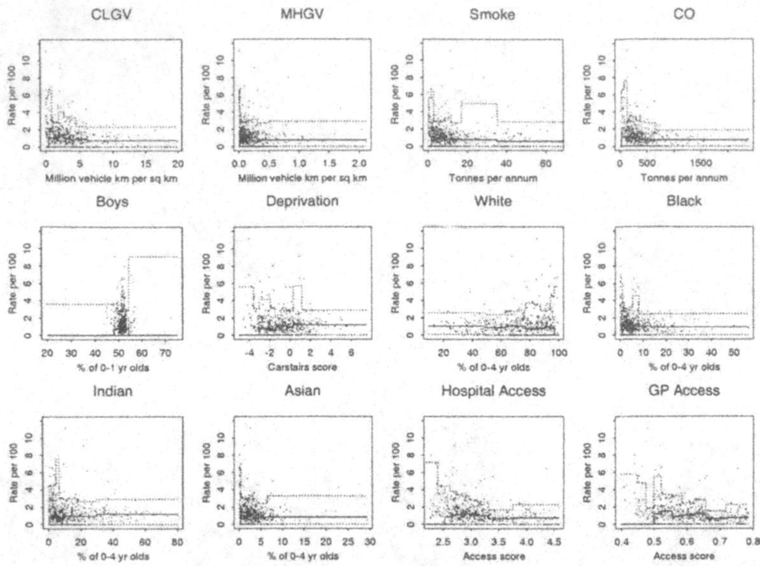

FIGURE 3.10. Raw admission rates for bronchiolitis versus covariates on the kilometer grid. Solid lines show the median admission rate, and dashed lines show the 5th and 95th percentiles, associated with each decile of exposure. For clarity, the vertical scale is truncated at a rate of 12 per 100; six areas had observed rates exceeding this threshold and are included in the calculations of the empirical rate percentiles shown by the dashed and solid lines.

There is some suggestion that the admission rate for bronchiolitis in deprived areas (high Carstairs score) is higher than in affluent areas (low Carstairs score). Otherwise, these plots show little evidence of strong associations between bronchiolitis admissions and most of the covariates. Similar plots were produced for non-respiratory and unspecified diagnoses (not shown). These showed some evidence that areas with high proportions of Indian and black children have relatively high admission rates for unspecified diagnoses, but otherwise revealed no clear patterns. However, interpretation is difficult or misleading due to the relatively sparse data in many grid cells leading to spuriously high or low estimates of disease risk in these areas. Discussions of the limitations and problems associated with mapping and interpreting raw disease rates in small geographic areas can be found elsewhere (Clayton and Bernardinelli, 1992; Mollié, 1996), and it is well recognized that such data typically exhibit overdispersion with respect to the assumed binomial or Poisson sampling distribution. To overcome this, the usual approach is to adopt some form of Bayesian hierarchical model to allow 'borrowing of strength' between areas and so smooth the raw disease rates and improve interpretation. As mentioned in Section 2.2 this is frequently done using the class of log-linear Markov random field models with CAR priors proposed by Besag et al. (1991). However, there are a number of potential limitations to this approach, some of which are fundamental to the model itself, and some of which arise within the context in which the models are usually applied. We expand on these in the next section in order to motivate the modeling approach we adopt for this Case Study.

5.1 Poisson regression analysis

To explore more closely the relationship of explanatory variables (gender, emissions data, area deprivation, *etc.*) to health effects, we use log-linear Poisson regression to model the hospital admissions numbers Y_i^B of bronchiolitis cases, Y_i^N of non-respiratory cases, and Y_i^U of unspecific diagnoses in the ith kilometer grid cell in the form

$$Y_i^* \sim \mathsf{Po}\big(w_i \exp(\textstyle\sum X_{ij}\phi_j^*)\big) \tag{3.1}$$

where $* = B, N$ or U, w_i is the estimated population of 0–1 year old children (in hundreds) and X_{ij} is the jth covariate (proportion of male births, smoke emission, Carstairs deprivation score, *etc.*) in the ith kilometer grid cell. (Note that $X_{i0} \equiv 1$, a vector of 1s, so that ϕ_0^* is an overall intercept.) We use a quasi likelihood approach (McCullagh and Nelder, 1989) to adjust for overdispersion and specify the variance function for the observed admission counts to be $\mathsf{V}(Y_i^*) = \kappa^*\mathsf{E}(Y_i^*)$. The unknown parameter $\kappa^* \geq 1$ measures the amount of overdispersion, and is equal to one in the conventional Poisson model. The effect on inference is simply to inflate the regression coefficient standard errors to account for the extra-Poisson variability; note that this approach assumes no parametric model for the form of the overdispersion.

We use the schematic representation of the hospital admission process in Figure 3.1 as the basis for building our statistical model. Primary interest concerns how the exposures in stage 1 influence the outcome at stage 6. However there are a number of factors that may influence intermediate stages which are unrelated to the true health status of the infant. Our approach is to first model these 'nuisance' factors to at least partially adjust for the possible biases they may introduce, and to then examine how the explanatory variables of interest (gender, traffic data, area deprivation, *etc.*) influence the adjusted admission rates.

Nuisance factors

We model the complex mix of factors influencing stages 4, 5 and 6 in Figure 3.1 by introducing separate 'hospital effects' for each of the hospital groups. These are defined by a (initially) 28-level factor X_{ij}, $j = 1, \ldots, 28$ taking values equal to the relative frequency of admission to hospital group j from grid cell i (see Figure 3.5); since the model includes an overall intercept, we alias X_{i1} during model fitting to maintain parameter identifiability. The effect of method of referral on admission rates (stage 4 in Figure 3.1) is modeled by a 3-level factor giving the relative frequency of referral from the hospital's A&E department, by a GP, or by other methods, for each grid cell; for this factor we alias A&E referral to ensure identifiability. The parents' decision to seek medical help and their choice of care (stages 2 and 3 in Figure 3.1) will be partly influenced by the accessibility of GP and hospital services in their area. This is modeled by separately including either GP or hospital access scores (Carr-Hill et al., 1994) as covariates in the model.

Explanatory variables

We regard the remaining covariates, *i.e.*, gender (proportion of male births in each grid cell), ethnicity (census proportions of 0-4 year old children in the following ethnic groups: white, black, Indian, Asian, other), area deprivation (Carstairs index), traffic density (MHGV and CLGV) and emission loads (CO and black smoke), as potential influences at stage 1b of Figure 3.1. These were entered simultaneously in the model and then the traffic density and pollutant emission variables were included one at a time (the latter were entered singly in the models to avoid problems with multicollinearity).

Results

The Poisson regression analysis described here may be seen as a preliminary screening analysis to eliminate any covariates which were clearly unimportant in explaining the risk of hospital admission for any of the outcomes (B, N and U). Our aim was to identify a reduced subset of potential nuisance factors and explanatory variables to explore in more detail using the Bayesian models presented in subsequent sections. We therefore chose to retain the same set of nuisance factors and explanatory variables in the model for each outcome (B, N and U). These were chosen to be any covariate whose 90% confidence interval for the relative risk of admission excluded one for at least one of the three outcomes (see Table 3.3).

Note that, having identified the subset of variables presented in this table, we then checked whether any of the remaining covariates met the 90% confidence interval criteria when added back into this model separately: none did. We then investigated interactions between the covariates, and threshold models for the traffic and emissions variables, but these did not significantly improve the model. One noteworthy finding is the significant but *negative* association between each of the traffic/emissions variables and risk of bronchiolitis. This is consistent with the exploratory plots in Figure 3.9; however it is unclear whether this is a genuine association, an artifact of data aggregation (ecological bias) or simply a result of residual confounding due to factors such as socioeconomic deprivation, or of systematic differences in the diagnostic coding policies of hospitals.

The results for κ shown at the bottom of Table 3.3 suggest that there is overdispersion in these data for all 3 outcomes. The residual deviance for each model is also larger than the degrees of freedom, suggesting that the Poisson regression models do not fit particularly well. However, the sparseness of our data in each grid cell casts doubt on the asymptotic chi-squared distribution of the deviance (Bithell et al., 1995). One reason for the apparent poor fit is that our problem is *hierarchical*, with individual, hospital, area and population levels. This structure should be acknowledged by explicit modeling of dependencies (including spatial dependence induced by geographic proximity) between variables at each level. In Sections 6 and 7 we discuss Bayesian hierarchical-spatial modeling approaches to address this issue.

6 Area-Level Hierarchical Models

The Poisson log-linear models described in Section 5.1 do not reflect the hierarchical structure or spatial dependence in the health or the traffic exposure data. The most common Bayesian hierarchical approach to ecological regression offering the opportunity to exploit spatial correlation in the data is a Markov Random Field (MRF) approach.

6.1 *Markov Random Field models*

Following Besag et al. (1991), we consider the hierarchical model given below for counts of hospital admissions Y_i^* in a set of geographical areas $R_i, i \in I$,

Covariate	Bronchiolitis		Non-Respiratory		Unspecified	
Hospital effects (relative to remaining 18 hospitals pooled together)						
H_3	1.01	(0.79,1.29)	0.93	(0.83,1.04)	0.22	(0.15,0.32)
H_6	1.04	(0.78,1.39)	1.25	(1.09,1.44)	0.38	(0.24,0.60)
H_7	0.86	(0.65,1.13)	1.22	(1.06,1.40)	0.46	(0.30,0.71)
H_8	1.75	(0.81,3.80)	1.30	(0.91,1.84)	0.03	(0.01,0.09)
H_{10}	0.61	(0.40,0.95)	0.71	(0.58,0.87)	6.05	(3.94,9.28)
H_{11}	0.94	(0.62,1.43)	2.10	(1.73,2.54)	0.95	(0.51,1.77)
H_{12}	0.50	(0.25,0.98)	0.83	(0.62,1.12)	2.72	(1.41,5.22)
H_{13}	0.04	(0.01,0.13)	0.56	(0.31,1.00)	0.00	(0.00,0.04)
H_{14}	0.01	(0.00,0.20)	0.19	(0.04,0.80)	0.07	(0.00,4.07)
H_{16}	8.91	(2.93,27.12)	0.82	(0.42,1.60)	0.12	(0.01,1.08)
Admission method (relative to referral from A&E)						
GP referral	1.18	(0.71,1.96)	2.75	(2.17,3.49)	2.65	(1.35,5.22)
Other	0.49	(0.34,0.72)	1.36	(1.15,1.62)	0.94	(0.59,1.49)
Ethnicity (relative to white/Asian/other)						
Black	0.46	(0.17,1.20)	1.41	(0.91,2.18)	4.15	(1.62,10.61)
Indian	0.38	(0.27,0.54)	0.63	(0.53,0.74)	1.43	(0.88,2.34)
Deprivation (most deprived vs least deprived)						
Carstairs	5.62	(3.78,8.35)	2.64	(2.20,3.17)	2.50	(1.50,4.16)
Gender						
Boys *vs* girls	0.93	(0.08,10.92)	5.23	(1.59,17.19)	2.42	(0.03,186.30)
Traffic pollution† *(most polluted vs least polluted)*						
MHGV	0.50	(0.35,0.71)	0.85	(0.73,0.98)	0.95	(0.64,1.41)
CLGV	0.54	(0.40,0.75)	0.96	(0.83,1.10)	0.95	(0.65,1.40)
CO	0.49	(0.33,0.71)	0.90	(0.77,1.06)	1.07	(0.69,1.67)
Smoke	0.58	(0.39,0.84)	1.00	(0.85,1.19)	1.05	(0.64,1.71)
Overdispersion and Residual Deviance						
κ	1.38		1.79		1.43	
Deviance	744.43		1045.12		661.28	
df	623		623		623	

† Pollution variables were included one at a time in the multiple regression model. Results for the non-pollution variables are reported for the model including MHGV; differences are negligible when MHGV is replaced by each of CLGV, CO or smoke.

TABLE 3.3. Relative risk estimates $e^{\phi_j^*}$ (90% confidence intervals) from the exploratory Poisson regression models.

$$Y_i^* \sim \text{Po}(w_i \exp(\sum_j X_{ij}\phi_j^* + V_i^* + U_i^*)) \tag{3.2}$$

$$V_i^* \sim \text{No}(0, \sigma_v^*)$$

$$U_i^* | U_{k,k\neq i}^* \sim \text{No}\left(\frac{\sum_{k\in I} q_{ik}u_k}{\sum_{k\in I} q_{ik}}, \frac{\sigma_u^*}{\sum_{k\in I} q_{ik}}\right) \tag{3.3}$$

As before, $* = B, N$ or U, and w_i and X_{ij} are the estimated population (in hundreds of 0–1 year old children) and covariate values in the ith area respectively. The terms V_i^* and U_i^* are area-specific random effects representing unobserved or unknown risk factors which display random and spatially correlated variation respectively. Information about the structure of the spatial variation of the U_i is expressed through the 'weights' q_{ik} in the MRF conditional autoregressive model (3.3): a common choice is $q_{ik} = 1$ if areas R_i and R_k share a common boundary, and $q_{ik} = 0$ otherwise. Within a Bayesian paradigm, prior distributions on the regression coefficients ϕ_j^* and the variance parameters σ_v^* and σ_u^* are also specified.

The usual choice of areas R_i for use with the above model are political or administrative boundaries such as EDs or wards in the UK, on which most official data are available. A preliminary analysis of our data at the ward level was described in Section 2.1. However, these areas tend to be highly irregular in size and shape and there is some concern over the appropriateness of using adjacency-based MRF conditional autoregressive models, which were developed originally for regular lattices in image analysis (Besag, 1974). In our present study the spatial resolution of the data permit us to overcome this problem by reapportioning our data onto a kilometer grid based on the National Grid. This grid, with 710 cells, is finer and far more regular than the partition into 269 wards, and matches the partition on which the minor roads traffic data are given.

We first fit model (3.2) with the intercept ϕ_0^* and random effects terms V_i^* and U_i^* only, and then repeat the analysis including the subset of potential explanatory covariates identified from the exploratory Poisson regression analysis (Table 3.3). We assume gamma prior distributions with shape parameters 0.5 and precisions 0.0005 for the inverses of the random effects variances $1/\sigma_v^*$ and $1/\sigma_u^*$; this choice represents a prior belief that the random effects variance is centered around 0.002, with about 1% chance each that it is less than 0.0002 or greater than 6.4 (Kelsall and Wakefield, 1999). For the vectors of coefficients ϕ^* we choose independent normal prior distributions with means equal to zero to reflect prior indifference to whether the relative risk factor X_{ij} increases or decreases the log-risk of hospital admission for bronchiolitis (or non-respiratory or unspecified conditions); prior variances are chosen such that the ratio between the 90th and the 10th percentile of the prior distribution for the relative risk associated with the maximum versus minimum exposure to X_{ij}, *i.e.*, $\exp\left(\phi_j^*(\max_i X_{ij} - \min_i X_{ij})\right)$, is 100. (We use the same prior distributions for the parameters ϕ^* described in the next section — see Table 3.6 for further details.) We model the vector of hospital effects as random rather than fixed as in the Poisson regression models of Section 5.1. This reflects the belief that the influences captured by the hospital term (*e.g.*, availability of beds, clincial staff, diagnostic coding) are likely to be similar across hospitals. We assume prior exchangeability, but allow for possibly outlying hospitals by choosing Student-t distributions with zero means, unknown scale parameters σ_h^* and 4 degrees of freedom for the population distributions of the hospital effects. Gamma prior distributions with shape and inverse scale parameters equal to 0.01 are chosen for the inverse scale parameters $1/\sigma_h^*$, reflecting the prior belief that between-hospital variation may be quite large.

Results

All results reported here and in subsequent sections are based on a single long run of the Markov Chain Monte Carlo (MCMC) sampler for each model. The models reported in Section 6 were estimated using the WinBUGS software (Spiegelhalter et al., 1999); programs to implement the models in Section 7 were written in ISO C. Convergence was assessed by visual inspection of the trace and autocorrelation plots: a burn-in of 1000–10000 samples was typically required. Posterior sample sizes were determined by running sufficient iterations to ensure that the Monte Carlo standard error of the mean (Roberts, 1996) was at least one order of magnitude smaller than the posterior standard deviation for each parameter of interest: 5000–20000 iterations were typically required to achieve this level of precision.

The first column of Figure 3.11 shows maps of the posterior mean admission rates $\exp(\phi_0^* + V_i^* + U_i^*)$ for each of the three outcomes $* = B, N$ or U before adjustment for observed covariates; the second column shows the same quantity (which we now term the residual or adjusted admission rate) after accounting for nuisance factors and potential explanatory variables. The strong spatial patterns of admission rates for each outcome are largely removed following covariate adjustment, although an area of high residual admission rates for unspecified diagnoses in the east of the study region remains unexplained.

We consider proportional admission rates for bronchiolitis relative to total admissions for bronchiolitis, non-respiratory and unspecified diagnoses by mapping the posterior mean of the ratio

$$\exp(\phi_0^B + V_i^B + U_i^B) / \sum_{* \in (B,N,U)} \exp(\phi_0^* + V_i^* + U_i^*),$$

both before and after adjustment for covariates (see Figure 3.12, top row). Also shown are corresponding maps of the posterior probabilities that the proportional admission rate in each grid cell exceeds the average proportional admission rate across the study region. These maps suggest that the high admission rates for bronchiolitis seen in the northwest of the study region in Figure 3.11 are consistent with higher overall admission rates in this area; there is a small cluster of cells with significantly elevated admission ratios for bronchiolitis in the southwest, some of which may be explained by the available covariates.

Table 3.4 shows the posterior variances of the random effects from the covariate-unadjusted and adjusted MRF models. We note that inclusion of covariates leads to a large reduction in total between-area variation ($\sigma_v^* + \sigma_u^*$): prior to covariate adjustment, the spatially-structured component of variation (σ_u^*) dominates; this is largely removed by inclusion of nuisance factors and explanatory covariates, although there continues to be unstructured heterogeneity. There is also considerably more spatially structured between-cell variation in admissions for unspecified diagnoses compared to bronchiolitis or non-respiratory illnesses.

Posterior means and 95% credible intervals (CI) for the relative risks (RR) associated with each explanatory covariate are shown in the top half of Table 3.5. The final two columns of this table give the ratio of relative risks of admission for bronchiolitis versus non-respiratory and unspecified diagnoses respectively associated with exposure to each covariate, *i.e.*, $\exp(\phi_j^B - \phi_j^N)$ and $\exp(\phi_j^B - \phi_j^U)$.

There was no evidence of significantly (*i.e.*, the 95% credible interval excludes one) elevated or reduced risk of bronchiolitis admission to any of the 11 hospital groups relative to the overall bronchiolitis admission rate. Non-respiratory admission rates to hospital 11 were significantly higher than average (RR = 1.69, 95% CI = 1.15–2.35), while admissions with unspecified diagnoses were higher than average to hospital 10 (RR = 6.28, 95% CI = 1.67–15.05) and lower than average to hospitals 3 (RR = 0.29, 95% CI = 0.10–0.65) and 13 (RR = 0.09, 95% CI = 0.00–0.14).

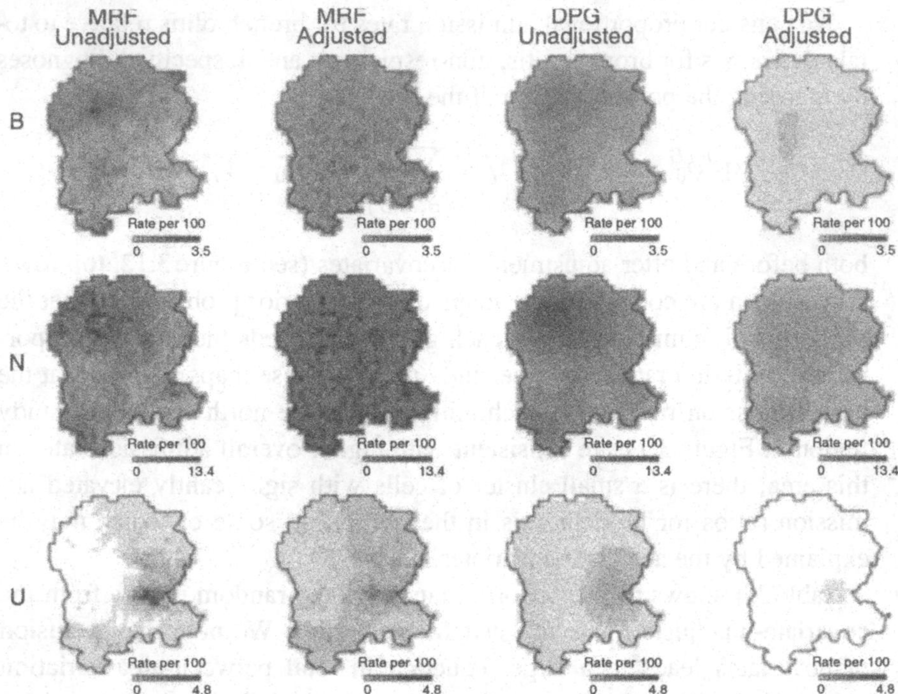

FIGURE 3.11. Posterior mean admission rates for bronchiolitis (*B*), non-respiratory illness (*N*) and unspecified diagnoses (*U*), before and after adjustment for covariates. First two columns relate to Markov Random Field (MRF) models of Section 6.1; last two columns relate to Discrete Poisson/gamma (DPG) models of Section 6.2.

GP-initiated hospital referrals were associated with a significantly higher risk of admission for bronchiolitis (RR = 1.91, 95% CI = 1.22–2.85) and non-respiratory diagnoses (RR = 2.86, 95% CI = 2.04–3.72), but not unspecified diagnoses, relative to referrals from the hospital's A&E department. Referrals by other methods were associated with a significantly lower risk of admission for bronchiolitis compared with A&E referrals (RR = 0.59, 95% CI = 0.33–0.96).

The MRF models have a number of advantages over the Poisson regression models reported in Section 5.1. In particular, they allow for 'borrowing of strength' across the units of analysis (1km^2 cells) and explicit modeling of spatial dependencies in the data. This leads to more reliable estimates of disease rates and of the magnitude of the risk associated with exposure to relevant covariates. However, as seen above, the MRF model

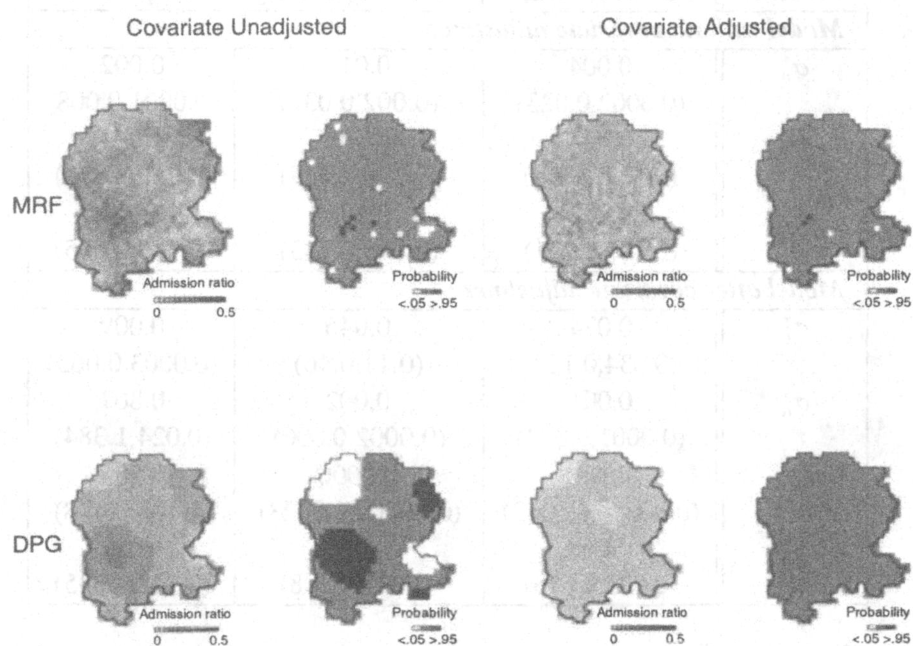

FIGURE 3.12. Posterior mean proportional admission rates for bronchiolitis before and after adjustment for covariates, and posterior probabilities of exceeding the corresponding overall proportional admission rate in the study region. Top row relates to Markov Random Field (MRF) models of Section 6.1; bottom row relates to Discrete Poisson/gamma (DPG) models of Section 6.2.

implicitly requires that all data be aggregated to a common spatial scale for analysis; this leads to a number of problems. First, the analysis will depend on the area units chosen. Second, the process of aggregation introduces approximation errors. Third, and most serious, the relationship between disease prevalence and average exposure at the area level may be quite different from the disease-exposure relationship for individuals. Different aspects of this problem have been described by a number of authors, under a number of different names: the terms *ecological bias*, *ecological fallacy*, *aggregation bias*, and the *modifiable areal unit problem* (MAUP) have all been used (Robinson, 1950; Greenland, 1992; Richardson, 1992; King, 1997). Carrying out a MRF analysis on kilometer grid cells rather than wards can only offer limited help in reducing the effects of the ecological fallacy.

Variance	Bronchiolitis	Non-Respiratory	Unspecified
Model without covariate adjustment			
σ_v^*	0.004	0.011	0.002
	(0.0002,0.023)	(0.002,0.032)	(0.0001,0.008)
σ_u^*	0.541	0.293	1.673
	(0.314,0.796)	(0.170,0.419)	(1.210,2.278)
$s_u^{*\dagger}$	0.162	0.105	0.630
	(0.109,0.221)	(0.075,0.133)	(0.480,0.815)
Model after covariate adjustment			
σ_v^*	0.074	0.046	0.009
	(0.034,0.122)	(0.11,0.46)	(0.0003,0.063)
σ_u^*	0.001	0.002	0.867
	(0.0001,0.004)	(0.0002,0.006)	(0.024,1.384)
$s_u^{*\dagger}$	0.0004	0.0006	0.285
	(0.00004,0.0013)	(0.00007,0.0023)	(0.106,0.478)
σ_h^*	0.078	0.055	1.082
	(0.005,0.613)	(0.012,0.208)	(0.142,4.635)

† s_u = empirical variance of the spatial random effects, $\sum_i U_i^2/(I-1)$; this may be interpreted as a *marginal* variance (whereas σ_u is a conditional variance) and so is directly comparable with the variance of the unstructured random effects, σ_v^*.

TABLE 3.4. Posterior random effects variances (95% credible intervals) from MRF models before and after adjustment for covariates.

6.2 *Discrete Poisson/Gamma models*

A principal reason why the MRF models may lead to ecological bias lies in the log-linear exposure/response relationship. The logarithmic link function leads to *products* rather than *sums* for exposure variables under aggregation of areal units, and thus leads to statistical models which are inconsistent under different levels of aggregation. In order to minimise the deleterious effects of ecological bias we now introduce a hierarchical Bayesian Poisson regression model with an *identity* link function leading to a linear exposure/response relationship (King, 1997, pp 249–255).

In this section we concentrate on a discrete version of the models employing our data at the 1km^2 grid cell level. As in the MRF approach we allow for modeling unmeasured confounders through a spatially varying random effect (latent) term. However, due to the identity link function we are not restricted to modeling these unmeasured or unobserved covariates on the same spatial scale as our data, but can choose an appropriate parti-

Covariate	*B*	*N*	*U*	*B vs N*	*B vs U*
Relative risk estimates from MRF model, and odds ratios for B versus N or U					
Ethnicity (relative to white/asian/other)					
Black	0.51 (0.25,0.99)	1.08 (0.72,1.58)	3.98 (1.19,8.94)	0.49 (0.21,1.03)	0.17 (0.04,0.49)
Indian	0.57 (0.40,0.81)	0.72 (0.57,0.89)	1.50 (0.67,2.67)	0.81 (0.52,1.21)	0.43 (0.19,0.90)
Gender					
Boys *vs* girls	2.70 (0.66,7.44)	1.23 (0.60,2.23)	0.71 (0.08,2.84)	2.46 (0.49,7.65)	8.81 (0.55,41.81)
Deprivation (most deprived vs least deprived)					
Carstairs	5.81 (3.32,9.55)	2.82 (2.05,3.79)	2.16 (0.89,4.66)	2.12 (1.09,3.68)	3.20 (1.06,7.44)
Traffic pollution (most polluted vs least polluted)					
MHGV	0.16 (0.01,0.82)	0.32 (0.06,0.97)	1.22 (0.04,7.22)	0.87 (0.02,5.44)	0.72 (0.01,4.90)
Relative risk estimates from DPG model					
Ethnicity (relative to white/asian/other)					
Black	0.62 (0.35,1.13)	1.06 (0.79,1.38)	3.18 (1.75,5.75)	0.59 (0.31,1.14)	0.20 (0.08,0.47)
Indian	0.56 (0.41,0.76)	0.66 (0.56,0.77)	1.61 (0.98,2.72)	0.86 (0.60,1.22)	0.35 (0.19,0.63)
Gender					
Boys *vs* girls	2.50 (0.42,13.88)	1.73 (0.77,4.13)	0.89 (0.12,8.80)	1.42 (0.18,10.10)	2.73 (0.16,43.05)
% Admissions attributed to each excess risk factor from DPG model, and % excess admissions for B vs N or U					
Baseline	12.8% (0.7%,19.8%)	23.4% (10.0%,31.5%)	31.9% (7.7%,50.4%)	−14.5% (−47.6%, 8.4%)	−22.6% (−43.1%, −8.9%)
Carstairs	53.8% (42.3%,69.0%)	33.1% (28.9%,39.0%)	28.1% (11.8%,57.1%)	11.9% (5.6%,17.1%)	15.6% (−2.3%, 31.6%)
MHGV	0.1% (0.0%,1.5%)	0.1% (0.0%,0.5%)	0.7% (0.0%,5.9%)	13.7% (−49.0%, 49.9%)	−31.4% (−50.0%, 48.8%)
Latent Spatial Risk Factors	34.0% (16.3%,49.3%)	44.9% (30.8%,53.1%)	37.8% (6.4%,74.8%)	−5.9% (−25.4%, 6.9%)	−4.6% (−20.1%, 33.4%)

B = bronchiolitis, *N* = non-respiratory illness and *U* = unspecified diagnoses

TABLE 3.5. Parameter estimates (95% credible intervals) from the MRF (top half) and $DPG_{8\times9;3}$ (bottom half) regression models.

tion of the study region to reflect their spatial structure (although we still need to specify this partition *a priori*). In Section 7 we will then extend this model class to a continuous random field model to allow for the finest level of aggregation for all parts of our data, including individual attributes, and eliminate the need to specify the structure of the latent spatial term in advance. These models should then further reduce the problem of the ecological fallacy.

Let $\mathcal{Y}$ denote our study region in the northwestern part of London, and, as before, let $\{y_i^B\}_{i\in I}$, $\{y_i^N\}_{i\in I}$, and $\{y_i^U\}_{i\in I}$ be the numbers of hospital admissions with a diagnosis of bronchiolitis, non-respiratory illness, and unspecified, respectively, in a region $R_i \subset \mathcal{Y}$, $i \in I$. The regions R_i could be EDs, wards, grid cells, or any other arbitrary partition of the study region. As for the MRF analysis, we take $\{R_i\}_{i\in I}$ to be part of the National Grid, *i.e.*, a collection of 710 grid cells of size 1km^2.

Again, we model the numbers of admissions in area R_i as independent random variables with Poisson distributions $\{Y_i^*\} \sim \mathsf{Po}(\Lambda_i^* \times w_i)$, $* = B, N$ or U, where, as before, w_i is an estimate for the population of 0–1 year old children (in hundreds) in region R_i.

On the next stage of the hierarchy we model the rates Λ_i^*, $* = B, N$ or U, in area R_i (in counts of admissions per 100 births) as a combination of additive (excess) and multiplicative (relative) risk factors. We discuss the rationale and interpretation of this model in the subsection *Specific discrete Poisson/gamma models* in the context of the specific risk factors considered for this Case Study; see also (Breslow and Day, 1980, pp 55–7). In this approach, we represent the effects of excess risk factors by the term $\sum_{j\in J_E} X_{ij}\beta_j^*$ of possibly spatially varying covariates X_{ij} in region R_i and regression coefficients β_j^* (with X_{i0} equal to a vector of 1s and β_0^* corresponding to the intercept term). The relative risk factors enter the regression model through the term $\exp(\sum_{j\in J_R} X_{ij}\phi_j^*)$ of possibly spatially varying covariates X_{ij} in region R_i and regression coefficients ϕ_j^*.

Thus, the intensity (w.r.t. to the population weights w_i) of our discrete Poisson regression models with identity link can generally be presented in the following form:

$$\Lambda_i^* = c(\phi^*) \times e^{\sum_{j\in J_R} X_{ij}\phi_j^*} \times \sum_{j\in J_E} X_{ij}\beta_j^*, \quad i \in I.$$

Here ϕ^* is the vector of regression coefficients ϕ_j^* and $c(\phi^*)$ is a normalising factor introduced both to maintain a mean relative risk factor term $c(\phi^*) \times \exp(\sum_{j\in J_R} X_{ij}\phi_j^*)$ of unity and to reduce correlations between the ϕ_j^* and β_j^* and thus to speed up convergence in our underlying MCMC scheme. We choose the inverse of this normalising factor to be

$$c(\phi^*)^{-1} = \exp(\sum_{j\in J_R} \bar{X}_j\phi_j^*) \quad \text{with} \quad \bar{X}_j = \sum_{i\in I} X_{ij}w_i / \sum_{i\in I} w_i \,,$$

or equivalently use centered relative risk covariates.

This class of Poisson regression models with identity link is consistent under different levels of aggregation if the models include only excess (*i.e.*, additive) risk factors; for an application see Best et al. (2000). Allowing for both excess and relative risk factors within the Poisson regression with identity link makes this class of models more flexible than Poisson models with log-linear link. Choosing which risk factors enter additively and which multiplicatively is then an additional step in our modeling task, and is usually based on *a priori* considerations (Breslow and Day, 1980, pp 122–4).

Our regression models (like the MRF models described earlier) include a spatially varying latent term which we model as additive (rather than multiplicative as in the MRF models). We represent this term by

$$X_{il}\beta_l^* = (\sum_{m\in M} k_{im}^*\gamma_m^*) \times \beta_l^*$$

where l is the last index of J_E. Each of the quantities $k_{im}^*\gamma_m^* \times \beta_l^*$, $i \in I$, $m \in M$, can be viewed as the influence that unmeasured or unobserved risk factors in area S_m, or at a fixed location s_m in S_m, exert in area R_i. Here the regions S_m are in northwestern London or nearby; they might again be EDs, wards, 1km^{2}grid cells or an arbitrary partition of Northwest London. In the following we take them to be small rectangles covering all of northwestern London plus a 4km surrounding buffer zone. The kernel matrix elements k_{im}^* are chosen to be Gaussian, *i.e.*,

$$k_{im}^* = \frac{1}{2\pi\rho^{*2}} e^{-|x_i - s_m|^2/2\rho^{*2}} \,,$$

with distance scale $\rho^* > 0$ which determines how rapidly the influence k_{im}^* declines with increasing distance $|x_i - s_m|$ from the center x_i of R_i to the center s_m of S_m. The latent risk factors γ_m^* are also taken to be uncertain.

Prior distributions

Inference about the parameters $\phi^* = \{\phi_j^*\}_{j\in J_R}$, $\beta^* = \{\beta_j^*\}_{j\in J_E}$, $\gamma^* = \{\gamma_m^*\}_{m\in M}$ and possibly the distance scale ρ^* of the Gaussian kernel and derived quantities such as the intensities Λ_i^*, is based on a Bayesian analysis using independent prior distributions obtained as follows.

Adopting the same approach as for the MRF models in Section 6.1, we choose independent normal prior distributions $\mathsf{No}(\mu_j^*, \sigma_j^*)$ for the vectors of (log) relative risk regression coefficients ϕ^*, where $\mu_j^* = 0$ and σ_j^* is chosen such that the ratio between the 90th and the 10th percentile of the prior distribution for the relative risk associated with the maximum versus minimum exposure to X_{ij} is 100. Hospital effects are again treated as random and are assumed to follow Student-t distributions with zero means, unknown scale parameters σ_h^* and 4 degrees of freedom. For the vectors of coefficients β^* of the excess risk factors we choose independent gamma distibutions $\mathsf{Ga}(\alpha_j^*, \tau_j^*)$ with means α_j^*/τ_j^* and variances $\alpha_j^*/(\tau_j^*)^2$ in order to maintain non-negativity. The parameters of these gamma distributions are taken such that with 80% prior probability the number of bronchiolitis admissions (or non-respiratory admissions or admissions with unspecified diagnoses) associated with each excess risk factor lies between one-tenth and ten times a nominal equal distribution. This is achieved by setting $\alpha_j^* = 0.575$, $j \in J_E$, $* \in \{B, N, U\}$, giving a factor of 100 for the ratio of the 90th to the 10th percentile of a gamma distribution, and by setting the prior means α_j^*/τ_j^* for each coefficient β_j^* such that *a priori* all excess additive risk factors contribute equally.

The prior distributions for the latent risk factors γ_m^*, $m \in M$, are also taken from the gamma family with shape parameters $\alpha_{\gamma_m}^*$ and precisions τ_γ^*. Their prior means are set to be $\alpha_{\gamma_m}^*/\tau_\gamma^* = |S_m|$, the area of region S_m in km^2, which leads to a spatially extensible model for the sum $\sum_{m \in M} k_{im}^*$ γ_m^* over any partition $\{S_m\}$. The prior uncertainty about the spatial distribution of the latent effects is expressed through τ_γ^*: our uncertainty is about the same as if we had to infer *a priori* the value of the latent spatially-varying function by observing locations of points randomly scattered with the same distribution, at a density of about τ_γ^* points per km^2. Thus, the smaller the values for τ_γ^*, the higher the spatial variation of the latent risk factor under our prior.

The distance scale ρ^* for the kernel we treat as certain but run our models for several fixed values (see below). Putting a discrete prior concentrated on a few values for ρ^* and calculating a posterior distribution would be more elegant but is computationally feasible only for a very small number M of rectangles in the latent partition $\{S_m\}$.

Specific discrete Poisson/gamma models

As mentioned above, the greater flexibility of the Poisson/gamma models allows us to choose whether covariate terms should enter multiplicatively or additively. However, there is often little scientific evidence to guide this choice; in the present application, our decision is based on the assumption that the following model represents a plausible mechanism by which the various risk factors for bronchiolitis may operate.

Bronchiolitis is a viral illness where the source of infection is often a family member with an apparently minor respiratory illness. Our model therefore assumes a baseline rate of bronchiolitis admissions which may be attributed to such infections. One possible route by which chronic exposure to high levels of traffic pollution may affect risk of bronchiolitis is by lowering an infant's 'tolerance threshold' to the extent that they develop clinical symptoms of bronchiolitis if exposed to an infection which would otherwise have led to only mild respiratory illness. Such a mechanism would result in a number of extra hospital admissions, attributable to chronic exposure to ambient air pollution, in addition to those 'baseline' cases of bronchiolitis which would have occurred anyway. We may envisage a similar mechanism for the effects of socioeconomic deprivation on bronchiolitis risk, and there may be other unknown or unmeasured risk factors which also operate by this route. We therefore treat the intercept (baseline), traffic related pollution variables, Carstairs deprivation index and the spatially varying random effects term representing latent unobserved covariates as additive excess risk factors in our model. A different mechanism is envisaged for risk factors such as gender and ethnicity, which we assume affect an infant's overall susceptibility to bronchiolitis. For example, exposure to the same underlying 'attributable' risk factor (baseline infection, traffic pollution *etc*) may be more likely to lead to clinical symptoms of bronchiolitis in boys compared to girls. We therefore regard gender and ethnicity as relative risk factors which enter our model multiplicatively. We also model the hospital effects and method of admission as multiplicative relative risk factors, since, although these do not directly affect the risk of an infant developing bronchiolitis, they are likely modify the overall rate of admission to hospital.

Figure 3.13 shows a schematic representation of the various mechanisms described above, and can be thought of as an expanded version of stages 1a and 1b shown in Figure 3.1. For comparability, we assume the same model for non-respiratory and unspecified admissions.

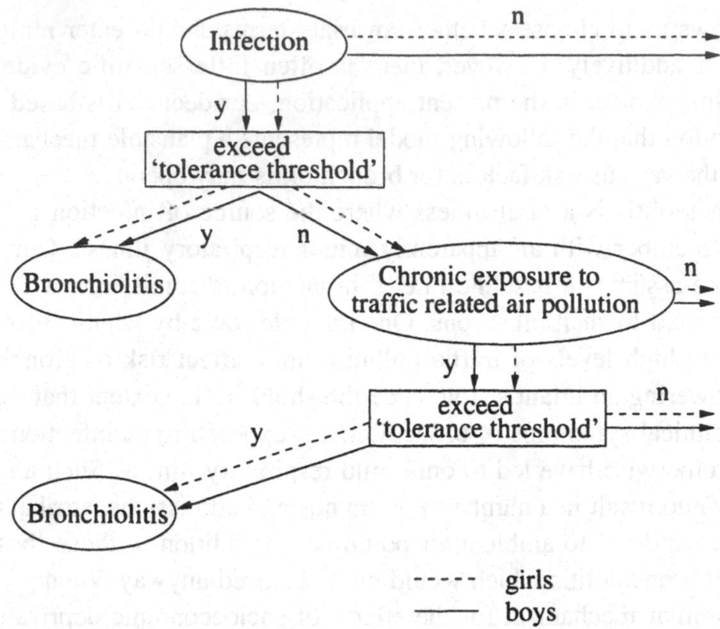

FIGURE 3.13. Schematic representation of the various mechanisms by which an infant may develop bronchiolitis. An underlying infectious ætiology is assumed for all cases. Excess cases may be generated if an infected infant has also been exposed to some other relevant risk factor, *e.g.*, ambient air pollution (other such exposures not shown include living in a deprived area or other unknown or unmeasured risk factors). Links between boxes have associated weights representing the probability that that particular pathway occurs. Risk factors such as gender (or ethnicity, which is not shown) may modify these probabilities, and so separate links are shown for boys and girls.

Before fitting the full model with additive and multiplicative risk factors as described above, we first consider a set of models that include only the intercept (representing the baseline admission rate) and spatially varying latent risk factors. As already noted, the latent term is expressed as a mixture of gamma-distributed random variables representing the influence of unmeasured risk factors in areas S_m. These areas are arbitrary, but must be pre-specified for these discrete Poisson/gamma models. Our choice is governed by a trade-off between the resolution required to detect local variations in disease risk and the computation time involved. We consider two alternatives: a partition of northwestern London plus a 4km surrounding buffer zone into (i) an 11×13 grid of approximately 11km^2 cells, and (ii) an 8×9 grid of approximately 22km^2 cells. For each latent partition, we

Regression coefficients			
Baseline	$\beta_0^B \sim \text{Ga}(\alpha_0^B, \tau_0^B)$	$\tau_0^B = 4\alpha_0^B / \overline{Y}^B = 1.93$	$\alpha_0^B = 0.575$
Carstairs	$\beta_1^B \sim \text{Ga}(\alpha_1^B, \tau_1^B)$	$\tau_1^B = 4\alpha_1^B \overline{X}_{1_E} / \overline{Y}^B = 9.489$	$\alpha_1^B = 0.575$
MHGV	$\beta_2^B \sim \text{Ga}(\alpha_2^B, \tau_2^B)$	$\tau_2^B = 4\alpha_2^B \overline{X}_{2_E} / \overline{Y}^B = 4.188$	$\alpha_2^B = 0.575$
Latent	$\beta_3^B \sim \text{Ga}(\alpha_3^B, \tau_3^B)$	$\tau_3^B = 4\alpha_3^B / \overline{Y}^B = 1.93$	$\alpha_3^B = 0.575$
GP ref.	$\phi_1^B \sim \text{No}(0, \tau_1^B)$	$\tau_1^B = \frac{\left(2.56(\max_i X_{i1_R} - \min_i X_{i1_R})\right)^2}{\log 100} = 0.31$	
Other ref.	$\phi_2^B \sim \text{No}(0, \tau_2^B)$	$\tau_2^B = \frac{\left(2.56(\max_i X_{i2_R} - \min_i X_{i2_R})\right)^2}{\log 100} = 0.31$	
Black	$\phi_3^B \sim \text{No}(0, \tau_3^B)$	$\tau_3^B = \frac{\left(2.56(\max_i X_{i3_R} - \min_i X_{i3_R})\right)^2}{\log 100} = 0.10$	
Indian	$\phi_4^B \sim \text{No}(0, \tau_4^B)$	$\tau_4^B = \frac{\left(2.56(\max_i X_{i4_R} - \min_i X_{i4_R})\right)^2}{\log 100} = 0.20$	
Male	$\phi_4^B \sim \text{No}(0, \tau_4^B)$	$\tau_5^B = \frac{\left(2.56(\max_i X_{i5_R} - \min_i X_{i5_R})\right)^2}{\log 100} = 0.10$	
Other parameters			
Latent risk factors	$\gamma_m^B \sim \text{Ga}(\alpha_{\gamma_m}^B, \tau_\gamma^B)$	$\alpha_{\gamma_m}^B = \tau_\gamma^B \lvert S_m \rvert = 22$	$\tau_\gamma^B = 1.0/\text{km}^2$
Distance scale	$\rho^B \equiv 3\text{km}$		
Hospital precision	$1/\sigma_h^B \sim \text{Ga}(0.5, 0.0005)$		

Notes:
$\overline{Y}^B$ = overall observed admission rate for bronchiolitis in the study region.
$\overline{X}_{j_E}$ = population weighted average value of jth excess risk factor.
X_{ij_R} = j^{th} relative risk factor.

TABLE 3.6. Prior distributions for the covariate-adjusted $\text{DPG}_{8\times9;3}$ regression model for bronchiolitis.

consider three distance scales $\rho^* = 1\text{km}$, $\rho^* = 3\text{km}$ and $\rho^* = 5\text{km}$ for $* = B, N$ or U. In what follows, we will use the notation $\text{DPG}_{x\times y;z}$ to denote the discrete Poisson/gamma model with latent partition on an $x \times y$ grid and kernel distance scale $z$km.

We then include socioeconomic deprivation and traffic related pollution as additional excess risk factors in the model, and introduce multiplicative terms for gender, ethnicity and the nuisance factors (hospital effects and admission method). Table 3.6 shows the specific prior distributions chosen for this model. Note that the original scale for the Carstairs deprivation score is centered about zero, but to ensure non-negative contributions from each excess risk factor term we add a positive constant (equal to the absolute value of the minimum) to the score for each area; this does not affect the interpretation of the deprivation index.

Results

Figure 3.14 shows the posterior mean of the admission rates $\beta_0^B + (\sum_{m \in M} k_{im}^B \gamma_m^B) \times \beta_l^B$ and latent influences γ_m^B for the six combinations of latent partition and kernel distance scale for the covariate-unadjusted DPG models for bronchiolitis. Not surprisingly, the largest distance scale (ρ^B=5km) produces the smoothest rate estimates for both the 8×9 and 11×13 latent partition; for a given value of ρ^B, both partitions lead to similar maps for the posterior mean bronchiolitis admissions rates. Results for non-respiratory and unspecified admissions followed a similar pattern for the six models. Partly for computational reasons, we restrict subsequent analyses to the $\text{DPG}_{8\times9;3}$ model, *i.e.*, with latent influences defined on an 8×9 grid of cells S_m and kernel distance scale $\rho^* = 3$km. The latter choice was based on results from the MRF models which suggest that much of the spatial variation in admission rates is removed after inclusion of the covariate effects. Hence the greater spatial variation expressed by the 1km kernel distance scale compared to larger scales may be unnecessary in the covariate-adjusted model. The 3km and 5km distance scales appear to result in quite similar amounts of spatial smoothing, and so the choice between these two values was arbitrary.

The third column of Figure 3.11 shows maps of the posterior mean admission rates $\beta_0^* + (\sum_{m \in M} k_{im}^* \gamma_m^*) \times \beta_l^*$ for each of the three outcomes estimated from the $\text{DPG}_{8\times9;3}$ model without adjustment for known covariates; the final column shows the same quantity after accounting for observed nuisance factors and explanatory covariates. These maps show similar spatial patterns of admission rates for each outcome compared to the MRF models, but are considerably smoother.

Maps of the posterior mean proportional admission rates for bronchiolitis relative to all admissions, given by the ratio

$$\frac{\beta_0^B + (\sum_{m \in M} k_{im}^B \gamma_m^B) \times \beta_l^B}{\sum_{* \in \{B,N,U\}} \beta_0^* + (\sum_{m \in M} k_{im}^* \gamma_m^*) \times \beta_l^*},$$

are shown in Figure 3.12 (bottom row) both before and after adjustment for observed covariates. Also shown are corresponding maps of the posterior probabilities that the proportional admission rate in each grid cell exceeds the average proportional admission rate across the study region. Again, the spatial patterns are similar to those estimated from the MRF model, although the grid cells with significantly elevated or reduced admission ratios for bronchiolitis cover a much larger geographical area due to the greater degree of smoothing induced by the DPG models.

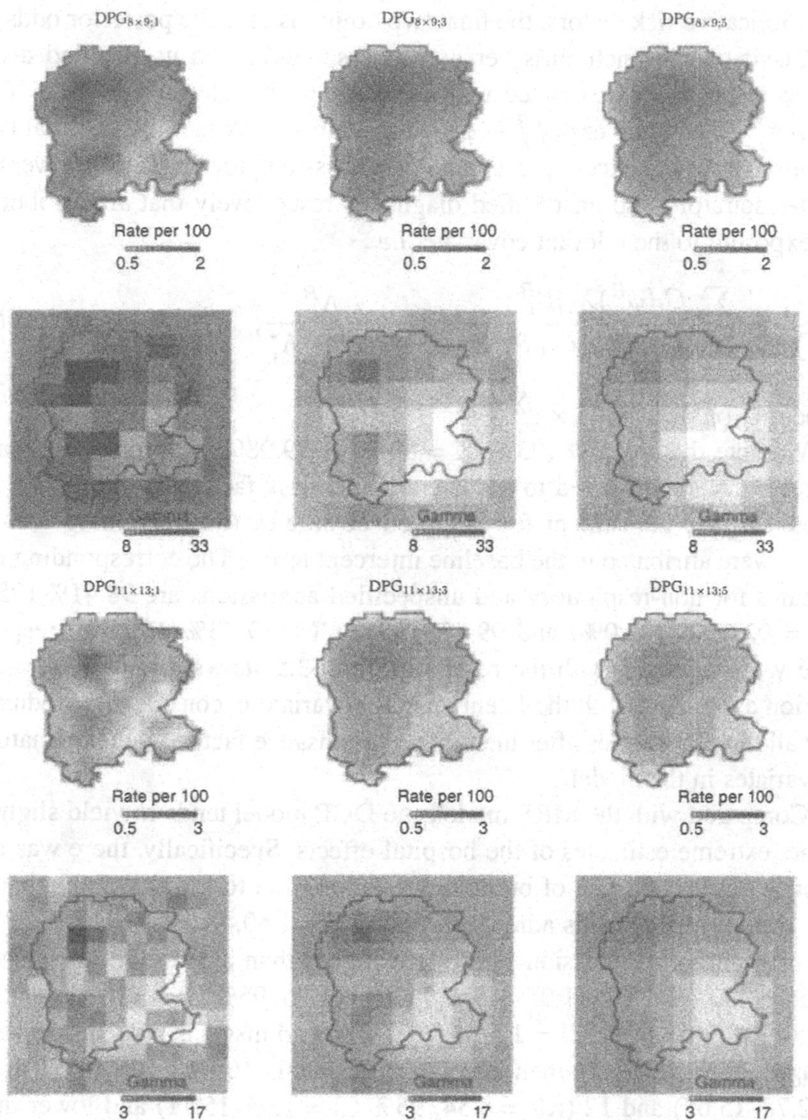

FIGURE 3.14. Posterior mean admission rates (rows 1 and 3) and latent influences (rows 2 and 4) for the covariate-unadjusted DPG models for bronchiolitis.

Posterior means and 95% credible intervals for the relative or percentage attributable risks associated with each multiplicative or additive explanatory covariate respectively are shown in the bottom half of Table 3.5. For multiplicative risk factors, the final two columns give the posterior odds of admission for bronchiolitis versus non-respiratory and unspecified diagnoses respectively associated with exposure to the relevant covariate, *i.e.*, $\exp(\phi_j^B - \phi_j^N)$ and $\exp(\phi_j^B - \phi_j^U)$; for additive risk factors, the final two columns give the excess percentage of admissions for bronchiolitis versus non-respiratory and unspecified diagnoses respectively that are attributed to exposure to the relevant covariate, *i.e.*,

$$\frac{\sum_i C_i(\phi^B) X_{ij}\beta_j^B}{\sum_i C_i(\phi^B) X_{ij}\beta_j^B + C_i(\phi^*) X_{ij}\beta_j^*} - \frac{\Lambda_i^B}{\Lambda_i^B + \Lambda_i^*}, \quad j \in J_E, * = N \text{ or } U$$

where $C_i(\phi^*) = c(\phi^*) \times e^{\sum_{j \in J_R} X_{ij}\phi_j^*}$.

We note that 96.71% (95% CI = 82.93%–99.98%) of bronchiolitis admissions were attributed to the latent spatial risk factors in the $\text{DPG}_{8\times9;3}$ model *before* adjustment for observed covariates (the remaining admissions were attributed to the baseline intercept term). The corresponding estimates for non-respiratory and unspecified admissions are 98.41% (95% CI = 92.67%–100.0%) and 99.5% (95% CI = 97.73%–100.0%) respectively. Comparison with the results in Table 3.5 shows that the risk attribution associated with the latent spatial covariate is considerably reduced for all three outcomes after including the nuisance factors and explanatory covariates in the model.

Compared with the MRF model, the DGP model tends to yield slightly more extreme estimates of the hospital effects. Specifically, there was evidence of elevated risk of bronchiolitis admission to hospital 7 relative to the overall bronchiolitis admission rate (RR = 1.50, 95% CI = 1.03–2.38). Non-respiratory admission rates were higher than average to hospitals 6 (RR = 1.30, 95% CI = 1.05–1.63), 7 (RR = 1.26, 95% CI = 1.03–1.59) and 11 (RR = 2.07, 95% CI = 1.56–2.75), while admissions with unspecified diagnoses were higher than average to hospitals 10 (RR = 6.70, 95% CI = 2.78–15.65) and 12 (RR = 5.54, 95% CI = 1.95–15.71) and lower than average to hospitals 3 (RR = 0.27, 95% CI = 0.13–0.55), 6 (RR = 0.37, 95% CI = 0.16–0.80), 8 (RR = 0.08, 95% CI = 0.01–0.40) and 13 (RR = 0.01, 95% CI = 0.00–0.06).

By contrast, the DPG model yields less extreme estimates for the effect of method of referral on admission rates compared to the MRF model. This is probably due to the fact that some hospitals tend to admit most patients via direct referral from their A&E department, while others admit a large proportion following GP referral, and so both the hospital effects and method of referral may offer similar explanations for the spatial pattern of admission rates seen in our data. The only effect of admission method for the DPG models was a higher risk of admission for bronchiolitis for GP-initiated hospital referrals compared to A&E referrals (RR = 1.98, 95% CI = 1.15–3.25).

6.3 *Model validation*

Model validation for complex hierarchical models is a topic of continuing research. Here we carry out a simple assessment of model adequacy by using each of the Poisson regression, MRF and DPG models reported in preceding sections to predict the number of admissions per grid cell for the validation dataset. Comparison of the distribution of deviance residuals for observed versus predicted counts for the various models suggests that the Poisson regression models lead to more extreme residuals than either the MRF or DPG models; the latter two modeling approaches show similar residual patterns.

7 Marked Point Process Regression Models

The Poisson regression models with identity link introduced in Section 6.2 already show some advantages over the MRF models: they allow for modeling the latent random effects term on a different spatial resolution than the cases and other covariates, and they are capable of modeling excess (additive) as well as relative risk factors. However, we still have to fix a partition $\{S_m\}_{m\in M}$ for modeling the unobserved covariates prior to our analysis (when we do not know on which spatial scale a latent spatial pattern might emerge) and we are still unable to include covariate information on an individual level. In this section we will extend the models of the previous section to a continuous random field version, eliminating all unnecessary aggregation by using all data at their finest spatial scale.

This is achieved by further refining both the areas $\{R_i\}_{i\in I}$ of $\mathcal{Y}$ on which the admission data are given and the areas $\{S_m\}_{m\in M}$ on which the latent gamma terms are given until in the limit both the admissions' locations and the random effects may be represented by continuous random fields. We can then introduce individual attributes to our model by going over to a marked random field.

Thus we view the locations $\{y_n^B\}_{n\in N_B}$, $\{y_n^N\}_{n\in N_N}$, and $\{y_n^U\}_{n\in N_U}$ of children admitted to hospitals for bronchiolitis, of the admissions for non-respiratory illnesses and of the admissions with an unspecified diagnosis, respectively, and their associated vectors $\{a_n^*\}_{n\in N_*}$, $* = B, N$ or U, of individual attributes (*e.g.*, gender) as *marked* points $\{x_n^*\}_{n\in N_*}$, $* = B, N$ or U in the product space $\mathcal{X} = \mathcal{Y} \times \mathcal{A}$ of pairs $x^* = (y^*, a^*)$ of locations $y^* \in \mathcal{Y}$ and attribute vectors (marks) $a^* \in \mathcal{A}$. We model the marked points $\{x_n^*\}_{n\in N_*}$, $* = B, N$ or U, as Poisson random fields $N^*(dx) \sim \mathsf{Po}(\Lambda^*(x)w^*(dx))$ on $\mathcal{X}$ with uncertain intensity $\Lambda^*(x)w^*(dx)$. Thus, $N^*(dx)$, $* = B, N$ or U, is a doubly-stochastic Poisson (or Cox) process on the product space. We take the intensity $\Lambda^*(x)w^*(dx)$ to have a density function $\Lambda^*(x)$ with respect to a reference weight measure $w^*(dx)$ intended to reflect the known distribution of the population at risk and their attributes. We model $w^*(dx) = w_Y^*(dy)w_A^*(da|y)$, $* = B, N$ or U, as a product of a non-negative spatial weight measure $w_Y^*(dy)$ and a location-specific conditional probability measure $w_A^*(da|y)$ of attributes.

On the next stage of the hierarchy we model the density $\Lambda^*(x)$, $* = B, N$ or U, as a combination of excess and relative risk factors. The excess (additive) risk factors, which here may include an intercept, traffic related pollution covariates, social deprivation and a spatially varying random effects term, are represented by the term $\sum_{j\in J_E} X_j(x)\beta_j$ with risk factors $X_j(x)$ which may depend on either or both the location y or the attribute vector a and regression coefficients β_j. The relative (multiplicative) risk factors, which here may include gender, ethnicity, admission method, and hospital effects, now enter the model as $\exp(\sum_{j\in J_R} X_j(x)\phi_j)$ with regression coefficients ϕ_j. In all $\Lambda^*(x)$, $* = B, N$ or U, is given by

$$\Lambda(x)^* = c(\phi^*) \times e^{\sum_{j\in J_R} X_j(x)\phi_j^*} \times \sum_{j\in J_E} X_j(x)\beta_j^* .$$

Again, $c(\phi^*)$ is a normalizing constant.

7.1 *A gamma random field model for the latent term*

Infinitely refining the areas $\{S_m\}_{m\in M}$ on which the latent gamma terms are given leads to a gamma random field model for the latent random effects. The gamma random field has a representation as a discrete sum of infinitely many point masses at random locations $s_m \in \mathcal{S}$ and corresponding magnitudes $\gamma_m > 0$. For a gamma random field $\Gamma(ds) \sim \mathsf{Ga}(\alpha(ds), \tau(s))$ with shape measure $\alpha(ds)$ and inverse scale function $\tau(s)$ the sum $\Gamma(A) =$

$\sum_m \{\gamma_m : s_m \in A\}$ of magnitudes for points in any small set A have approximately the gamma probability distribution with shape parameter $\alpha(A)$ and mean $\alpha(A)/\tau(a)$ for any $a \in A$ (if $\tau(a)$ is constant in A this is exact). Furthermore, $\Gamma(A)$ and $\Gamma(B)$ for non-overlapping sets A and B are independent. For a discussion on gamma random fields including their construction see Wolpert and Ickstadt (1998).

The gamma random field can therefore be interpreted as a sum of point sources at random locations s_m with positive magnitudes γ_m. We now regard our unmeasured or unobserved covariates as arising from these point sources, and falling off with increasing distance from these sources at rates described by a kernel $k(x, s)$. The aggregate effect at a location y for an individual with attribute vector a is then given by the kernel mixture $\sum k(x, s_m)\gamma_m$ (where $x = (y, a)$) or in integral form by $\int_S k(x, s)\Gamma(ds)$. The latent random effects term which we again treat to be excess additive can thus be written as

$$X_l(x)\beta_l^* = \left(\sum k^*(x, s_m)\gamma_m^*\right) \times \beta_l^* = \beta_l^* \times \int_S k^*(x, s)\Gamma^*(ds),$$

where l is the last index of set J_E and $* = B, N$ or U; as in the discrete versions of the models we use a Gaussian kernel $k^*(x, s) = 1/(2\pi\rho^{*2}) \times \exp(-|y - s|^2/(2\rho^{*2}))$.

7.2 *Specific point process regression models*

We illustrate the continuous Poisson/gamma approach with a model that includes as excess risk factors an intercept, a traffic pollution measure (MHGV), and a spatially varying latent random field. It includes a single relative risk factor, gender, which enters as an individual attribute. We only consider admissions for bronchiolitis. Prior distributions for the regression coefficients β^B and ϕ^B were chosen in the same manner as described earlier for the DPG model (Table 3.6). Prior distributions for the shape measure $\alpha(ds)$ and inverse scale function $\tau(s)$ of the latent gamma random field were chosen to ensure that each of the three excess risks (intercept, MHGV, and spatial random effect) would have equal prior expectations (so that about 1/3 of the bronchiolitis cases might be associated with each risk factor, *a priori*), with about 90% prior probability that the effect would be between 1/10 and 10 times this amount. In contrast to the DPG models, we treat the distance scale ρ here as uncertain, with a lognormal prior distribution chosen to ensure a prior median interaction distance of 400m with prior 95% interval of 50-3000m.

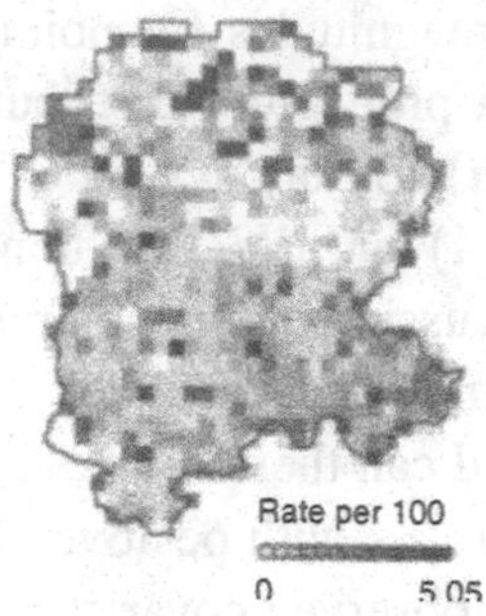

FIGURE 3.15. Posterior mean admission rates for bronchiolitis attributed to the 'unmeasured' sources (*i.e.*, the baseline and latent covariates), estimated using the point process regression models.

7.3 Results

Figure 3.15 shows the posterior mean of the admission rate $\beta_0^B + \beta_l^B \times \int_S k(x,s)\Gamma(ds)$ attributed to the 'unmeasured' sources (*i.e.*, the baseline and latent covariates). We present these rates integrated over the 1km grid for comparability with the results from the discrete MRF and DPG models; however, the continuous random field model allows risk associated with the latent spatial covariates to be presented consistently at arbitrary spatial scales, and so other partitions could be chosen for presentation if desired. The overall spatial pattern is similar to that estimated using the covariate-unadjusted MRF and DPG models (Figure 3.11), although it is less smooth than either of the discrete models. This is consistent with the fact that the posterior mean of the distance scale ρ was 728m with 95% credible interval (489m, 1070m), which suggests that spatial correlation in the unexplained (*i.e.*, latent) risk extends over a smaller neighborhood than was assumed for either the MRF or DPG models.

The posterior percentage attributed risks for each excess risk factor are summarized as follows: Baseline 12.02% (95% CI 7.49%–16.22%); Latent spatial covariates 87.46% (95% CI 83.17%–92.05%); MHGV 0.52% (95% CI 0.03%–2.00%). These are broadly consistent with our findings for the DPG model. In particular, only a very small fraction of cases are attributed to traffic related pollution exposure, with the majority of admissions being attributed to the latent spatial covariate. Note that since we do not include social deprivation in this model, the results are not directly comparable with those reported for the covariate-adjusted DPG model in Table 3.5; however, since social deprivation shows a strong spatial pattern, the effects of this covariate are likely to be captured by the latent spatial random field.

The posterior mean relative risk of bronchiolitis admission for boys compared to girls was 1.61 (95% CI = 1.07–2.58). This agrees with the findings from the discrete MRF and DPG models that areas with a higher proportion of boys tend to have higher admission rates for bronchiolitis; however, the relative risk parameter in the present model is directly interpretable in terms of the *individual* level risk associated with being a boy, rather than the *ecological* level risk associated with living in areas with a high proportion of boys. The risk estimate is also somewhat smaller than we obtained from the MRF or DPG models (1.61 compared to 2.70 and 2.50 respectively). This suggests that the discrete model estimates based on aggregate covariates may be subject to ecological bias, although the models are not adjusted for the same set of covariates and so cannot be compared directly.

8 Discussion

This Case Study attempts to address an important public health question using a variety of routinely collected data sources, many of which have only recently been made available for epidemiological research. As in any observational study one must be careful in making inference from these non-experimental data. The specific concerns inherent in the HES data are not yet well understood, but some valuable insights into their quality, interpretation and usefulness have emerged from this study. For example, it is now clear that London hospitals differ widely in their diagnostic coding policies and in the referral process by which their patients are admitted. These factors may lead to large apparent geographical variations in the ICD-specific hospital admission rates that are unrelated to variations in the incidence and severity of bronchiolitis across the study region. Careful interpretation of the substantive findings of this study is therefore imperative in order to avoid misleading conclusions.

We now return to the four questions originally posed in Section 1.

1. Do infants living in areas of high traffic density have higher rates of respiratory illness?

Analysis based on our Poisson regression and MRF models does not reveal higher respiratory illness rates in areas of high traffic density; if anything, it suggests that rates might be lower there. This probably reflects confounding (due to multicollinearity) and bias (due to systematic differences in diagnostic coding between inner city and outer city hospitals) in the HES data. This interpretation is supported by the discovery of a similar negative association between MHGV and risk of admission for non-respiratory diagnoses, suggesting that the effect of traffic exposure is non-specific. Furthermore, since admission rates for all three diagnoses (B, N and U) are expected to be affected similarly by any biases in the HES data, the *ratios* of admission rates for the different diagnoses should be less affected. We found no 'significant' difference between the relative risk estimates for the effect of MHGV on admission for bronchiolitis compared with either non-respiratory or unspecified diagnoses, in the sense that the 95% credible intervals for the relative risk ratios include one (see final two columns of Table 3.5). This again suggests that exposure to high traffic density does not have a specific effect on admissions for bronchiolitis in infants.

To check whether the true association between traffic density and bronchiolitis might be masked by correlation between MHGV and other covariates such as deprivation or hospital effects, we re-ran the MRF model including only MHGV and a spatial random effect as linear predictors. The estimated association with MHGV remained negative, although there was considerable posterior uncertainty and the 95% credible interval for the relative risk included one (RR = 0.24; 95% CI = 0.01–1.22).

The discrete and continuous Poisson/gamma (DPG and CPG) models were proposed as more flexible alternatives to the MRF analysis. They allow both excess and relative risk factors to be included in the same model (the MRF models allow only relative risks). Our choices between additive and multiplicative forms for the covariate effects in these models were based on *a priori* considerations rather than being data driven, since the data rarely offer conclusive support in favor of one or other model (Breslow and Day, 1980). However, we may view the contrast between the log linear MRF models on the one hand, and the linear DPG/CPG models on the other, as a form of sensitivity analysis to the assumed form of the relationship between traffic pollution and risk of bronchiolitis. Despite different interpretations for the regression coefficients associated with the relative versus excess risk factor models, and with different prior assumptions, both approaches lead us to the same *qualitative* conclusion: these data provide no evidence to support concerns that chronic exposure to traffic pollution poses a significant risk of hospitalization for bronchiolitis in infants living in northwest London. To this extent, the methodologies appeared robust, although other applications may show more sensitivity (particularly in applications where the data show a stronger positive association between exposure and response).

This conclusion appears to conflict with the results of our preliminary ward-level analysis of bronchiolitis admissions in the whole of Greater London (Section 2.1). However, the present study used much finer levels of spatial aggregation (indeed the CPG model requires no aggregation at all), and so may suffer less from the possible effects of ecological bias. Our handling of admissions with unspecified diagnoses also differs between studies: in the original analysis, we treated these cases as missing at random, and reassigned them as bronchiolitis, other respiratory or non-respiratory admissions according to the observed admission rates for the three diagnoses. This would lead to more apparent cases of bronchiolitis in central London (the southeast corner of the present study region) where many of the unspecified cases are clustered and where traffic density is also highest. However, it is not clear that the true diagnosis for these admissions is really missing at random (it may be that cases presenting with complicated conditions, or infants with non-specific symptoms, are more likely to be assigned an ICD 799 code), and so we chose to model unspecified diagnoses as a separate outcome in the present study. This analysis also focuses on a subset of the original study region (*i.e.*, the northwest quadrant). Preliminary examination of the association between bronchiolitis admission rates and traffic density in other quadrants shows inconsistencies across London: there is some suggestion that the northeast and southeast quadrants show a slight positive relationship, while the southwest shows no relationship. However, these findings may well be driven by a few outlying areas and further detailed follow-up analysis is required.

2. Which pollutants or vehicle types are associated with the greatest risk?

It proved impossible to distinguish among the effects of different pollutants on risk of bronchiolitis admission due to the extremely high correlations between the emissions data available for this study. Since we found no evidence of an increase in risk of bronchiolitis admissions associated with exposure to high traffic density, we did not pursue this question further.

3. Do factors such as socioeconomic deprivation of the area of residence, or individual characteristics such as gender increase or modify this risk?

All our models showed a consistent and large positive relationship between deprivation and admission rates for all three diagnoses, suggesting that infants living in socioeconomically disadvantaged areas are more likely to be admitted to hospital, irrespective of the cause. The Carstairs index used to measure deprivation in this study is a relatively crude proxy for a range of factors such as poverty, smoking, poor diet, poor housing conditions and overcrowding, which may explain this finding, although more detailed individual level data on such covariates would be needed in order to further identify the specific risk factors involved. It may also reflect such non-ætiological factors as access to health care, use of A&E (emergency room) services and thresholds for admission to hospital. Comparison of relative admission rates for bronchiolitis versus non-respiratory or unspecified diagnoses revealed that deprivation had a significantly greater impact on the former. Under the MRF model, infants living in the most deprived area were over 2.5 times more likely to be admitted to hospital for bronchiolitis than for either non-respiratory or unspecified diagnoses compared with infants living in the most affluent area; under the DPG model, approximately 12%–15% more admissions for bronchiolitis were attributed to living in a deprived rather than affluent area compared to non-respiratory or unspecified diagnoses.

Investigation of *individual* covariate effects is not feasible using the area-level MRF and DPG models. Instead it was necessary to treat attributes such as gender as aggregated covariates, which complicates interpretation. This problem is avoided by using the CPG model, which has the great advantage of allowing all data to enter at their natural level of (dis)aggregation and so allows regression coefficients for individual attributes to be interpreted directly as *individual* level risks rather than as *ecological* level risks. The gender effect in this model was consistent with that in the area-level models (all show that boys experience elevated risk), but much better-defined (the 95% credible intervals were more than five times narrower).

4. Can we account for residual confounding due to unknown or unmeasured risk factors?

Both the MRF and Poisson/gamma models addressed this issue through the inclusion of spatial random effects, although the distributional form varied between models. One difficulty with the area-level MRF and DPG models is that the spatial structure for these random effects must be specified *a priori*; however, there is generally very little scientific evidence to inform this choice. By contrast, the location and shape of the latent spatial random field in the CPG model is estimated from the data and so avoids the need to impose arbitrary prior assumptions about this aspect of the model.

In summary, this Case Study revealed no association between chronic exposure to traffic pollution and hospital admissions for viral bronchiolitis strong enough to overcome other sources of variation in the observational data. It revealed the need for a systematic strengthening of diagnostic coding standards across hospitals before the HES database will support detailed epidemiological analyses using sophisticated modern statistical modeling. However, without the detailed statistical modeling carried out for this study it would remain unclear whether the unexpected negative results for the association between traffic density and bronchiolitis admissions were indicative of fundamental inadequacies in the data or to the failure of simple models to account fully for the multiple sources of variation. In this respect the Bayesian models presented in this paper have allowed us to gain considerable insight into the structure and nature of the available data and the potential sources of bias, have raised a number of questions which require further study, and have pointed the way to needed improvements in HES data collection.

Acknowledgments

The authors gratefully acknowledge the support of the following grants for this work: US NSF grant DMS-9626829; MRC Career Establishment Grant No. G9803841; Pan-Thames Environmental R&D Initiative Grant No. 339. The authors are grateful to the Office for National Statistics and the Department of Health and the London Research Centre for providing data. The authors would also like to thank Lucy Sadler from the London Research Centre for helpful discussions concerning the road traffic data used in this project, the London School of Hygiene and Tropical Medicine for the joint linking of the road networks, and Richard Arnold for helpful discussions. The work is based on data provided with the support of the ESRC and JISC and uses census and boundary material which is copyright of the Crown, the Post Office and the ED-LINE consortium. The computing facilities used for the work were partially supported by a Wellcome equipment grant.

References

Anderson, H., Butland, B., and Strachan, D. (1994). Trends in prevalence and severity of childhood asthma. *British Medical Journal* **308**, 1600–4.

Anderson, H. R., Ponce de Leon, A., Bland, J. M., Bower, J. S., Emberlin, J., and Strachan, D. P. (1998). Air pollution, pollens, and daily admissions for asthma in London 1987–92. *Thorax* **53**, 842–8.

Bernardinelli, L., Pascutto, C., Best, N., and Gilks, W. (1997). Disease mapping with errors in covariates. *Statistics in Medicine* **16**, 741–52.

Besag, J. (1974). Spatial interaction and the statistical analysis of lattice systems. *Journal of the Royal Statistical Society, Series B (Methodological)* **36**, 192–236. (With discussion).

Besag, J., York, J., and Mollié, A. (1991). Bayesian image restoration, with two applications in spatial statistics. *Annals of the Institute of Statistical Mathematics* **43**, 1–59. (With discussion).

Best, N., Bennett, J., Cockings, S., Falconer, S., Maheswaran, R., Wakefield, J., Smith, A., and Elliot, P. (1998). Small area mapping and statistical analyses of variations in disease risk and possible explanatory factors in the Thames Regions. Tech. rep., Department of Epidemiology & Public Health, Imperial College, London.

Best, N. G., Ickstadt, K., and Wolpert, R. L. (1998b). Spatial Poisson regression for health and exposure data measured at disparate resolutions. *Journal of the American Statistical Association* **95** 1076–88.

Bithell, J. F., Dutton, S. J., Neary, N. M., and Vincent, T. J. (1995). Controlling for socio-economic confounding using regression methods. *Journal of Epidemiology and Community Health, Supplement 2* **49**, S15–S19.

Breslow, N. E. and Day, N. E. (1980). *Statistical methods in cancer research. Volume 1 - The analysis of case-control studies.* No. 32 in IARC Scientific Publications. International Agency for Research on Cancer, Lyon.

Briggs, D. J., Collins, S., Elliott, P., Fischer, P., Kingham, S., Lebret, E., Pryl, K., van Reeuwijk, H., Smallbone, K., and van der Veen, A. (1997). Mapping urban air pollution using GIS: A regression-based approach. *International Journal of Geographic Information Science* **11**, 699–718.

Brunekreef, B., Dockery, D. W., and Krzyanowski, M. (1995). Epidemiologic studies on short-term effects of low levels of major ambient air pollution components. *Environmental Health Perspectives* **103 (Supplement 2)**, 3–13.

Brunekreef, B., Janssen, N. A. H., de Hartog, J., Harssema, H., and Knape, M. (1997). Air pollution from truck traffic and lung function in children living near motorways. *Epidemiology* **8**, 298–303.

Buckingham, C., Clewley, L., Hutchinson, D., Sadler, L., and Shah, S. (1997). London Atmospheric Emissions Inventory. Tech. rep., London Research Centre.

Burney, P. (1988). Asthma deaths in England and Wales 1931-85: evidence for a true increase in asthma mortality. *Journal of Epidemiology and Community Health* **42**, 316–20.

Carr-Hill, R., Hardman, G., Martin, S., Peacock, S., Sheldon, T., and Smith, P. (1994). A formula for distributing NHS revenues based on small area use of hospital beds. Tech. rep., Occasional Paper Series, Centre for Health Economics, University of York.

Carstairs, V. and Morris, R. (1991). *Deprivation and health in Scotland.* Aberdeen University Press, Aberdeen.

Chestnut, L. G., Schwartz, J., Savitz, D. A., and Burchfiel, C. M. (1989). Pulmonary function and ambient particulate matter: epidemiological evidence from NHANES I. *Arch Environ Health* **46**, 135–44.

Clayton, D. G. and Bernardinelli, L. (1992). Bayesian methods for mapping disease risk. In Elliott et al. (1992), chap. 18, 205–20. Paperback edition published 1996.

Committee of Medical Effects of Air Pollution (1995a). *Asthma and outdoor air pollution.* Her Majesty's Stationary Office, London.

Committee of Medical Effects of Air Pollution (1995b). *Non-biological particles and health.* Her Majesty's Stationary Office, London.

Committee of the Environmental and Occupational Health Assembly of the American Thoracic Society (1996). Health effects of outdoor pollution. *American Journal of Respiratory Critical Care Medicine* **153**, 3–50.

Cummins, P. and Rathwell, T. (1991). Geographical Information Systems and the National Health Service. Tech. rep., Association for Geographic Information Education, Training and Research Publication No.3.

Department of Health (1998). *How HES Data is Processed 1996/97 Datayear v.3.0. SD2HES.*

Dixon, J., Sanderson, C., Elliott, P., Walls, P., Jones, J., and Petticrew, M. (1998). Assessment of the reproducibility of clinical coding in routinely collected hospital activity data: a study of two hospitals. *J Pub Health Med* **20**, 63–69.

Dockery, D., Pope, III, C., Xiping, X., Spengler, J., Ware, J., Fay, M., Ferris, B., and Speizer, F. (1993). An association between air pollution and mortality in six US cities. *New England Journal of Medicine* **329**, 1753–1759.

Dockery, D. W., Spengler, J. D., Reed, M. P., and Ware, J. (1981). Relationships among personal, indoor and outdoor NO_2 measurements. *Environment International* **5**, 101–7.

Duhme, H., Weiland, S. K., Keil, U., Kraemer, B., Schmid, M., and Stender, M. (1996). The association between self-reported symptoms of asthma and allergic rhinitis and self-reported traffic density on street of residence in adolescents. *Epidemiology* **7**, 578–82.

Edwards, J., Walters, S., and Griffiths, R. (1994). Hospital admissions for asthma in preschool children: relationship to major roads in Birmingham, United Kingdom. *Archives of Environmental Health* **49**, 223–7.

Elliott, P., Cuzick, J., English, D., and Stern, R., eds. (1992). *Geographical and environmental epidemiology: Methods for small-area studies.* Oxford University Press, Oxford. Paperback edition published 1996.

Gardner, M. T., Winter, P. D., and Barker, D. J. P. (1984). *Atlas of mortality from selected diseases in England and Wales 1968–1978.* Wiley, Chichester.

Gatrell, A. C., Dunn, C. E., and Boyle, P. J. (1991). The relative utility of the Central Postcode Directory and Pinpoint Address Code in applications of Geographical Information Systems. *Environment and Planning A* **23**, 1447–1458.

Gatrell, A. C. and Senior, M. (1999). Health and healthcare applications. In *Geographical Information Systems: Principles, Techniques, Applications and Management*, eds. P. Longley, M. Goodchild, D. Maguire, and D. Rhind, vol. 2, 925–938. John Wiley & Sons.

Greenberg, M. R. (1983). *Urbanization and cancer mortality.* Oxford University Press, Oxford.

Greenland, S. (1992). Divergent biases in ecologic and individual-level studies. *Statistics in Medicine* **11**, 1209–1223.

HMSO (1994). *Royal Commission on Environmental Pollution. Eighteenth Report. Transport and the Environment.* HMSO, London.

Hoek, G., Brunekreef, B., and Hofschreuder, P. (1989). Indoor exposure to airborne particles and nitrogen dioxide during an air pollution episode. *J Air Poll Contr Assoc* **39**, 1348–9.

Howe, H. L., Keller, J. E., and Lehnherr, M. (1993). Relation between population density and cancer incidence, Illinois, 1986–90. *Am J Epidemiol* **138**, 29–36.

Jolley, D., Jarman, B., and Elliott, P. (1992). Socio-economic confounding. In Elliott et al. (1992), chap. 11, 115–124. Paperback edition published 1996.

Kelsall, J. E. and Wakefield, J. C. (1999). Discussion of 'Bayesian models for spatially correlated disease and exposure data' by Best et al. In *Bayesian Statistics 6*, eds. J. M. Bernardo, J. O. Berger, A. P. Dawid, and A. F. M. Smith, 131–56. Oxford University Press, Oxford.

King, G. (1997). *A solution to the ecological inference problem.* Princeton University Press, Princeton.

Kleinschmidt, I., Hills, M., and Elliott, P. (1995). Smoking behaviour can be predicted by neighbourhood deprivation measures. *J Epidemiol Community Health* **49 (Supplement 2)**, S72–S77.

Knorr-Held, L. and Besag, J. (1998). Modelling risk from a disease in time and space. *Statistics in Medicine* **17**, 2045–60.

Livingstone, A. E., Shaddick, G., Grundy, C., and Elliott, P. (1996). Do people living near inner city main roads have more asthma needing treatment? Case-control study. *British Medical Journal* **312**, 676–7.

Mage, D. T. and Buckley, T. J. (1995). The relationship between personal exposures and ambient concentrations of particulate matter. Presented at the 88th Annual Meeting of the Air and Waste Management Association, San Antonio, TX.

McCullagh, P. and Nelder, J. A. (1989). *Generalized Linear Models*. Chapman and Hall, London.

McKee, M. and Petticrew, M. (1993). Disease staging — a case-mix system for purchasers? *J Pub Health Med* **15**, 25–36.

Mollié, A. (1996). Bayesian mapping of disease. In *Markov Chain Monte Carlo in Practice*, eds. W. R. Gilks, S. Richardson, and D. J. Spiegelhalter, 359–379. Chapman & Hall, New York, NY, USA.

Murakami, M., Ono, M., and Tamura, K. (1990). Health problems of residents along heavy-traffic roads. *Journal of Human Ergology* **19**, 101–6.

MVA Limited (1999). London Transportation Studies Technical Note 25: B1.10 Base Forcast Validation. Tech. rep., MVA Limited, 115 Shaftesbury Avenue, London.

Noy, D., Brunekreef, B., Boleij, J. S. M., Houthuijs, D., and De Koning, R. (1990). The assessment of personal exposure to nitrogen dioxide in epidemiological studies. *Environment International* **12**, 407–11.

Osterlee, A., Drijver, M., Lebret, E., and Brunekreef, B. (1996). Chronic respiratory symptoms in children and adults living along streets with high traffic density. *Occupational Environment Medicine* **53**, 241–7.

Ozkaynak, H. and Spengler, J. (1996). The role of outdoor particulate matter in assessing total human exposure. In *Particles in Outdoor Air: Concentrations and Health Effects*, eds. R. Wilson and J. Spengler, 63–84. Harvard University Press, Cambridge, MA.

Parkin, D. M., Whelan, S. L., Ferlay, J., Raymond, L., and Young, J. (1997). *Cancer incidence in Five Continents*. IARC Scientific Publications No. 143, Lyon.

Pershagen, G., Rylander, E., Norberg, S., Eriksson, M., and Nordell, S. (1995). Air pollution involving nitrogen dioxide exposure and wheezing bronchitis in children. *International Journal of Epidemiology* **24**, 1147–53.

Pope, C., Thun, M., Namboodiri, M., Dockery, D., Evans, J., Speizer, F., and Heath, C. (1995). Particulate air pollution as a predictor of mortality in a prospective study of US adults. *American Journal of Respiratory Critical Care Medicine* **151**, 669–674.

Quackenboss, J. J., Krzyanowski, M., and Lebowitz, M. D. (1991). Exposure assessment approaches to evaluate respiratory health effects of particulate matter and nitrogen dioxide. *Journal of Exposure Analysis and Environmental Epidemiology* **1**, 83–106.

Quackenboss, J. J., Spengler, J., Kanarek, M. S., Letz, R., and Duffy, C. P. (1986). Personal exposure to nitorgen dioxide: relationship to indoor/outdoor air quality and activity patterns. *Environ Sci Tecnol* **20**, 775–83.

Raper, J. F., Rhind, D. W., and Shepherd, J. W. (1992). *Postcodes: the new geography*. Longman Group UK Limited, Essex.

Richardson, S. (1992). Statistical methods for geographical correlation studies. In Elliott et al. (1992), chap. 17, 181–204. Paperback edition published 1996.

Richardson, S., Monfort, C., Green, M., Draper, G., and Muirhead, C. (1995). Spatial variation of natural radiation and childhood leukaemia incidence in Great Britain. *Statistics in Medicine* **14**, 2487–2501.

Roberts, G. O. (1996). Markov chain concepts related to sampling algorithms. In *Markov chain Monte Carlo Methods in practice*, 45–58. Chapman & Hall, New York, NY, USA.

Robinson, W. S. (1950). Ecological correlations and the behaviour of individuals. *American Sociological Review* **15**, 351–357.

Roemer, W., Hoek, G., and Brunekreef, B. (1993). Effect of ambient winter air pollution on respiratory health of children with chronic respiratory symptoms. *Am Rev Respir Dis* **147**, 118–24.

Schwartz, J. (1989). Lung function and chronic exposure to air pollution: a cross-sectional analysis of NHANES II. *Environ Research* **50**, 309–21.

Schwartz, J. (1993). Particulate air pollution and chronic respiratory disease. *Environmental Research* **62**, 7–13.

Schwartz, J. (1994). Air pollution and daily mortality: a review and meta-analysis. *Environmental Research* **64**, 36–52.

Sheldon, T. A., Smith, P., Borowitz, B., Martin, S., and Carr-Hill, R. (1994). Attempt at deriving a formula for setting general practitioner fundholding budgets. *British Medical Journal* **309**, 1059–64.

Shima, M. and Adachi, M. (1998). Indoor nitrogen dioxide in homes along trunk roads with heavy traffic. *Occup Environ Med* **55**, 428–33.

Smith, R. and Jarvis, C. (1998). Just the medicine. *Mapping Awareness* **12**, 30–33.

Spiegelhalter, D., Thomas, A., and Best, N. (1999). On-line User Manual. WinBUGS user manual, version 1.2. URL `http://www.mrc-bsu.cam.ac.uk/bugs`.

Strachan, D. P. (1991). Damp housing, mould allergy and childhood asthma. *Proc R Coll Physicians Edinb* **21**, 140–6.

Strachan, D. P., Anderson, H. R., Limb, E. S., O'Neill, A., and Wells, N. (1994). A national survey of asthma prevalence, severity, and treatment in Great Britain. *Arch Dis Child* **70**, 174–8.

Urell, M. and Samet, J. (1993). Particulate air pollution and health — new evidence on an old problem. *Am Rev Respir Dis* **147**, 1334–5.

Vostal, J. J. (1994). Physiologically based assessment of human exposure to urban air pollutions and its significance for public health risk. *Environmental Health Perspectives* **102 (Supplement 4)**, 101–6.

Waldron, G., Pottle, B., and Dod, J. (1995). Asthma and motorways — one district's experience. *Journal of Public Health Medicine* **17**, 85–9.

Wallace, L. (1996). Indoor particles: a review. *J Air Waste Manage Assoc* **46**, 98–126.

Walshe, K., Harrison, N., and Renshaw, M. (1993). Comparison of the quality of patient data collected by hospital and departmental computer systems. *Health Trends* **25**, 105–8.

Weiland, S., Mundt, K., Ruckmann, A., and Keil, U. (1994). Self-reported wheezing and allergic rhinitis in children and traffic density on street of residence. *AEP* **4**, 243–7.

Williams, B. (1997). Utilisation of national health service hospital in England by private patients 1989-95. *Health Trends* **29**, 21–5.

Williams, D. R. R., Anthony, P., Young, R. J., and Tomlinson, S. (1994). Interpreting hospital admissions data across the Korner divide: the example of diabetes in the North Western region. *Diabetic Med* **11**, 166–9.

Wjst, M., Reitmeir, P., Dold, S., Wulff, A., Nicolai, T., Freifrau von Loeffelholz-Colberg, E., and von Mutius, E. (1993). Road traffic and adverse effects on respiratory health in children. *British Medical Journal* **307**, 596–600.

Wolpert, R. L. and Ickstadt, K. (1998). Simulation of Lévy random fields. In *Practical Nonparametric and Semiparametric Bayesian Statistics*, eds. D. Dey, P. Müller, and D. Sinha, vol. 133 of *Lecture Notes in Statistics*, 227–242. Springer-Verlag, New York, NY, USA.

Discussion

Francesca Dominici and Jonathan Samet, Johns Hopkins University

Best et al. address questions of scientific and public health significance on the adverse health effects of traffic air pollution in children. This analysis is motivated by recent concern that air pollution generated by road traffic adversely affects children. The authors have developed and implemented elegant and complex statistical models to assess association between air pollution and hospital admission rates in childhood, offering a superb case-study. But will the authors be able to convince policy-makers with their the findings?– The ultimate goal of the analysis.

Infectious bronchiolitis, the outcome of interest, is a clinical syndrome, most common in the first years of life. It reflects inflammation of the small airways of the lung, the bronchioles, by respiratory viruses. Respiratory syncytial virus, a predominant causal agent, produces epidemics that take place sporadically, mostly during the winter months. The clinical hallmark of bronchiolitis is wheezing, but cough is also a frequent symptom (Samet et al. 1993). While bronchiolitis often necessitates medical evaluation and treatment, many mild cases, probably constituting the majority, go untreated and are not captured in health care data systems.

Ambient air pollution has long been considered as possibly contributing to the occurrence of acute lower respiratory illnesses in childhood such as bronchiolitis (American Thoracic Society,1996). Animal models have shown that air pollution exposure worsens experimental infections and air pollution exposure affects elements of lung defenses (Smith, Samet, and Romieu 1999). Thus, the hypothesis explored by Best and colleagues is grounded in plausibility offered by biomedical evidence. Their findings have potential public and regulatory implications.

How could air pollution affect hospitalization rates for bronchiolitis (or other respiratory diagnoses) in childhood? An effect on hospitalization rates would be anticipated if vehicle-related pollution either increased the incidence or the severity of bronchiolitis. Diminished host defenses associated with pollution exposure might either increase the likelihood of infection or of more severe infection, or perhaps air pollution associated airways inflammation at the time of viral infection could also increase severity of infections. The outcome measure in the present study, hospitalization, only captures more severe cases, and consequently may be insensitive to increased incidence associated with vehicle traffic.

These biomedical considerations further suggest that the relevant exposure window immediately precedes the illness event and perhaps spans only days. For one key traffic pollutant, NO_2, exposure to brief, high peaks has been hypothesized as more relevant to risk for respiratory infection than longer-term exposures (Samet and Utell 1990). To interpret the findings of Best and colleagues, it will be necessary to characterize the relationship between their indicators of typical exposures, generated from source considerations, and the likelihood of spikes of exposure in time, corresponding to peaks of traffic.

For estimating exposure, the investigators develop a complex system that transfers estimates of the total annual traffic flows from the London Research Center are transferred onto a set of 1 km × 1 km squares. How valid are these estimated exposure levels? The estimation procedure of exposure is heavily model-based and a validation study should be a key component of these analyses. How does vehicle traffic play into this complex web in London?

The inherent limitation of measurement error is recognized and might even be addressed head on, using a validation study with relatively inexpensive, passive CO or NO_2 monitors. Is the second question of the investigators reachable in this study: "...which pollutants or vehicle types are associated with the greatest risk?" Common sources, correlated pollutant concentrations, and measurement error may be insurmountable limitations. Prior hypotheses are abundant and we are pessimistic about finding strong evidence for one credible hypothesis against equally credible alternatives.

Several models of increased complexity are introduced: a Poisson regression model, a hierarchical Markov random field model (with the inclusion of area-specific random effects), a discrete Poisson/Gamma model (adding flexibility in modelling excess risk and including relative risks, analysis at smaller aggregation level), and a continuous Poisson/Gamma model (proposing an analysis at the finest spatial scale of the data).

Poisson regression and Markov random field models provide the puzzling result that hospital admissions for bronchiolitis are negatively associated with traffic load (MHGV) and Black Smoke. Area deprivation scores, hospital effects, admission methods, gender and ethinicity are included in the models to control for counfouding factors. For respiratory infections in infancy, the causal web is complex, involving environmental factors (outdoor and indoor air pollution, crowding, family structure, housing quality, child care contact, and nutrition) and host factors (gender, atopy, risk for asthma, and history of prematurity). Socioeconomic indicators, as considered by Best and colleagues, are surrogates for several of these factors, depending on the population. The concept of "confounding" implies sufficient prior knowledge of these relationships. Do the investigators have a basis for sufficient certainty as to the underlying causal relations?

The estimated map of the unadjusted admission rates for bronchiolitis under the discrete Poisson/Gamma model is similar to the one provided by the Markov random field model, but considerably smoother. In addition, model validation results suggest that the discrete Poisson/Gamma model does not perform substantially better in predicting the admission rates, compared to Markov random field model. What is the "right" model and how is it to be selected?

Estimating the adverse health effects of air pollution is an obviously complicated problem, and complex statistical models are needed for the task. The modeling strategies developed by the authors are state-of-the-art. However it is important not to draw conclusions that are too heavily model-based; no direct answers to the scientific questions are provided; public policy implications of a possible protective effect of traffic air pollution are difficult to interpret.

Let us project ourselves forward in time. Best and colleagues are presenting their findings to an erudite committee charged with reviewing air quality guidelines for the United Kingdom. Some possible scenarios are:

- Traffic and traffic pollution are associated with hospitalizations and the findings are robust to model specification.

- Traffic and traffic pollution are associated with hospitalizations in some models and for some pollutants.
- Traffic and traffic pollution are not associated with hospitalizations.

The appropriate response is self-evident for the first scenario. What does the evidence tell us in the second and third?

References

American Thoracic Society, Committee of the Environmental and Occupational Health Assembly, Bascom, R., Bromberg, P.A., Costa, D.A., Devlin, R., Dockery, D.W., Frampton, M.W., Lambert, W., Samet, J.M., Speizer, F.E. and Utell, M. Health effects of outdoor air pollution. Part 1. Am J Resp Crit Care Med 1996;153:3-50.

American Thoracic Society, Committee of the Environmental and Occupational Health Assembly, Bascom, R., Bromberg, P.A., Costa, D.A., Devlin, R., Dockery, D.W., Frampton, M.W., Lambert, W., Samet, J.M., Speizer, F.E. and Utell, M. Health effects of outdoor air pollution. Part 2. Am J Resp Crit Care Med 1996;153:477-98.

Samet, J.M., Cushing, A.H., Lambert, W.E., Hunt, W.C., McLaren, L.C., Young, S.A. and Skipper, B.J. Comparability of parent-reported respiratory illnesses to clinical diagnoses. Am Rev Respir Dis 1993;148:441-6.

Samet, J.M. and Utell, M.J. The risk of nitrogen dioxide: What have we learned from epidemiological and clinical studies? Toxicol Indust Health 1990; 6: 247-62.

Smith, K.R., Samet, J.M., Romieu, I. and Bruce, N. Air pollution and acute lower respiratory infections in children. Thorax 1999 (submitted).

Rejoinder

We would like to thank Professors Dominici and Samet for their careful study and thoughtful discussion of our work. Indeed as they suggest the ultimate goal of our analysis is to guide and support policy-makers in making well-informed choices affecting public health; we are grateful for the opportunity this Workshop has afforded us to share these ideas with a wider audience.

We agree that traffic pollution may plausibly affect children's respiratory ailments in two different ways, either by increasing the severity of existing cases or by initiating additional cases; and that the present study, through its focus on cases severe enough to warrant hospitalization, is perhaps more sensitive to the first of these plausible effects than to the second. Our negative conclusion must be interpreted in this light— we have shown that there is little evidence that traffic pollution causes a marked increase in respiratory hospital admissions, but have not shown whether or not there is an observable effect on the incidence of milder cases not requiring hospitalization. Our choice of *hospital admissions* as the incidence criterion was motivated by the recent availability of the HES dataset, providing a population-based register of all such events as described in §3.2; we know of no comparable dataset in the UK reporting incidence rates of mild cases of children's respiratory disease on a population-wide basis.

The discussants' concerns about nonlinearity of response to traffic pollution, and their resulting concern about the acute effects of air pollution, are well taken but do not concern us here— in §2 we review the substantial literature on the acute effects of air pollution, but in the present study we explore the question of whether there is an additional (or separate) effect of *chronic* exposure to air pollution on respiratory illness. Once again our findings need to be interpreted in this light— our finding of no evidence of association with chronic exposure does not necessarily conflict with the generally positive association between acute air pollution episodes and respiratory illness that is widely recognized in the literature. Rather, it suggests that, while the acute effects may be due to "harvesting" (*i.e.*, precipitating or exacerbating already existing respiratory illness cases), any chronic effects of exposure are more likely to produce *new* cases of disease that may not be detectable in our study due to the insensitivity of our outcome measure to milder cases, as discussed above.

Work describing and validating data from the London Transportation Studies (LTS) model of Buckingham et al. (1997) appears in MVA Limited (1999). We should emphasize that this extrapolation of traffic flows is neither new nor attributable to the present authors; our novel contribution is to improve the grid extrapolation accuracy by using GIS techniques to map traffic flows directly to the actual road network rather than to the linearized "stick" network approximation, before averaging over 1km $\times$ 1km grid cells. Note that it is *emissions* and not concentrations that are extrapolated to the grid, and related through our model to hospitalization rates. The discussants' suggestion of validating this extrapolation through measured concentrations is well taken, but would entail additional measurement and modeling to relate emissions to concentrations, adding new sources of uncertainty, variability and measurement error.

The discussants' question "Is the second question of the investigators reachable in this study" is answered (in the negative) in the Discussion section of the paper, §8; the high correlations among the emissions levels, and the low level of association between any of them and disease incidence, prevent us from reaching the goal of distinguishing which of the pollutants may be most strongly associated with disease incidence.

No, alas, we do not have a deep enough understanding of the "causal web" leading to respiratory illness to be confident that our model includes all the important covariates in exactly the right way. We do, however, construct models that allow for both contributory ("excess risk") factors and modifying ("relative risk") factors, while the more usual log-linear approach includes only the latter. We do also consider a possible mechanism (in Figure 3.1) for how a child may become a case and use this to suggest which variables to include as contributing and which as modifying factors, but there are undoubtedly other important effects that we omitted. Our spatial random effect (RE) is intended to act as a surrogate for all such unmodeled sources; the nearly flat posterior distribution for this RE suggests that no unmeasured spatially varying covariate has a particularly strong influence on respiratory illness rates.

In the spirit of George Box's aphorism that *All models are wrong, but some are useful* (Box, 1979) we resist trying to answer the question of which model is "right;" indeed, it is notoriously difficult to do sensible model criticism in complex high-dimensional models. The Poisson/gamma model with its additive spatial RE is intended to reveal an unmodeled spatially-varying influence, if one is present; the fact that we found none does not lessen the value of such an exploratory tool.

We agree that our analyses, on their own, do not offer the "direct answers to scientific questions" on which we would like to base public-health policy decision-making. Indeed, our study revealed important issues for the analysis of the HES data that went beyond the initial focus of the work and that seriously complicate interpretation of the scientific questions initially considered. A less elaborate and less intensive modeling effort would have left the lingering doubt that our routinely collected HES data might hold the keys, if only we could be imaginative enough to exploit them more fully; we find instead that specific and targeted experiments and data collection will be required to make progress in resolving the public-health policy implications of the scenarios posed by the discussants.

We thank the discussants for their role in clarifying all these points, and in helping to make the Workshop a spirited and exciting exercise for us all.

References

Box G.E.P., (1979). Robustness in the strategy of scientific model building. In *Robustness in Statistics*, 201–236. Academic Press, New York, NY, USA

Buckingham, C., Clewley, L., Hutchinson, D., Sadler, L., and Shah, S. (1997). London Atmospheric Emissions Inventory. Tech. rep., London Research Centre.

MVA Limited (1999). London Transportation Studies Technical Note 25: B1.10 Base Forcast Validation. Tech. rep., MVA Limited, 115 Shaftesbury Avenue, London.

CONTRIBUTED

PAPERS

A Hierarchical Model for Estimating Distribution Profiles of Soil Texture

Pamela J. Abbitt
F. Jay Breidt

ABSTRACT The MLRA (Major Land Resource Area) 107 pilot project involved implementation of a multi-phase probability sampling design to update the soil surveys for two counties in western Iowa. We consider estimation of distribution profiles of soil texture using a hierarchical model and data from the pilot project. Soil texture measurements are recorded for each horizon (or layer) of soil. Soil horizon profiles are modeled as realizations of Markov chains. Conditional on the horizon profile, transformed field and laboratory determinations of soil texture are modeled as a multivariate mixed model with normal errors. The posterior distribution of unknown model parameters is numerically approximated using a Gibbs sampler. The hierarchical model provides a comprehensive framework which may be useful for analyzing many other variables of interest in the pilot project.

1 Introduction

The National Cooperative Soil Survey (NCSS) is a collaborative program involving the U.S. Department of Agriculture and a state agency, often the state's Agricultural Experiment Station. The NCSS program is charged with constructing county reports which contain soil maps and descriptions of all soils found within the county. These reports are periodically updated through the NCSS program to provide current information on soil characteristics. Updates are based on surveys involving extensive field work. This information is used by contractors, farmers and others to guide land use planning purposes and by scientists to develop models based on soil characteristics.

Texture is a vector of the proportions of clay, sand and silt making up a soil. Soil texture is an important consideration in land use and management so most soil surveys describe the texture of soils found within the county

(Natural Resources Conservation Service (henceforth NRCS), 1999). Soil texture measurements are typically recorded for each horizon (or layer of soil) since the texture of the soil may change with depth. In this study, horizon-based texture measurements are recorded to a depth of 48 inches. Horizon-based data present challenges in combining data across sites because the types and depths of horizons differ from site to site.

We assume there is a soil texture profile underlying the observed horizon-based measurements at a particular site. By *soil texture profile*, we mean a collection of soil texture values indexed by inch. The main objective of this paper is to investigate the (48×3 dimensional) distribution of soil texture profiles using horizon-based soil texture measurements. In particular, we estimate marginal distribution profiles for each component of texture. A *distribution profile* refers to a collection of distributions indexed by inch, which describes how the distribution of a soil characteristic changes across depth. We may construct mean profiles, variance profiles or quantile profiles to summarize features of a distribution profile.

Details of the data collection protocol for the MLRA 107 pilot project are given in Section 2. Section 3 introduces a hierarchical model for horizon-based soil texture data. A Bayesian analysis for the model is described in Section 4 and concluding discussion is in Section 5.

2 Data structure

The study population in the MLRA 107 pilot project is a county. Sampling units are dimensionless points on the land referred to as sites. The sampling design used in the pilot project consists of three phases. In all phases, variables are recorded by horizon. A *horizon* is a layer of soil which differs from the adjacent layers in physical, biological or chemical properties (NRCS, 1999). The thickness, labels and order of horizons differ from site to site.

For the first phase sample sites, information is collected on the physical characteristics that are easily determined from the *surface horizons*. The surface horizons are the top one or two horizons at the site. For second phase sample sites, field-observable data are collected on all horizons to a depth of 48 inches, where possible. In the third phase sample, laboratory determinations are made on soil samples taken from each horizon. For more details of the sampling design, see Abbitt and Nusser (1995).

We will use $\mathcal{L}$ to refer to the set of sites where field and laboratory measurements are available to a depth of 48 inches. The set of sites with full profiles of only field data will be denoted $\mathcal{F}$. We will use $\mathcal{S}$ to refer to sites where field data are available only for the surface horizon. Let $\mathcal{D}$ denote $\mathcal{S} \cup \mathcal{F} \cup \mathcal{L}$.

Let H_g be the number of observed horizons for site g and let I_g be the lower boundary of the last observed horizon. For $g \in \mathcal{F} \cup \mathcal{L}$, I_g will usually be 48 inches, although some exceptions exist in the data. For $g \in \mathcal{S}$, I_g will usually be in the range of 6 to 12 inches. The value of H_g is limited to a maximum of 10 by the data collection protocol. For $g \in \mathcal{S}$, H_g is 1 or 2. In general, we use $h = 1, \ldots, H_g$ to index horizons and $i = 1, \ldots, I_g$ to index inches for site g.

For each observed horizon, the data collector records the horizon name. The horizon name includes the master horizon designation, denoted by capital letters and modifying prefixes or suffixes. For simplicity, we assume the possible values of master horizon designation are A, B and C. In general, A horizons are mineral horizons formed at or near the surface. B horizons form below A horizons. C horizons are layers that have not been strongly affected by soil-forming processes. Deviations may occur in this ordering (e.g., an A horizon may be buried below a C horizon) (NRCS, 1999).

Table 4.1 shows an example of the horizon data collected at a site. Note that a site may have multiple occurrences of A horizons and B horizons as demonstrated in this example. Multiple C horizons are also possible. For simplicity, when the master horizon designation consists of two capital letters, we use only the first. Let m_{gh} and d_{gh} represent the master horizon designation and the horizon depth, respectively, of horizon h at site g. Horizon depth refers to the lower boundary of the horizon in inches.

Horizon index h	Horizon name	Master horizon m_{gh}	Horizon depth d_{gh}	Inch index i
1	Ap	A	7	$1, \ldots, 7$
2	A	A	15	$8, \ldots, 15$
3	AB	A	24	$16, \ldots, 24$
4	B1	B	30	$25, \ldots, 30$
5	B2	B	36	$31, \ldots, 36$
6	BC	B	42	$37, \ldots, 42$
$7 = H_g$	C	C	$48 = I_g$	$43, \ldots, 48$

TABLE 4.1. Example horizon data for site g.

Texture is the vector $\boldsymbol{c} = (c_1, c_2, c_3)$ of proportions of clay, sand and silt present in the soil. These three proportions must sum to 1.0. Figure 4.1 contains a ternary diagram which displays all possible values of soil texture. The ternary diagram is labeled with different texture classes used by soil scientists. This diagram is known as the texture triangle (NRCS, 1999).

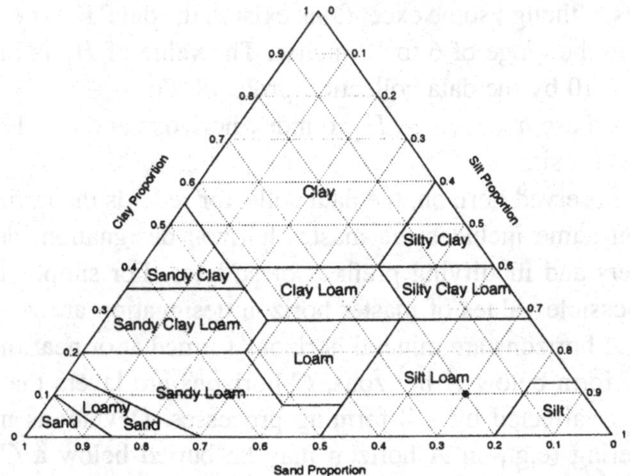

FIGURE 4.1. Ternary diagram of soil texture values. Clay content is constant along each horizontal line; silt content is constant along each 60° line; sand content is constant along each −60° line. For example, the proportions of clay, silt and sand for the point labeled "Silt Loam" are 0.1, 0.2 and 0.7.

Due to the multi-phase sampling design used for data collection, the amount of soil texture data collected varies from site to site. Let the field and laboratory determinations of texture for horizon h of site g be denoted

$$\boldsymbol{c}_{gh}^{(f)} = \left(c_{gh,1}^{(f)}, c_{gh,2}^{(f)}, c_{gh,3}^{(f)}\right) \text{ and } \boldsymbol{c}_{gh}^{(l)} = \left(c_{gh,1}^{(l)}, c_{gh,2}^{(l)}, c_{gh,3}^{(l)}\right),$$

respectively. Figure 4.2 shows field and laboratory texture data from the pilot project for a group of soils called Old Alluvium (OA) soils.

For a given horizon, the field and laboratory measurements are both estimates of the true soil texture. Investigating the distribution of true soil texture is difficult because the bias and variance of the measurement errors in laboratory and field measurements are unknown. We suspect that measurement error is a significant source of variability in both sets of measurements. The measurement error variance may be larger in the laboratory measurements than in the field measurements, but it is expected that the laboratory measurements are less biased than the field measurements.

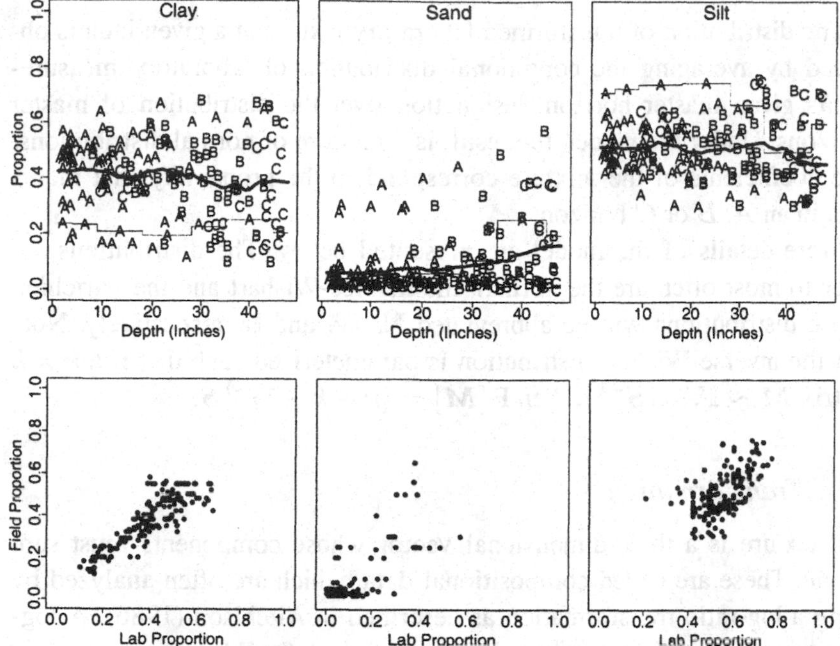

FIGURE 4.2. Texture data for Old Alluvium soils. Top row of plots shows laboratory measurements versus depth for clay, sand and silt. Letters indicate master horizon designation; horizontal plotting coordinate is the midpoint of the horizon. Light solid line is an example of a point profile; dark solid line is empirical mean profile obtained by averaging point profiles inch by inch. Bottom row of plots shows field measurements versus laboratory measurements for clay, sand and silt.

Thus, we will derive and estimate the distribution profile of laboratory determinations of soil texture.

3 Hierarchical model

Soil texture must lie in the two-dimensional space represented in Figure 4.1. A transformation of texture will be used to create a two-dimensional vector with components which no longer have a sum constraint. Transformed field texture measurements are assumed to follow a normal linear model, conditional on the transformed laboratory measurements. Conditional on the master horizon designations, transformed laboratory texture measurements follow a normal mixed linear model, with site-specific random effects.

The master horizon profile (indexed by inch) can be obtained from the horizon data collected. Horizon profiles are assumed to be realizations of a Markov chain. Some of the transition probabilities of the Markov chain are unknown parameters.

The distribution of transformed laboratory texture at a given inch is obtained by averaging the conditional distribution of laboratory measurements given master horizon designation over the distribution of master horizons. Under the model, the result is a mixture of normal distributions. The coefficients of the mixture correspond to the probability that inch i falls in an A, B or C horizon.

More details of the model are presented below. The distributions we refer to most often are the normal, the inverse-Wishart and the Dirichlet. These distributions will be abbreviated $\mathbb{N}$, $\mathbb{IW}$ and $\mathbb{D}$, respectively. Note that the inverse-Wishart distribution is parameterized such that if a $k \times k$ matrix $\boldsymbol{M} \sim \mathbb{IW}_\nu\left(\boldsymbol{S}^{-1}\right)$, then $\mathrm{E}[\boldsymbol{M}] = (\nu - k - 1)^{-1}\boldsymbol{S}$.

3.1 *Transformation*

Soil texture is a three-dimensional vector whose components must sum to one. These are called compositional data, which are often analyzed by using a log-ratio transformation as described in Aitchison (1986). A log-ratio transformation of a texture, (c_1, c_2, c_3), is defined by

$$\mathbb{L}(c_1, c_2, c_3) = \left(\log\left(\frac{c_1}{c_3}\right), \log\left(\frac{c_2}{c_3}\right)\right),$$

for $c_1 > 0$, $c_2 > 0$ and $c_3 > 0$. The log-ratio transformation maps a three-dimensional vector of positive values to $\mathbb{R}^2$. Other transformations could be used to accomplish this mapping. The choice of c_3 as the denominator is arbitrary. Because the transformed measurements are assumed to follow a normal distribution in Section 3.2, the results will be invariant to this choice (Aitchison, 1986, Section 6.6).

Transformed laboratory and field measurements are denoted $\boldsymbol{l}_{gh}$ and $\boldsymbol{f}_{gh}$, respectively. That is, for horizon h of site g, we have

$$\boldsymbol{l}_{gh} = (l_{gh,1}, l_{gh,2}) = \mathbb{L}\left(c_{gh,1}^{(l)}, c_{gh,2}^{(l)}, c_{gh,3}^{(l)}\right)$$
$$\text{and } \boldsymbol{f}_{gh} = (f_{gh,1}, f_{gh,2}) = \mathbb{L}\left(c_{gh,1}^{(f)}, c_{gh,2}^{(f)}, c_{gh,3}^{(f)}\right).$$

For any vector $(x_1, x_2) \in \mathbb{R}^2$, the inverse of the log-ratio transformation is

$$\mathbb{L}^{-1}(x_1, x_2) = (\exp(x_1) + \exp(x_2) + 1)^{-1}(\exp(x_1), \exp(x_2), 1).$$

By construction, $\mathbb{L}^{-1}$ creates a vector that satisfies the sum constraint.

3.2 *Field and laboratory measurements*

We assume the transformed field measurements are normally distributed with homoskedastic errors and a mean that depends on the corresponding transformed laboratory measurements. Specifically,

$$\begin{aligned} \boldsymbol{f}_{gh} = & \begin{pmatrix} \psi_{01} \\ \psi_{02} \end{pmatrix} + \begin{pmatrix} \psi_{11} & 0 \\ 0 & \psi_{22} \end{pmatrix} \begin{pmatrix} l_{gh,1} \\ l_{gh,2} \end{pmatrix} + \begin{pmatrix} \omega_{gh,1} \\ \omega_{gh,2} \end{pmatrix} \\ = & \boldsymbol{\psi}_0 + \boldsymbol{\psi}_1 \boldsymbol{l}_{gh} + \boldsymbol{\omega}_{gh}, \end{aligned} \tag{4.1}$$

where

$$\{\boldsymbol{\omega}_{gh}\} \,\big|\, \boldsymbol{\Sigma}_\omega \sim \mathrm{N}(\boldsymbol{0}, \boldsymbol{\Sigma}_\omega),$$

for a positive-definite matrix $\boldsymbol{\Sigma}_\omega$. We assume that

$$\boldsymbol{\Sigma}_\omega \sim \mathrm{IW}_{\nu_\omega}\left(\boldsymbol{S}_\omega^{-1}\right)$$

and that

$$\boldsymbol{\psi} = (\psi_{01}, \psi_{11}, \psi_{02}, \psi_{22})' \sim \mathrm{N}(\boldsymbol{b}, \boldsymbol{V}_\psi).$$

The transformed laboratory measurements are assumed to be normally distributed with a mean and variance which depend on the master horizon designation. That is,

$$\boldsymbol{l}_{gh} = \boldsymbol{\mu}_{m_{gh}} + \boldsymbol{\alpha}_g + \boldsymbol{\zeta}_{gh},$$

where

$$\boldsymbol{\mu}_m \sim \mathrm{N}(\boldsymbol{\mu}_{0m}, \boldsymbol{V}_m)$$

for $m \in \{A, B, C\}$ and the $\{\boldsymbol{\zeta}_{gh}\}$ are distributed independently with

$$\boldsymbol{\zeta}_{gh} \,\big|\, m_{gh}, \boldsymbol{\Sigma}_A, \boldsymbol{\Sigma}_B, \boldsymbol{\Sigma}_C \sim \mathrm{N}\left(\boldsymbol{0}, \boldsymbol{\Sigma}_{m_{gh}}\right)$$

for positive-definite matrices $\boldsymbol{\Sigma}_A$, $\boldsymbol{\Sigma}_B$ and $\boldsymbol{\Sigma}_C$. Finally, for $m \in \{A, B, C\}$,

$$\boldsymbol{\Sigma}_m \sim \mathrm{IW}_{\nu_m}\left(\boldsymbol{S}_m^{-1}\right).$$

The model also includes a random effect $\boldsymbol{\alpha}_g$ for each site. We assume the $\{\boldsymbol{\alpha}_g\}$ are independent and identically distributed (iid) with

$$\boldsymbol{\alpha}_g \,\big|\, \boldsymbol{\Sigma}_\alpha \sim \mathrm{N}(\boldsymbol{0}, \boldsymbol{\Sigma}_\alpha)$$

for a positive-definite matrix $\boldsymbol{\Sigma}_\alpha$, where

$$\boldsymbol{\Sigma}_\alpha \sim \mathrm{IW}_{\nu_\alpha}\left(\boldsymbol{S}_\alpha^{-1}\right).$$

3.3 *Horizon profiles*

The distribution of l_{gh} is specified as a conditional distribution given the master horizon designation, m_{gh}. The values of (m_{gh}, d_{gh}) can be used to obtain the (inch-indexed) horizon profile at site g which is a collection of master horizon designations indexed by inch. The horizon profile is assumed to be a realization of an inhomogeneous Markov chain which evolves across inches. A first-order Markov chain with six states is selected; this chain is a simple approximation to reality, but is quite flexible due to its depth-varying transition probabilities. The model is expected to capture much of the dependence in the master horizon sequence.

In the six-state chain, states 1 through 6 correspond to continuing an A horizon, beginning a new A horizon, continuing a B horizon, beginning a new B horizon, continuing a C horizon and beginning a new C horizon, respectively. These states are selected so that multiple A, B or C horizons can be generated and distinguished; this would not be possible in a chain with only three states A, B, C.

The value of the Markov chain for inch i at site g is denoted T_{gi}. The values of $\{T_{gi}\}_{g\in\mathcal{D};i=1,\ldots,I_g}$ are observed. Note that I_g is random; it is likely to be around 6 to 12 for $g \in \mathcal{S}$ and should be 48 for all $g \in \mathcal{F} \cup \mathcal{L}$. However, this is not always the case because of minor deviations from the data collection protocol. If a horizon ended "near" 48 inches, I_g is often the lower boundary of the last horizon. When $I_g \neq 48$, we assume that I_{g+1} is the beginning of a new horizon. That is, $T_{g,I_g+1} \in \{2, 4, 6\}$.

At inch i, the Markov chain is governed by transition matrix $\mathbf{\Delta}_i$. Each profile is assumed to begin with a new A horizon, i.e. $T_{g1} = 2$, for all g. Define $\delta(j, k, i)$ to be the (j, k)th element of $\mathbf{\Delta}_i$. That is,

$$\delta(j, k, i) = \Pr(T_{gi} = k \mid T_{g,i-1} = j)$$

for $i = 2, \ldots, 48$. Each matrix $\mathbf{\Delta}_i$ has the form

$$\begin{pmatrix} \delta(1,1,i) & \delta(1,2,i) & 0 & \delta(1,4,i) & 0 & \delta(1,6,i) \\ 1 & 0 & 0 & 0 & 0 & 0 \\ 0 & \delta(3,2,i) & \delta(3,3,i) & \delta(3,4,i) & 0 & \delta(3,6,i) \\ 0 & 0 & 1 & 0 & 0 & 0 \\ 0 & \delta(5,2,i) & 0 & \delta(5,4,i) & \delta(5,5,i) & \delta(5,6,i) \\ 0 & 0 & 0 & 0 & 1 & 0 \end{pmatrix}$$

For example, $\delta(1, 3, i) = 0$ means that the probability of continuing a B horizon in inch i given that the previous inch was a continuation of an A

horizon is zero. On the other hand, $\delta(1,4,i)$ is not necessarily zero. This parameter represents the probability of beginning a B horizon in inch i given that the previous inch was a continuation of an A horizon. Because $\delta(3,2,i)$, $\delta(5,2,i)$ and $\delta(5,4,i)$ are not necessarily zero, buried horizons are possible. Transitions from new horizons to continued horizons occur with probability one, so the minimum horizon thickness is two inches in this model. The unconditional probability for each state j at inch i is given by

$$\Pr(T_{gi} = j) = \left[\prod_{i'=2}^{i} \mathbf{\Delta}_{i'} \right]_{2j},$$

where $[M]_{jk}$ is the (j,k)th element of the matrix M.

Under this model, the conditional transition probability profiles are not constant. However, not enough data are available to support estimation of $\mathbf{\Delta}_i$ at each inch. Thus, we consider simplifying assumptions on the form of $\{\mathbf{\Delta}_i\}$. Consideration of empirical transition probabilities and subject matter lead us to model $\mathbf{\Delta}_i$ as a piecewise constant matrix across inches. Let $\mathbb{I}_n$ denote an interval on which $\mathbf{\Delta}_i$ is assumed constant. Intervals of six inches were chosen so that $n = 1, \ldots, 8$. The distribution of the unknown transition probabilities is assumed to be

$$\begin{aligned}(\delta(j,j,i),\ \delta(j,2,i),\ \delta(j,4,i),\ \delta(j,6,i))& \\ \sim \mathbb{D}\left(d(j,j,n), d(j,2,n), d(j,4,n), d(j,6,n)\right)&,\end{aligned}$$

for $j \in \{1,3,5\}$, $n = 1, \ldots, 8$, and $i \in \mathbb{I}_n$.

3.4 Marginal distribution profiles

The unconditional joint distribution profile is the collection of distributions of the random variables $\mathbf{c}_{\cdot\cdot i}$ for $i = 1, \ldots, 48$. The distribution of $\mathbf{c}_{\cdot\cdot i}$ represents the distribution of laboratory determinations of soil texture across all sites and all horizons containing inch i. (The subscript $\cdot\cdot i$ is used because the first and second subscripts of $\mathbf{c}$ have been used for site and horizon indices, respectively.) The corresponding transformed random variable will be denoted $\mathbf{l}_{\cdot\cdot i}$. We wish to estimate the corresponding marginal distribution profiles for each component of texture.

Under the model, the distribution of $\mathbf{l}_{\cdot\cdot i}$ is a mixture of normal distributions. The coefficients in the normal mixture depend on $\{\mathbf{\Delta}_i\}$. Let

$$\boldsymbol{\theta} = \left(\boldsymbol{\psi}, \mathbf{\Sigma}_\omega, \mathbf{\Sigma}_\alpha, \boldsymbol{\mu}_A, \boldsymbol{\mu}_B, \boldsymbol{\mu}_C, \mathbf{\Sigma}_A, \mathbf{\Sigma}_B, \mathbf{\Sigma}_C, \{\mathbf{\Delta}_i\}_{i=2}^{48}\right)$$

be the vector of all unknown parameters in the model. Then the distribution of $\boldsymbol{l}_{\cdot\cdot i}$ is

$$\boldsymbol{l}_{\cdot\cdot i} \,\big|\, \boldsymbol{\theta} \sim \sum_{m\in\{A,B,C\}} \mathcal{M}_{mi} \times \mathrm{N}(\boldsymbol{\mu}_m, \boldsymbol{\Sigma}_m + \boldsymbol{\Sigma}_\alpha), \tag{4.2}$$

where $\mathcal{M}_{mi}$ represents the probability that master horizon m occurs at inch i:

$$\begin{aligned} \mathcal{M}_{Ai} &= \left[\prod_{i'=2}^{i} \boldsymbol{\Delta}_{i'}\right]_{21} + \left[\prod_{i'=2}^{i} \boldsymbol{\Delta}_{i'}\right]_{22}, \\ \mathcal{M}_{Bi} &= \left[\prod_{i'=2}^{i} \boldsymbol{\Delta}_{i'}\right]_{23} + \left[\prod_{i'=2}^{i} \boldsymbol{\Delta}_{i'}\right]_{24} \\ \text{and } \mathcal{M}_{Ci} &= \left[\prod_{i'=2}^{i} \boldsymbol{\Delta}_{i'}\right]_{25} + \left[\prod_{i'=2}^{i} \boldsymbol{\Delta}_{i'}\right]_{26} \end{aligned} \tag{4.3}$$

for $i = 2, \dots, 48$ and $\mathcal{M}_{A1} = 1$, $\mathcal{M}_{B1} = 0$ and $\mathcal{M}_{C1} = 0$.

The collection of distributions in (4.2) for $i = 1, \dots, 48$ is the joint distribution profile of transformed soil texture. Next, we demonstrate how this joint distribution profile can be used to obtain marginal distribution profiles on the original scale. The distribution function of clay content at the ith inch, $c_{\cdot\cdot i,1}$, is

$$\begin{aligned} \Pr\left(c_{\cdot\cdot i,1} \le q\right) &= \Pr\left(\frac{\exp(l_{\cdot\cdot i,1})}{\exp(l_{\cdot\cdot i,1}) + \exp(l_{\cdot\cdot i,2}) + 1} \le q\right) \\ &= \Pr\left(l_{\cdot\cdot i,1} \le \log\left[\frac{q}{q-1}(\exp(l_{\cdot\cdot i,2}) + 1)\right]\right) \\ &= \sum_{m\in\{A,B,C\}} \mathcal{M}_{mi} \mathrm{E}\left[\Phi\left(\frac{\log\left[\frac{q}{q-1}(\exp(l)+1)\right] - \mathrm{E}_m[l_{\cdot\cdot i,1} \mid l_{\cdot\cdot i,2} = l]}{\sqrt{\mathrm{Var}_m\left(l_{\cdot\cdot i,1} \mid l_{\cdot\cdot i,2} = l\right)}}\right)\right], \end{aligned} \tag{4.4}$$

where $l \sim \mathrm{N}(\mu_{m,2}, [\boldsymbol{\Sigma}_m]_{22})$, $\Phi(\cdot)$ denotes the cumulative distribution function of a standard normal random variable and

$$\begin{aligned} \mathrm{E}_m\left[l_{\cdot\cdot i,1} \mid l_{\cdot\cdot i,2} = l\right] &= \mu_{m,1} + \frac{[\boldsymbol{\Sigma}_m]_{12}}{[\boldsymbol{\Sigma}_m]_{22}}(l - \mu_{m,2}), \\ \mathrm{Var}_m\left(l_{\cdot\cdot i,1} \mid l_{\cdot\cdot i,2} = l\right) &= [\boldsymbol{\Sigma}_m]_{11} - \frac{([\boldsymbol{\Sigma}_m]_{12})^2}{[\boldsymbol{\Sigma}_m]_{22}}, \end{aligned}$$

for i in master horizon m. For a given inch, marginal clay quantiles can be obtained by numerically inverting the distribution in (4.4) for a desired level of probability. A quantile profile is a sequence of such quantiles (with a fixed level of probability) indexed by inch.

4 Estimation under a Bayesian approach

In the case in which the random effect $\boldsymbol{\alpha}_g$ is identically zero, closed forms for maximum likelihood estimates (MLEs) of some parameters can be obtained following the results of Anderson (1957). However, evaluating the quality of these estimates depends on the asymptotic normality of the MLEs. It is not clear how good this approximation would be for the sample sizes in the soil texture data. The MLEs do not have closed forms if the site-specific random effect has non-zero variance. Further, future work on the soils project will involve small area estimation for small groupings of soils. For these reasons, a Bayesian approach has considerable appeal.

The prior distribution of $\boldsymbol{\theta}$ is assumed to be of the form

$$\begin{aligned} p(\boldsymbol{\theta}) =& p(\boldsymbol{\psi})p(\boldsymbol{\Sigma}_\omega)p(\boldsymbol{\mu}_A)p(\boldsymbol{\mu}_B)p(\boldsymbol{\mu}_C)p(\boldsymbol{\Sigma}_A)p(\boldsymbol{\Sigma}_B)p(\boldsymbol{\Sigma}_C) \\ &\times p(\boldsymbol{\Sigma}_\alpha)p(\{\boldsymbol{\Delta}_i\}_{i=2}^{48}). \end{aligned} \tag{4.5}$$

For each factor, a vague, conjugate prior distribution was chosen. All variance components ($\boldsymbol{\Sigma}_\omega, \boldsymbol{\Sigma}_\alpha, \boldsymbol{\Sigma}_A, \boldsymbol{\Sigma}_B, \boldsymbol{\Sigma}_C$) are assigned inverse-Wishart prior distributions. The parameters $\boldsymbol{\psi}, \boldsymbol{\mu}_A, \boldsymbol{\mu}_B, \boldsymbol{\mu}_C$ are each assigned a Normal prior distribution. The unknown parameters of the horizon profile model are assigned Dirichlet prior distributions.

To force conjugacy, the parameter vector $\boldsymbol{\theta}$ is augmented with some unobserved quantities. If $I_g \neq 48$, we assume that $T_{g,I_g+1} = 2, 4$ or 6. That is, we have a "partially" observed value of T_{g,I_g+1}. Similarly, when $\boldsymbol{f}_{gh}$ is observed, but $\boldsymbol{l}_{gh}$ is not, we have partial information about the value of $\boldsymbol{l}_{gh}$. We augment the parameter vector with these "partially" observed values. Let $\mathcal{Z}$ represent all observed data. The full posterior distribution is

$$\begin{aligned} &p(\boldsymbol{\theta}, \{\boldsymbol{l}_{gh}\}_{g\in\mathcal{S}\cup\mathcal{F};h=1,\dots,H_g}, \{\boldsymbol{\alpha}_g\}_{g\in\mathcal{D}}, \{T_{g,I_g+1}\}_{g\in\mathcal{D}:I_g<48} \,\big|\, \mathcal{Z}) \qquad (4.6) \\ &\propto p(\{\boldsymbol{f}_{gh}\}_{g\in\mathcal{D};h=1,\dots,H_g} \,\big|\, \{\boldsymbol{l}_{gh}\}_{g\in\mathcal{D};h=1,\dots,H_g}, \boldsymbol{\psi}, \boldsymbol{\Sigma}_\omega) \\ &\times p(\{\boldsymbol{l}_{gh}\}_{g\in\mathcal{D};h=1,\dots,H_g} \big| \{m_{gh}\}_{g\in\mathcal{D};h=1,\dots,H_g}, \{\boldsymbol{\mu}_m, \boldsymbol{\Sigma}_m\}_{m\in\{A,B,C\}}, \{\boldsymbol{\alpha}_g\}_{g\in\mathcal{D}}) \\ &\times p(\{\boldsymbol{\alpha}_g\}_{g\in\mathcal{D}} \,\big|\, \boldsymbol{\Sigma}_\alpha) \\ &\times p(\{T_{gi}\}_{g\in\mathcal{D};i=1,\dots,I_g}, \{d_{gh}\}_{g\in\mathcal{D};h=1,\dots,H_g} \{T_{g,I_g+1}\}_{g\in\mathcal{D}:I_g<48} \,\big|\, \{\boldsymbol{\Delta}_i\}_{i=2}^{48}) \\ &\times p(\boldsymbol{\theta}), \end{aligned}$$

where $p(\boldsymbol{\theta})$ is given in (4.5).

4.1 Gibbs sampler

In order to obtain numerical summaries of the posterior distribution in (4.6), Gibbs sampling is used (Geman and Geman, 1984, Gelfand and Smith, 1990). The augmented parameter vector is divided into subvectors. The subvectors and prior distributions have been chosen in such a way that each of the conditional posterior distributions is a recognizable distribution that is easily sampled.

To demonstrate the application of this model, we use data from the pilot project for a group of soils called Old Alluvium (OA) soils. For OA soils, we have full profiles of field data and laboratory data for 22 sites, full profiles of field data for 25 sites and surface horizon field data for 135 sites. Thus, we have data from a total of 182 sites.

The convergence of the Gibbs sampler is assessed using the potential scale reduction statistic for each scalar parameter (Gelman and Rubin, 1992). The values of the statistics indicate that 2000 iterations are sufficient for the Gibbs sampler to converge. Posterior predictive assessment was used to judge the adequacy of the proposed model (Gelman et al, 1996). The assumption of linearity and the choice of one parameter, ψ, in the field measurement model for OA soils appears adequate, despite having different data collectors in the field. There is little evidence of heteroskedasticity among the residuals. The normality assumptions in both the field and laboratory measurement models appear adequate, although other transformations might improve the fit of the model in this respect. Posterior predictive assessment also suggests that the horizon profile model provides an adequate fit, despite the somewhat arbitrary choice of intervals for the piecewise constant function.

4.2 Posterior summaries

Certain parameters of the model are themselves of interest. For example, $\boldsymbol{\mu}_A$, $\boldsymbol{\mu}_B$ and $\boldsymbol{\mu}_C$ are the means of $\boldsymbol{l}_{gh}$ for each master horizon designation. Figure 4.3 shows contours of the posterior marginal distributions of $\boldsymbol{\mu}_A$, $\boldsymbol{\mu}_B$ and $\boldsymbol{\mu}_C$ mapped back to the original scale of the data and superimposed on part of the texture triangle from Figure 4.1. In this plot, we can see that the posterior distributions of the three means are all very tight in the sand dimension, but show more variability in the other two components (clay and silt). Also, the amount of variability in these means is small relative

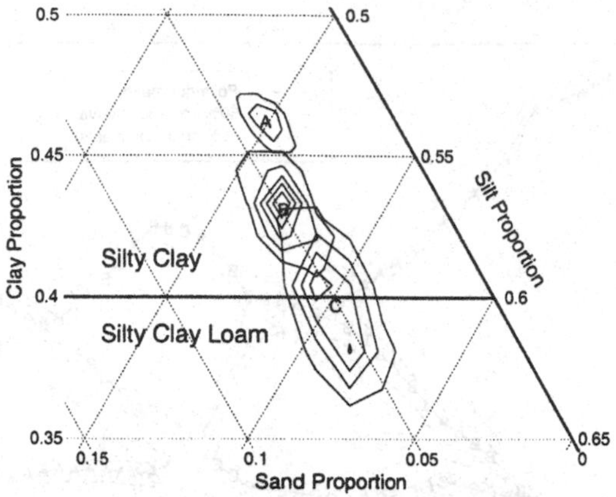

FIGURE 4.3. Contours of the marginal posterior distributions of μ_A, μ_B and μ_C mapped back to the original scale of the data. Contours are at fixed heights for each posterior distribution. The outermost contour contains approximately 88%, 95% and 93% of the mass of the posterior distribution for μ_A, μ_B and μ_C, respectively.

to the "size" of the texture classes. (Refer to Figure 4.1.) Finally, there is more variability in the estimation of B and C horizon parameters than in the estimation of A horizon parameters due to the greater number of observed A horizons.

We can also investigate the posterior distribution of transition probabilities. One meaningful way to summarize simultaneously all elements of $\{\boldsymbol{\Delta}_i\}$ is to consider posterior profiles of the coefficients of the mixture in (4.4). A posterior draw of $\{\mathcal{M}_{mi}\}_{m=A,B,C;i=1,\ldots,48}$ is obtained by substituting appropriate values from a posterior draw of $\boldsymbol{\theta}$ in (4.3). Recall that these coefficients represent the probability of each master horizon designation occuring at inch i. Figure 4.4 shows posterior means of the coefficients together with pointwise 50% and 95% credible intervals. Note that the probability of finding an A horizon near the top of the profile is close to 1. This probability decreases to nearly zero by the bottom of the profile. The probability for a B horizon starts out close to zero. This coefficient reaches a maximum around 35 inches, where B horizons are the most likely master horizon. The probability for C horizons stays near zero for over half of the profile and then begins to increase until C horizons become the most likely near the bottom of the profile.

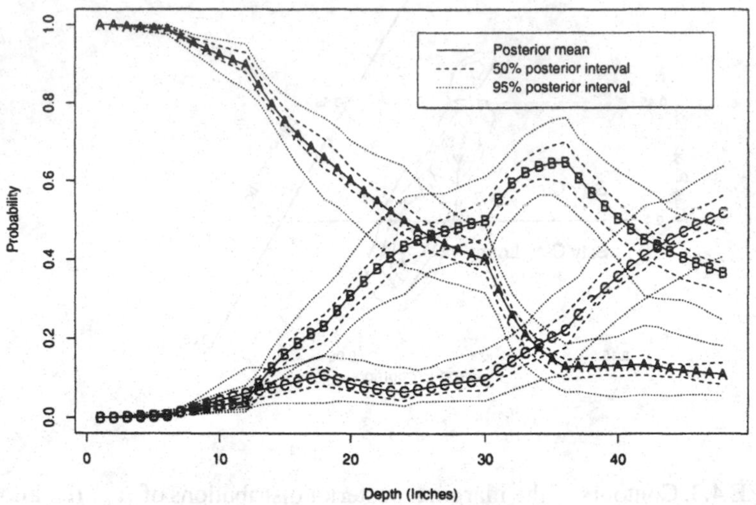

FIGURE 4.4. Posterior means of coefficients of the mixture in (4.4) with 50% and 95% posterior credible intervals.

The main objective of the analysis is to estimate the marginal distribution profiles of soil texture. Quantile profiles can be used to summarize these distribution profiles. A posterior draw of a quantile profile can be obtained by substituting a posterior draw of all appropriate parameters in (4.4) and numerically inverting the distribution. In this way, we can investigate the posterior distribution of the quantile profiles. A point estimate for a quantile profile can be obtained by using posterior means in (4.4) and numerically inverting the distribution. Examples of such point estimates are shown in Figure 4.5a.

In addition to distribution profiles, we can also estimate the distribution for a component of texture for each master horizon designation. Figure 4.5b shows the marginal density function for clay content when posterior means are used. Note that Figure 4.5a represents profiles of quantiles of distributions obtained by mixing the three components shown in Figure 4.5b. Near the top of the profile, the dominant component of the mixture is A. Further down the profile, there are increasing contributions from B and C which have lower average clay content and higher variability than A. These features are reflected in the decreasing median profile and the increasing separation of the other quantiles. Both summaries (by inch and by master horizon designation) are of interest to soil scientists. The computational Bayesian approach adopted here allows us to have both by simply considering different functions of posterior draws of the model parameters.

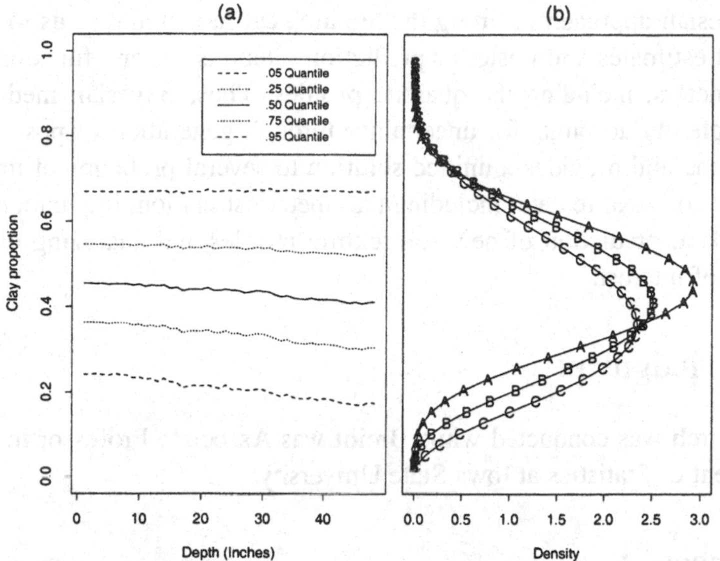

FIGURE 4.5. (a) Quantile profiles for clay evaluated at the posterior means of all parameters. (b) Density of clay content for each master horizon designation evaluated at posterior means of all parameters.

5 Conclusions

Soil texture profiles are an important consideration for land use and management. Quantile profiles can be used to summarize the marginal distribution profiles of the components of soil texture. A hierarchical model allows us to incorporate relevant features of the data structure in the estimation process. Under the hierarchical model presented, we can obtain quantiles at each inch by numerically inverting a distribution which is a function of the model parameters.

Future research will include incorporating small area features into the current model, since estimates are desired for smaller groups of soils than have been used for illustration here. Scientists are also interested in analysis of other variables, some of which may be expressed as functions of parameters in the horizon transition model. For example, the expected thickness of the A horizon is a function of the model transition probabilities. However, analyzing these variables may require a formulation of the horizon profile model which allows extrapolation beyond 48 inches. Other improvements to the model, including consideration of alternative transformations in the field and laboratory measurement models, will also be investigated.

A Bayesian approach to fitting the hierarchical model allows us to obtain point estimates and posterior prediction intervals for any function of the parameters, including the quantile profiles. Thus, Bayesian methodology explicitly accounts for uncertainty through generation of posterior distributions and provides a unified solution to several problems of interest for the soil texture data, including parameter estimation, imputation of missing data, prediction of new soil texture profiles and analyzing other variables of interest.

Acknowledgment

This research was conducted while Breidt was Associate Professor in the Department of Statistics at Iowa State University.

References

Abbitt, P. J. and Nusser, S. M. (1995). Sampling approaches for soil survey updates. *ASA Proceedings of the Section on Statistics and the Environment*, p. 87–91.

Aitchison, J. (1986). *The Statistical Analysis of Compositional Data*. Chapman and Hall, London.

Anderson, T. W. (1957). Maximum likelihood estimates for a multivariate normal distribution when some observations are missing. *Journal of the American Statistical Association*, 52:200–203.

Gelfand, A. E. and Smith, A. F. M. (1990). Sampling-based approaches to calculating marginal densities. *Journal of the American Statistical Association*, 85:398–409.

Gelman, A., Meng, X., and Stern, H. S. (1996). Posterior predictive assessment of model fitness via realized discrepancies. *Statistica Sinica*, 6:733–807.

Gelman, A. and Rubin, D. B. (1992). Inference from iterative simulation using multiple sequences (disc:P483–501, 503–511). *Statistical Science*, 7:457–472.

Geman, S. and Geman, D. (1984). Stochastic relaxation, Gibbs distributions, and the Bayesian restoration of images. *IEEE Transactions on Pattern Analysis and Machine Intelligence*, 6:721–741.

Natural Resources Conservation Service, Soil Survey Staff. (1999). *National Soil Survey Handbook*, title 430-VI, U.S. Government Printing Office, Washington, D.C. http://www.statlab.iastate.edu/soils/nssh

Assessing the Homogeneity of Three Odds Ratios: A Case Study in Small-Sample Inference

John B. Carlin

ABSTRACT In an experiment on the effects of different types of high-frequency ventilation on lung damage in rabbits, six groups of 6 to 8 animals each were compared, using a factorial treatment structure of 3 frequency values crossed with 2 amplitudes. The resulting data were reduced to binary outcomes for each animal, producing a $3 \times 2 \times 2$ contingency table. Although the numbers were small, there appeared to be a large effect of amplitude at the two extreme frequency levels, but there were no failures at either amplitude in the middle frequency group. The question of interest was whether the data provided evidence that the effect of amplitude differed between the 3 frequencies and in particular whether the effect in the middle group was lower than in the two extreme groups. Various models were considered for the 3 odds ratios in question, all seeking to incorporate minimally informative prior assumptions. Because of the small numbers, sensitivity to prior distribution specifications was considerable, and we compared the effect of assuming independent prior distributions on each cell in the 3×2 factorial with that of using a more structured prior distribution incorporating exchangeable row, column and interaction effects. Only under rather strong prior assumptions could it be concluded that there was substantial evidence of non-homogeneity. The analysis provides a case study of the sensitivity of inferences in small samples, in an example where the popular exact frequentist approach, based on a null hypothesis of equality of the odds ratios, breaks down.

1 Introduction

Much biomedical research is performed on animals, and for ethical reasons the numbers of animals used in experiments are often very limited. A popular convention seems to be to run experiments using groups of about 6 animals each. From a statistical point of view, it is difficult to draw confident conclusions from the results of such experiments, unless the results are very clearcut. Where the results are analysed in terms of discrete outcomes, they may be summarized in the form of a contingency table. Often, many of the cells contain very small numbers of animals. Since standard asymptotic inference methods break down with such data, common practice now relies on frequentist methods of "exact" inference, based on the sampling distributions of all possible configurations of the data, conditional on marginal totals (e.g. Mehta et al., 1985; Cytel, 1999). Such methods produce classical significance tests, which give an indication of the "extremeness" of the observed pattern of data relative to a particular null hypothesis, and some extensions provide parameter estimates. From a Bayesian point of view, these methods are of doubtful value, since the results rarely have a convincing Bayesian interpretation—that is, an interpretation conditional on the data. From a purely pragmatic point of view, the methods break down completely in certain problems where there are zero marginal totals.

This article reports a detailed Bayesian analysis of an experiment in neonatology where a striking pattern of results emerged but the "significance" of the observed pattern was difficult to discern, and the "exact" frequentist methods were not useful. Given that the clinical investigators were unwilling to introduce genuine prior information into the analysis, we explored the effect of a range of "noninformative" prior distributions in the analysis. In particular, emphasis is placed on the importance of structural assumptions in the prior distribution, concerning whether or not parameters are assumed to be independent *a priori*. Although the results essentially only relate to the particular dataset analyzed, they give some general insight on possible approaches to similar problems, and illustrate that currently available Bayesian software can readily handle such small-sample inference problems. The methods used are not technically novel (for a review of Bayesian approaches to the analysis of contingency tables, see Leonard and Hsu, 1994), but there are few published discussions of the practical aspects of such applications of Bayesian methodology.

I begin in the next section by briefly describing the background to the problem and presenting the data. The following section considers the important problem of how to translate the researchers' experimental questions into formal statistical terms, and the interplay between approaches to this problem and the choice of the Bayesian paradigm for inference. The fourth section considers prior distributions and computational methods, while the fifth presents a range of results. I conclude with some general observations that arise from considering these results.

	Frequency (Hz)		
Amplitude	5	10	15
20	1/6	0/6	0/6
60	4/6	0/6	4/8

TABLE 5.1. Data from an experiment on lung damage in rabbits experimentally exposed to 6 different regimes of high-frequency mechanical ventilation. Each cell shows [number of animals with lung damage]/[total number].

2 The problem and data

Babies born prematurely often suffer severe respiratory problems due to underdeveloped lungs. Over the course of the past few decades much progress has been made in reducing mortality and longterm morbidity in such infants by the use of sophisticated mechanical ventilation devices. The data to be considered here arose from laboratory experiments on the effects of varying the settings of a recently developed ventilation method that uses high frequency oscillation.

In particular, the lungs of newborn rabbits were manipulated to simulate lung disease in preterm infants, and a factorial experiment was conducted to assess the effect of changing the amplitude and frequency of ventilation. Two amplitude settings were used at each of three frequencies, giving rise to 6 experimental combinations. For each combination, 6 rabbits were used, except for one where 8 were included (by happenstance rather than plan). The outcome of interest was whether, according to subsequent pathological examination, the rabbit suffered lung damage (which would indicate that the ventilation regime was too aggressive). The resulting data, indicating the number of rabbits with lung damage in each of the 6 experimental combinations, is displayed in Table 5.1.

It can be seen that no rabbits in either of the 10Hz (middle frequency) groups suffered lung damage, while in both the 5 and 15Hz groups more than half the animals in the high amplitude groups recorded the poor outcome. The researchers did not have strong prior hypotheses about which frequency would be preferable, but were intrigued by the suggestion in the data that at 10Hz a higher amplitude could apparently be used without incurring a risk of increased lung damage. Could statistical analysis provide some assessment as to how strongly this suggestion was really supported by the evidence in the data?

3 Statistical formulation of the problem

A concise summary of the question of primary interest is whether or not the relative risk of lung damage due to the higher (compared with the lower) amplitude is lower in the middle frequency group compared with the high and low frequencies. A natural parametrization is to consider the odds ratio comparing the risk of lung damage at high amplitude with that at low amplitude. This can be defined separately for each of the three frequencies as

$$\lambda_j = \frac{\pi_{1j}/(1-\pi_{1j})}{\pi_{0j}/(1-\pi_{0j})} \tag{5.1}$$

where π_{ij} is the probability of lung damage in the ij^{th} group, with i indexing amplitude ($i = 0, 1$ for 20 and 60 respectively), and j indexing frequency (=1,2,3 for 5, 10, 15Hz respectively).

An initial exploratory analysis began by examining simple point estimates of the λ_js. If 0.5 is added to each of the event and no-event totals (Cox, 1970), point estimates of the 3 odds ratios are 6.6, 1.0 and 13.0 for $j = 1, 2, 3$ respectively. Standard (one-sided) Fisher exact p-values for the null hypotheses $\lambda_1 = 0$ and $\lambda_3 = 0$ are 0.12 and 0.07, which might lead one to conclude that the evidence for amplitude effects was weak even at the extreme frequencies.

Analyzing the three 2×2 tables separately does not, however, address the question of whether there is evidence of substantial variation in the strength of association across the 3 tables. This question has traditionally been assessed in a frequentist framework by using a significance test of the null hypothesis

$$H_0 : \lambda_1 = \lambda_2 = \lambda_3. \tag{5.2}$$

Such a test, conditional on the marginal totals of all of the tables, was proposed by Zelen (1971) and is available in StatXact (Cytel, 1999). In the current problem, however, this test is of no value. Because of the zero marginal totals in the middle table ($j = 2$), the test only uses information from the two extreme tables, ignoring the middle table, which is of course the one of primary interest. There are also well-known asymptotic tests, which do not condition on marginal totals (Cox, 1970; Breslow and Day, 1980), but these are clearly not appropriate with such small numbers. Furthermore, from a Bayesian perspective, the null hypothesis (5.2) is not a natural focus for considering the real underlying question. Prior belief is unlikely to give strong support to such a sharp point hypothesis and a more relevant approach is to try and quantify the strength of support for interesting departures from this hypothesis.

The approach chosen here was to focus on the posterior distribution of the ratio

$$\rho = \frac{\lambda_2}{\min(\lambda_1, \lambda_3)}. \tag{5.3}$$

In particular, the probability that ρ is less than one is the probability that the odds ratio at 10Hz is less than that at either 5Hz or 15Hz. The probability that ρ is less than 0.5 reflects the posterior evidence that the middle frequency has an amplitude effect (in the odds scale) no greater than half that at either of the more extreme frequencies.

4 Probability model and computation

To perform Bayesian inferences for ρ we require a model for the observed data along with prior distributions for unknown parameters. Letting y_{ij} represent the number of animals with lung damage at amplitude i and frequency j, and given that rabbits within each group were exchangeable, the natural model for y_{ij} given π_{ij} is binomial with index n_{ij} (= the number of animals in the ij^{th} cell):

$$y_{ij} \sim \text{Bin}(n_{ij}, \pi_{ij}). \tag{5.4}$$

We specified two families of prior distribution for π_{ij}. Both of these attempted to be as "noninformative" as possible; in particular, under all prior distributions that were considered, the prior probability that ρ was less than 1 was 0.5.

4.1 Independent beta prior

The only prior distribution that allows closed-form calculation of posterior distributions for the π_{ij} is one that specifies independent beta distributions for each π_{ij}. This may be expressed as

$$\pi_{ij} \sim \text{Beta}(\eta, \zeta), \quad \text{for all } i, j, \tag{5.5}$$

where we assumed the same values of η and ζ for all values of i and j (there being no reason to vary this choice). To be impartial about the relative probabilities of failure and success, we set $\eta = \zeta$, and to be "noninformative", standard settings are (i) $\eta = \zeta = 0.5$, the Jeffreys' prior, or

Beta distributions			
	η	ζ	
B1	0.5	0.5	
B2	1.0	1.0	
Hierarchical logistic models			
	τ_α	τ_β	τ_γ
H1	1.5	1.5	1.5
H2	1.5	1.5	1.0
H3	1.5	1.5	0.5
H4	0.5	0.5	1.5
H5	0.5	0.5	1.0
H6	0.5	0.5	0.5

TABLE 5.2. Summary of prior distributions.

(ii) $\eta = \zeta = 1$, the uniform prior. The improper prior distribution with $\eta = \zeta = 0$, which corresponds to a uniform prior in the logit scale, cannot be used here because it does not produce a proper posterior distribution when either y_{ij} or $n_{ij} - y_{ij}$ are 0. Settings (i) and (ii) led to two distinct priors (Table 5.2), with the key assumption of prior independence between the six π_{ij}s.

4.2 *Exchangeable row, column and interaction effects prior*

The assumption of prior independence cannot in fact be regarded as "non-informative" with respect to the question of homogeneity of effect across the values of j. An alternative is to create a prior distribution based on a logistic model for π_{ij}. We defined row, column and interaction effects by expressing π_{ij} as:

$$\pi_{ij} = \text{logit}^{-1}(\mu + \alpha_i + \beta_j + \gamma_j \mathbf{1}_{\{i=1\}}) \tag{5.6}$$

and specifying independent normal prior distributions for the α_is, β_js and γ_js:

$$\begin{aligned} \alpha_i &\sim \text{N}(0, \tau_\alpha^2) \\ \beta_i &\sim \text{N}(0, \tau_\beta^2) \\ \gamma_i &\sim \text{N}(0, \tau_\gamma^2). \end{aligned}$$

This model was introduced by Leonard (1975) but it appears to have been little used in practice. In principle it would be possible to estimate the variance parameters introduced in the prior distributions, by introducing a (hyperprior) distribution for them. In practice, with data as sparse as in the present example, there is very little if any information in the data on these parameters. The approach taken was therefore to specify a range of reasonably "noninformative" fixed settings for these values and to assess the consequent variation in posterior distributions.

A first step in specifying the variance parameters is to observe that, ignoring interaction effects for the moment, the standard deviations τ_α and τ_β can be interpreted as determining the likely range in the odds of lung damage between rows (amplitudes) and columns (frequencies), respectively. At one extreme it may be argued that setting $\tau_\alpha = \tau_\beta = 1.5$ corresponds to allowing very large variations among both the amplitude and frequency effects, since it corresponds to a relative standard deviation in the odds scale of $e^\tau = 4.5$ or a 95% range of effects covering a ratio (maximum to minimum odds) of $e^{2\times 1.96\times\tau} = 360$. As another extreme, which might be considered a relatively informative specification in this context, we set $\tau_\alpha = \tau_\beta = 0.5$, which corresponds to a relative standard deviation of 1.6 and a 95% range of 7.1-fold.

For the variance of the interaction effects, the same two "extreme" settings were used, along with an intermediate value, $\tau_\gamma = 1.0$ (relative SD = 2.7, 95% range 50-fold). The resulting 6 combinations of hyperparameter settings are labeled H1-H6 in Table 5.2. It is easy to see that the interaction variance, τ_γ^2, is the critical parameter in determining posterior inferences about homogeneity of the three odds ratios. On the one hand, a prior judgment might be made that interaction effects are likely to be smaller than main effects, suggesting that τ_γ should be smaller than τ_α and τ_β. On the other hand, if it could be argued that across this range of frequency settings, amplitude effects are likely to vary considerably, it might be reasonable to set τ_γ^2 to be as high as or even higher than the row and column variances.

A standard, nearly uniform, noninformative prior distribution was used for μ: $\mu \sim \mathrm{N}(0, 1000^2)$.

4.3 Computation

Under the conjugate independent beta prior distributions, the posterior distribution of the π_{ij}s is also a product of independent beta distributions. Using direct simulation from these beta distributions, it is straightforward to generate samples from the posterior distributions of λ_j ($j = 1, 2, 3$) and

ρ, and thus to estimate posterior probabilities such as $\Pr(\rho < 1|\mathbf{y})$ (e.g. see Gelman et al., 1995). Results were based on 100,000 independent draws, with simulations performed using S-PLUS.

The hierarchical logistic regression model does not allow closed-form expressions for posterior distributions. The model can, however, be estimated readily using Gibbs sampling as implemented, using adaptive-rejection sampling, in the BUGS or WinBUGS packages (MRC Biostatistics Unit, 1999). (Code available from the author.) Preliminary analysis explored convergence of the Gibbs sampler, using 5 parallel chains with dispersed starting values, and it was clear that the sampler converged rapidly after 1-2,000 iterations. Results reported are from 2,000 iterations obtained after a 2,000 update burn-in, pooling over 5 parallel chains. The estimated Monte Carlo standard error of posterior means was about 1%.

5 Results

Figure 5.1 displays interval estimates of the three odds ratios under each of the 8 different prior distributions, and summary inferences about ρ are shown in Table 5.3. Note that there is really no frequentist-based analogue for these Bayesian inferences about ρ; perhaps a point estimate could be made, from the point estimates of the 3 odds ratios, as 1/6.6 = 0.15, but an asymptotically based standard error would be difficult to derive as well as clearly unreliable with such small numbers.

The results indicate firstly that the two prior distributions based on independent beta specifications give similar results, except that the inferences for λ_2 and λ_3 are considerably more uncertain under the less informative specification ($\eta = \zeta = 0.5$). Under these priors, the posterior probability that λ_2 is the smallest of the 3 odds ratios is just over 70%, but these highly uninformative, unstructured priors also give substantial posterior probability to some very extreme values of ρ.

The structured hierarchical prior distributions give a wider range of results with respect to the comparison of the 3 odds ratios. Predictable effects of altering the priors are apparent in the posterior distributions. In particular, as the prior variance of the interaction effects, τ_γ^2, is reduced, holding the main effect variances constant (H1 $\rightarrow$ H3 and H4 $\rightarrow$ H6), the evidence that λ_2 is substantially less than λ_1 and λ_3 weakens considerably. If one assumed that interaction effects were likely to be smaller than main effects (priors H2, H3), the evidence for heterogeneity is very slim; in particular, note that under prior H3 the posterior distribution for λ_2 overlaps almost

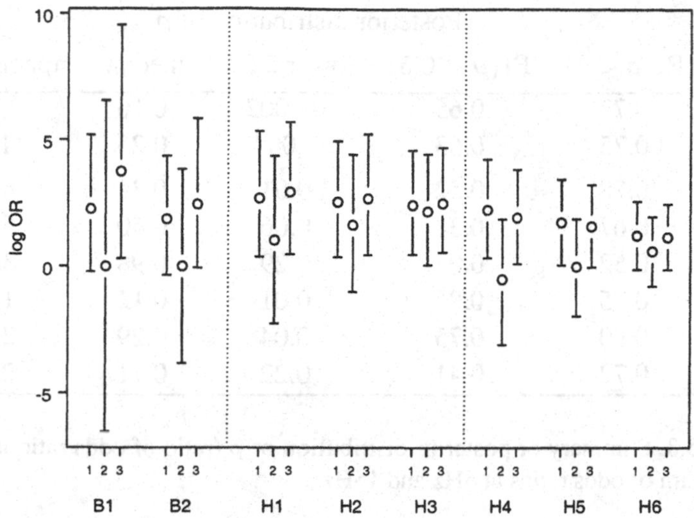

FIGURE 5.1. 95% posterior probability intervals for λ_j ($j = 1, 2, 3$) under the range of prior distributions discussed in text (see Table 5.2 for labels).

entirely with those for λ_1 and λ_3 (Figure 5.1). It is only under the relatively implausible assumption that interaction effects are likely to be larger than main effects (priors H4, H5) that the evidence for lack of homogeneity is strong.

6 Conclusions

In this article I have explored the extent of variation in posterior inferences for whether one of 3 odds ratios could be concluded to be smaller than the other two, where the data relating to the odds ratio in question contained no events in either the "exposed" (high amplitude) or "unexposed" (low amplitude) groups. It was shown that assuming independent beta prior distributions for binomial probabilities in this problem resulted in very diffuse posterior distributions that provided slight evidence of nonhomogeneity. Given the factorial structure of the experiment, the assumption of prior independence between the event probabilities was unrealistic, and with such a small amount of data no prior distribution is truly noninformative.

Prior	Posterior distribution of ρ				
	$\Pr(\rho < 1)$	$\Pr(\rho < 0.5)$	lower 2.5%	median	upper 2.5%
B1	0.73	0.65	0.0002	0.16	153
B2	0.75	0.63	0.004	0.25	18
H1	0.78	0.52	0.01	0.31	6.0
H2	0.67	0.31	0.06	0.60	5.7
H3	0.52	0.07	0.29	0.98	3.4
H4	0.95	0.86	0.01	0.12	1.5
H5	0.90	0.75	0.04	0.29	2.0
H6	0.72	0.41	0.22	0.71	2.3

TABLE 5.3. Summary of posterior distribution of ρ (ratio of odds ratio at 10Hz to minimum of odds ratios at 5Hz and 15Hz).

Analyses using a hierarchical logistic regression model, with a range of minimally informative (but fixed) variances used for each of the structural "random effects" in the model, produced a wider and much more informative range of results, where the implications of different prior specifications could be clearly seen in the resulting range of posterior distributions. It was shown that the only prior distributions that gave substantial support to the hypothesized non-homogeneity were ones that (unreasonably?) assumed greater prior uncertainty about interactions than about main effects.

It is worth noting that an approach that used continuous prior distributions for the variances (rather than point values) would produce a weighted average of the results obtained here, so in general more diffuse conclusions. The purpose of the present analysis was to explore the extent to which more or less confident conclusions *could* be reached if specific prior assumptions were made about parameters for which the data were uninformative. To sharpen conclusions, an alternative to introducing strong prior assumptions would of course be to collect more data, one version of which would be to attempt to combine these data with those from other experiments using meta-analysis. Although meta-analysis has been widely adopted for combining data from clinical trials, it does not seem to have been used much in examining laboratory studies, perhaps because of the difficulty of defining common estimands across such studies (Carlin, 2000).

Some tricky issues remain with respect to this case study. Our analysis ignored the ordering of the three frequency values, and it would often be held that a monotonic relationship of effect size to frequency is *a pri-*

ori more plausible than a pattern with the smallest effect at the middle frequency. On the other hand, the clinical investigators were willing to suggest possible mechanisms for a non-monotonic relationship, so it did not seem appealing to enforce any such structure on the model. A further concern is that it could be argued that the framing of the question about non-homogeneity in terms of the ratio of λ_2 to the minimum of λ_1 and λ_3 represented a *post hoc* choice of effect measure. Certainly, the evidence for other forms of departure from homogeneity would clearly be weaker than that found for the chosen alternative hypothesis, which proposed a diminished effect at the middle frequency. From a Bayesian viewpoint, however, it is perhaps best to deal with concern about overinterpretation of the apparent effect of interest by incorporating appropriate skepticism into the model, for instance by preferring the prior distributions that give smaller variance to the interaction than to the main effects.

Statisticians who work in consulting environments are often asked to provide answers to essentially unanswerable questions. One variety of such questions is exemplified by the example discussed in this article, where the experimental data are simply not ample enough on their own to provide strong evidence either way about the substantive research question. There is a great deal of pressure to provide answers to questions such as this, and a certain seduction in popular frequentist methods that generate a single p-value as an answer (especially when the methods are labeled "exact"!). The example presented here suggests that frequentist analyses of small datasets implicitly overstate the definitiveness of their conclusions.

This case study shows that it is possible to analyse small contingency tables in a Bayesian manner using available software. Exploring the range of inferences obtained under a range of relatively noninformative prior distributions provides a realistic assessment of the strength of evidence in the data for the substantive question(s) of interest. In this problem, it was important to use suitably structured prior distributions rather than convenient conjugate forms that are only feasible with strong assumptions about prior independence of parameters. Estimates for the hierarchical model were readily obtained using the WinBUGS package, currently available as freeware. These sorts of analyses could be used to promote greater realism about the limitations of small datasets in addressing scientific questions.

Acknowledgments

Thanks to Drs Jag Ahluwahlia and Peter Loughnan for providing the problem, and to David Spiegelhalter for helpful correspondence. I am also grateful to Hendricks Brown and the Department of Epidemiology and Biostatistics, University of South Florida, for sabbatical support during the writing of this paper.

References

Breslow, N.E. and Day, N.E. (1980). *Statistical Methods in Cancer Research I: Analysis of Case-Control Studies.* Lyon: IARC.

Carlin, J.B. (2000) Meta-analysis: formulating, evaluating, combining, and reporting by S-L. T. Normand [letter]. *Statistics in Medicine* **19**, 753-759.

Cox, D.R. (1970). *Analysis of Binary Data.* London: Methuen. [See also 2nd ed. (1989) with E.J. Snell. Chapman & Hall: London.]

Cytel Software Corporation (1999). StatXact software package. Cambridge, Massachusetts. (Web address: http://www.cytel.com/)

Gelman, A., Carlin, J.B., Stern, H.S. and Rubin, D.B. (1995). *Bayesian Data Analysis.* Chapman & Hall: London.

Leonard, T. (1975) Bayesian estimation methods for two-way contingency tables. *JRSS-B* , 23-37.

Leonard, T. and Hsu, J.S.J. (1994) The Bayesian analysis of categorical data–a selective review. In Freeman, P.R. and Smith, A.F.M. (eds) *Aspects of Uncertainty.* Chichester: John Wiley.

Mehta, C.R., Patel, N.R., Gray, R. (1985) On computing an exact confidence interval for the common odds ratio in several 2×2 contingency tables. *JASA* **80**, 969-973.

MRC Biostatistics Unit (1999). WinBUGS software package, version 1.2 (beta). Cambridge, U.K. (Web address http://www.mrc-bsu.cam.ac.uk/bugs)

Zelen, M. (1971). The analysis of several contingency tables. *Biometrika* **58**, 129-137.

Modeling Rates of Bone Growth and Loss Using Order-Restricted Inference and Stochastic Change Points

Richard B. Evans
J. Sedransk
Siu L. Hui

ABSTRACT The bone mineral density (BMD) of the spine and hip (femoral neck) is central to the management of osteoporosis, because spinal fractures are the most prevalent and hip fractures are the most debilitating. BMD has been shown to predict fractures, and the rates of bone growth and loss with age have been identified as key determinants of osteoporosis in old age. To develop therapies for osteoporosis, we need to better understand the pattern of bone growth and loss with aging. In this paper, we use data from 449 female subjects measured at the spine to provide inference for the age parameters t_1 and t_2 corresponding to changepoints demarking the stages in the pattern of age-specific mean rates of change of BMD, $\mu(t), t = 8, \ldots, 80$, where t is age in years. We assume the condition that the $\mu(t)$ behave according to $\mu(8) \leq \ldots \leq \mu(t_1) \geq \ldots \geq \mu(t_2) \leq \ldots \leq \mu(80)$. Two statistical problems arise using order-restricted parameters with unknown change points. First, proper, seemingly noninformative prior distributions may overwhelm the likelihood (Gelman, 1996). Using a simple example, we demonstrate this problem and propose a solution. Second, we discuss problems with two straightforward Markov chain methods that either converge very slowly or fail to sample the space of order restrictions.

1 Introduction

The bone mineral density (BMD) of the spine and hip are central to the management of osteoporosis, because spinal fractures are the most prevalent and hip fractures are the most debilitating. BMD has been shown to predict fractures, and the rates of bone growth and loss with age have been identified as key determinants of osteoporosis in old age (Hui, Slemenda, and Johnston, 1990).

To develop therapies for osteoporosis, we need to better understand the pattern of bone growth and loss with aging. The problem is to provide inference for the ages t_1 and t_2 that correspond to the changepoints demarking the stages in the pattern of the age-specific mean rates of change of BMD, $\mu(t)$, $t = 8, \ldots, 80$, where t is age in years. In other words, $\mu(t)$ is the population mean for people of age t.

Let $\beta_j(t)$, $j = 1, \ldots, n_t$, $t = 8, \ldots, 80$, represent the rate of bone growth for person j of age t (in years). If $\beta_j(t) < 0$ then bone is losing mass, i.e., $\beta_j(t)$ is the rate of bone loss.

It is generally accepted that the rate of bone growth increases until puberty, then decreases with age, that is, bone gains mass but at a lower rate each year. The rate of growth eventually ceases, then becomes negative (bone begins to lose mass) at an increasing rate until an older age, when the negative rate of growth slows. The negative rate of growth (bone loss) may decrease in old age because bone has lost much of its density.

To pattern the rate of bone growth (loss) we developed a strategy based on the assumption that the age-specific mean bone growth (loss) rates $\mu(t)$ (defined as $E(\beta_j(t)\,|\mu(t)) = \mu(t)$), and conditional on t_1 and t_2, follow the order restriction

$$\begin{aligned}\mu(8) \leq \ldots \leq \mu(t_1 - 1) \leq \mu(t_1) \geq \mu(t_1 + 1) \geq \ldots \\ \geq \mu(t_2 - 1) \geq \mu(t_2) \leq \mu(t_2 + 1) \leq \ldots \leq \mu(80). \qquad (6.1)\end{aligned}$$

The order restriction (6.1) depends on t_1 and t_2. The parameter t_1 is interpreted as the age of the maximum rate of bone growth, and t_2 is interpreted as the age of the maximum rate of bone loss. Our goal is to provide inference for t_1 and t_2 through the posterior distributions $\pi(t_1\,|y)$ and $\pi(t_2\,|y)$ where y represents the available data.

Previous approaches to inference for t_1 and t_2 have focused on fitting a curve $h(t)$ to the least squares estimates $\widehat{\beta}_j(t)$ of the $\beta_j(t)$, then estimating t_1 and t_2 from $h(t)$ (Matkovic, Jelic, Wardlaw, Ilich, Goel, Wright, Andon, Smith, and Heaney 1990). Using $h(t)$ appears to work when the range of ages is small, but our data show that the choice of the parametric form of $h(t)$ dramatically affects the point estimates of t_1 and t_2.

Another approach to inference for t_1 and t_2 uses nonparametric curves and the parametric bootstrap (Hui, Zhou, Evans, Slemenda, Peacock, Weaver, McClintock, and Johnston, 1999). Demonstrating the method of Hui et al. (1999) in the context of our problem, we first draw a bootstrap variate

$\underline{\mu}^{(\ell)} = (\mu(8)^{(\ell)}, \ldots, \mu(80)^{(\ell)})$ from the joint distribution

$$\prod_{t=8}^{80} N\left(\widehat{\mu}(t), \widehat{\eta}^2(t)\right),$$

where $\widehat{\mu}(t)$ and $\widehat{\eta}^2(t)$ are estimated from the data (see Hui et al., 1999, for details). Then the variate $t_1^{(\ell)}$ is defined as the age at which the maximum value of $\underline{\mu}^{(\ell)}$ occurs, and $t_2^{(\ell)}$ is defined as the age at which the minimum value of $\underline{\mu}^{(\ell)}$ occurs. Monte Carlo methods applied to $t_1^{(\ell)}$ and $t_2^{(\ell)}$, $\ell = 1, \ldots, L$, are used to provide inference for t_1 and t_2.

Nandram, Sedransk and Smith (1997) give a Bayesian approach for order-restricted inference with a stochastic change point. Their objects of inference are 8 order-restricted proportions with an unknown modal position. The number of order-restricted parameters is important because, as Gelman (1996) observed, the number of order-restricted parameters can influence the effect of the prior distribution of the order-restricted parameters on the posterior distribution. That is, when using an improper uniform prior distribution for a large number of constrained parameters, the prior will overwhelm the data. Gelman (1996) suggests, as a general remedy, replacing the "improper uniform distribution by a hierarchical family of proper prior distributions." For our data set, a Markov random field prior distribution on the $\mu(t)$ adequately modeled the data.

Section 2 describes the data and how they were collected. Section 3 presents the likelihood and prior distribution for the $\mu(t)$. Section 4 gives the form of the MCMC method that we use, and discusses problems with two alternative methods. Section 5 gives our inference for t_1, t_2, and $E(\mu(t)\,|y)$. Concluding comments are in section 5. The appendix has two examples and a discussion why it may be unsatisfactory to use noninformative prior distributions together with (6.1).

2 The data

The spine data in this study were collected as part of several longitudinal studies performed at Indiana University (Hui et al., 1999). All of these studies enrolled normal subjects who did not have bone disease and had not taken estrogen or other medications known to affect bone metabolism. Only white females were used in the analyses and data were excluded once postmenopausal women initiated estrogen replacement therapy. Some studies were observational, while others were randomized controlled trials from which only the control subjects were included here. For each of several age spans, there was a primary source of study subjects. The children and teenagers were mostly in the placebo arm of a randomized trial of calcium supplement. The young women were primarily in the control group of a randomized trial of exercise. From age 30 through age 70, most of the women were in ongoing observational studies of sex hormones and bone loss. The oldest age group (70-80 years) came primarily from the placebo arm of a randomized trial of calcium and vitamin D supplements in the elderly.

Each subject had her BMD measured at least 3 times over a period of less than 6 months. For this brief period, least squares regression for BMD against time is an acceptable model (Hui and Berger, 1983). The observed rate of change of BMD, $\widehat{\beta}_j(t)$, is the fitted slope of the least squares line for subject j of age t, $t = 8, \ldots, 80$, $j = 1, \ldots, n_t$. The set consisting of subjects that have the same age t is called age group t. Each subject is in exactly one age group (i.e., there is one $\widehat{\beta}_j(t)$ for each subject), and the number of subjects in age group t is n_t.

The data are 449 observations $\left(\widehat{\beta}_j(t), \widehat{\sigma}_j^2(t)\right)$, where $\widehat{\sigma}_j^2(t)$ is the square of the sample standard error of $\widehat{\beta}_j(t)$. Figure 6.1 is a plot of the $\widehat{\beta}_j(t)$ against t in years. The $\widehat{\beta}_j(t)$ do not follow an order restriction, but do follow an increasing trend to approximately age 13, and then a decreasing trend, with an upturn around age 50.

Two observations are clearly outliers, a large positive value at age 49 and a large negative value at age 50. These observations have relatively large values of $\widehat{\sigma}_j^2(t)$, so they have little impact on inference.

There are no observations for ages 15, 17, 18, 21, and 29, but $\mu(15)$, $\mu(17)$, $\mu(18)$, $\mu(21)$, and $\mu(29)$ must be in the model because a change-point may occur at one of those ages. The lack of data is no problem because $\mu(15)$, $\mu(17)$, $\mu(18)$, $\mu(21)$, and $\mu(29)$ are the parameters of Markov random field prior distributions for the $\mu(t)$ that have corresponding data (see section 3 for a description of the prior). The remaining age groups have $1 \leq n_t \leq 17$.

The Kolmogorov-Smirnov test of normality was applied to the $\widehat{\beta}_j(t)$ in the 14 age groups that have 10 or more observations. In each case, the null hypothesis that the $\widehat{\beta}_j(t)$ follow a normal distribution was not rejected.

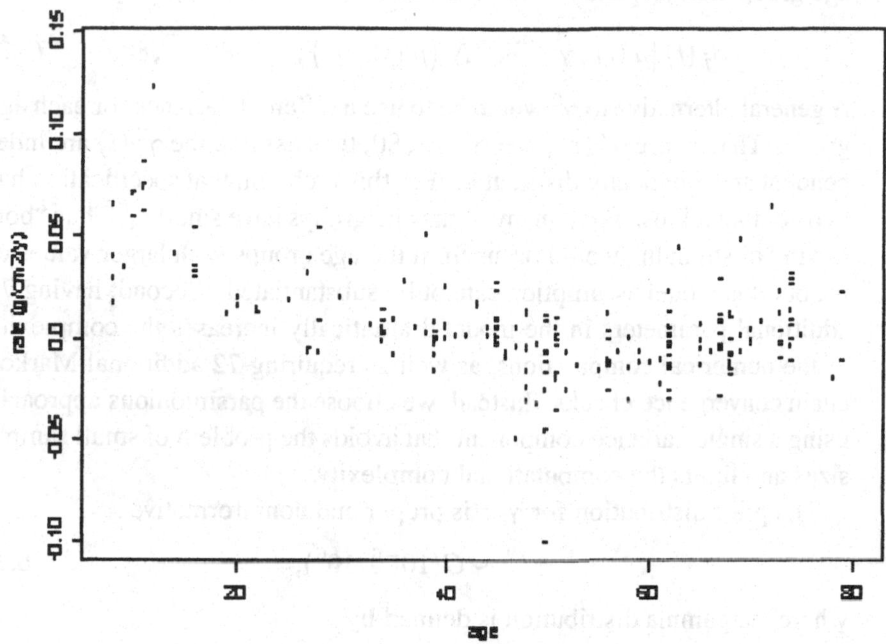

FIGURE 6.1. Scatter plot of Individual Rates of Bone Growth (Loss) by Age.

3 The Model

3.1 Introduction

The general approach to Bayesian order-restricted inference (Gelfand, Smith, and Lee, 1992) is to first determine the unconstrained posterior distribution of the parameters of interest. The posterior distribution of the constrained parameters is proportional to the kernel of the unconstrained posterior, but supported only on the constrained region. Following this general approach, we begin the description of our model with the sampling distribution and the prior distributions without consideration of the order restrictions. Let

$$\widehat{\beta}_j(t)\,|\beta_j(t) \overset{indep}{\sim} N\left(\beta_j(t), \sigma_j^2(t)\right), \quad j = 1, \ldots, n_t, \; t = 8, \ldots, 80. \tag{6.1}$$

In (6.1) $\sigma_j^2(t)$ is estimated by $\widehat{\sigma}_j^2(t)$, because we do not have the original BMD measurements that would permit modeling the components of $\widehat{\beta}_j(t)$.

Next, we model the prior distribution of the $\beta_j(t)$. Individuals within an age group are independent, and (6.1) motivates the conjugate specification

$$\beta_j(t)\,|\mu(t),\gamma^2 \overset{indep}{\sim} N\left(\mu(t),\gamma^2\right),\ t=8,\ldots,80. \tag{6.2}$$

A general alternative to γ^2 would be to use a different variance for each age group. That is, use $\gamma^2(t)$, $t=8,\ldots,80$, then assume the $\gamma^2(t)$ are independent and identically distributed. For this problem, that specification has two complications. First, many of the age groups have small n_t. Thus "borrowing of strength" would occur from the age groups with larger values of n_t, but this model assumption cannot be substantiated. Second, having 72 additional parameters in the model dramatically increases the complexity of the numerical computations, as well as requiring 72 additional Markov chain convergence checks. Instead, we choose the parsimonious approach, using a single variance component that avoids the problem of small sample sizes and limits the computational complexity.

The prior distribution for γ^{-2} is proper and noninformative

$$\gamma^{-2} \sim G(10^{-6},10^{6}), \tag{6.3}$$

where the gamma distribution is defined by

$$g(x\,|\alpha,\beta) \propto x^{\alpha-1}e^{-\frac{x}{\beta}},\ x>0,\ \alpha>0,\ \beta>0. \tag{6.4}$$

The prior distribution for the $\mu(t)$, without the order restriction, is motivated by pairwise difference priors (Besag, York, and Mollié, 1991); i.e.,

$$p\left(\mu(80)\,|\nu^2\right) \propto e^{-\frac{1}{2\nu^2}\mu(80)^2},$$

and

$$p\left(\mu(t)\,|\mu(t+1),\mu(t+2),\ldots,\mu(80),\tau_t^2\right) \propto e^{-\frac{1}{2\tau_t^2}(\mu(t)-\mu(t+1))^2},\quad t=8,\ldots,79 \tag{6.5}$$

where $\tau_t^2=\tau_1^2$ when $t\le 17$ and $\tau_t^2=\tau_2^2$ when $t>17$. We refer to τ_1^2 and τ_2^2 as smoothing parameters (Bernardinelli, Pascutto, Best, and Gilks, 1997). When τ_1^2 and τ_2^2 are small, the plot of the $\mu(t)$ against t is smooth when compared to a plot of the $\mu(t)$ against t for large values of τ_1^2 and τ_2^2. Originally, we used one smoothing parameter in (6.5), but it failed to capture the variation by age that is evident in the data (see Figure 6.1); after age 17, the data are smooth compared to the swift increase and decrease around age 13. For this example we chose $\nu^2=1$, a value that emphasizes the data in inference for $\mu(80)$. The prior distributions for τ_1^2 and τ_2^2 are developed in section 3.2.

3.2 Priors for the smoothing parameters

Preliminary work indicated that noninformative prior distributions for τ_1^2 and τ_2^2 would provide inferences that are inconsistent with the data (e.g., see example 2 in the appendix). Because we cannot use noninformative prior distributions, the data are used as a source of information about τ_1^2 and τ_2^2. The approach is to use (6.1)-(6.5) and proper, noninformative gamma priors on τ_1^{-2} and τ_2^{-2}, without (6.1), to obtain the posterior distributions $\pi(\tau_1^2|y)$ and $\pi(\tau_2^2|y)$. Then $\pi(\tau_1^2|y)$ and $\pi(\tau_2^2|y)$ are used as prior distributions for τ_1^2 and τ_2^2 in (6.5) for the *constrained* model. Ideally, we would use a random subset of the data like prior information to estimate $\pi(\tau_1^2|y)$ and $\pi(\tau_2^2|y)$. However, sparse data in the young age groups preclude that approach. Although the data are used in the prior distributions, $\pi(\tau_1^2|y)$ and $\pi(\tau_2^2|y)$, they come from a different model (the $\mu(t)$ unconstrained) than the one used to provide inference for t_1 and t_2 (the $\mu(t)$ constrained).

Numerical methods may be more easily implemented if the parameters have distributions of common form. We tested $\pi(\tau_1^2|y)$ and $\pi(\tau_2^2|y)$ against gamma distributions using the Kolmogorov-Smirnov test and determined that

$$\tau_1^{-2} \sim G(3.11, 1342.704) \tag{6.6}$$

and

$$\tau_2^{-2} \sim G(3.8, 63141.42) \tag{6.7}$$

are acceptable prior distributions (where the gamma distribution is defined in (6.4)). The prior mean of τ_1^{-2} is 4,175.8, and the prior mean of τ_2^{-2} is 239,937.4 which suggests that there is considerable smoothing of the $\mu(t)$.

3.3 The prior for t_1 and t_2

The maximum rate of bone growth occurs around puberty, so t_1 is assigned uniform positive probability on the ages 11, 12, 13, and 14. The maximum rate of bone loss occurs in older people so we assigned uniform positive probability to the ages 45, 46,...,60. The prior distribution on the ages corresponding to change points is

$$p(t_1, t_2) = 1/64,\ t_1 = 11, 12, 13, 14,\ t_2 = 45, 46, ..., 60. \tag{6.8}$$

Because the pairs (t_1, t_2) can be enumerated, we will sometimes index the set of order-restrictions with the parameter g. For example, $g = 1$ is the order restriction defined by $(t_1 = 11, t_2 = 45)$, and $g = 2$ is the order restriction defined by $(t_1 = 11, t_2 = 46)$. In this context, g can be treated as a model parameter.

3.4 *Summary*

The model that we used is given by (6.1)-(6.3) and (6.5)-(6.7), but for given t_1 and t_2, (6.5) is constrained by (6.1). Finally the prior distribution for t_1 and t_2 is (6.8). The appendix has two examples and a discussion why a prior distribution on $\mu(t)$ such as (6.5) is needed; that is, why a locally uniform distribution on the $\mu(t)$ together with (6.1) and (6.8) is insufficient.

4 Computational methods

In this section, we describe the numerical method used to generate the Markov chains $\underline{\mu}^{(\ell)}$, $\underline{\tau}^{2(\ell)}$, and $\gamma^{2(\ell)}$, $\ell = 1, \ldots, L$, from the joint distribution $f\left(\underline{\mu}, \underline{\tau}^2, \gamma^2, t_1, t_2, y\right)$, defined by (6.1) and (6.1)-(6.8). These chains are then used to give Monte Carlo estimates, $\widehat{\pi}(t_1|y)$ and $\widehat{\pi}(t_2|y)$, of the posterior distributions $\pi(t_1|y)$ and $\pi(t_2|y)$.

An appealing approach would be to use the Metropolis algorithm to sample over the space of order-restrictions. That is, to consider t_1 and t_2 as discrete model parameters with a domain of 64 (4×16) pairs of values. The Metropolis algorithm could generate a Markov chain $(t_1^{(\ell)}, t_2^{(\ell)})$, $\ell = 1, \ldots, L$, from $f\left(\underline{\mu}, \underline{\tau}^2, \gamma^2, t_1, t_2, y\right)$,with $\widehat{\pi}(t_1 = t^*|y)$ proportional to the number of times t^* occurs in the chain $t_1^{(\ell)}$, $\ell = 1, \ldots, L$, and with $\widehat{\pi}(t_2 = t^{**}|y)$ proportional to the number of times t^{**} occurs in the chain $t_2^{(\ell)}$, $\ell = 1, \ldots, L$. Unfortunately, this method converges very slowly because $f\left(\underline{\mu}, \underline{\tau}^2, \gamma^2, t_1, t_2, y\right)$ has a local maximum at every pair (t_1, t_2). One method of circumventing this problem is to use multiple chains with different initial values.

An alternative is the Gibbs sampler (Gelfand, Smith, and Lee, 1992). A straightforward application of the Gibbs sampler would be to generate Markov chains $g^{(\ell)}$, $\underline{\mu}^{(\ell)}$, $\underline{\tau}^{2(\ell)}$, and $\gamma^{2(\ell)}$, $\ell = 1, \ldots, L$ from the complete conditionals $\left[g|\underline{\mu}, \underline{\tau}^2, \gamma^2, y\right]$, $\left[\mu(t)|g, \underline{\mu}(-t), \underline{\tau}^2, \gamma^2, y\right]$, $\left[\underline{\tau}_i^2|g, \underline{\mu}(t), \underline{\tau}^2_{-i},\right.$ $\left.\gamma^2, y\right]$, and $\left[\gamma^2|g, \underline{\mu}, \underline{\tau}^2, \gamma^2, y\right]$, where $\underline{\mu}(-t)$ is $\underline{\mu}$ without $\mu(t)$ and where g represents a value of (t_1, t_2) (see section 3.3). The chain $g^{(\ell)}$, an approximately random sample from the distribution $\pi(t_1, t_2|y)$, is used for inference for t_1 and t_2. However, the $\mu(t)$ are sampled conditional on g, so the $\mu(t)$ follow the order restriction given by g. The complete conditional $\left[g|\underline{\mu},\underline{\tau}^2,\gamma^2,y\right]$ is a point mass on the order restriction that the $\mu(t)$ follow (i.e., g). Therefore, the Markov chain $g^{(\ell)}$ never moves from its initial value.

We circumvent this problem by using 64 Gibbs samplers, each conditional on an order restriction, and then apply an identity from Nandram, Sedransk and Smith (1997) to obtain the marginal distribution $m\left(y\,|g\right)$. The idea is to use the Gibbs sampler to estimate

$$\frac{1}{m\left(y\,|g\right)}=\int_0^\infty\int_{\Omega_1}\int_{\Omega_g}\frac{1}{f\left(y\,|\underline{\mu},\underline{\tau}^2,\gamma^2\right)}\pi\left(\underline{\mu},\underline{\tau}^2,\gamma^2\,|y,g\right)d\underline{\mu}d\underline{\tau}^2d\gamma^2, \tag{6.9}$$

where the inner integral is a 73-dimensional integral over Ω_g, which is R^{73} constrained by g, and the middle integral is a 2-dimensional integral over $\Omega_1 = R^+ \times R^+$.

Conditional on the order restriction g, we generate samples $\underline{\mu}_g^{(\ell)}$, $\underline{\tau}_g^{2(\ell)}$, and $\gamma_g^{2(\ell)}$, $\ell = 1, \ldots, L$ from $\pi\left(\underline{\mu},\underline{\tau}^2,\gamma^2\,|y,g\right)$, using the Gibbs sampler (Gelfand, Smith, and Lee, 1992) with the complete conditionals $\left[\mu(t)|g,\underline{\mu}(-t),\underline{\tau}^2,\gamma^2,y\right]$, $\left[\tau_i^2|g,\underline{\mu}(t),\underline{\tau}^2_{-i},\gamma^2,y\right]$, and $\left[\gamma^2|g,\underline{\mu},\underline{\tau}^2,\gamma^2,y\right]$. The first complete conditional (for $\mu(t)$) is a constrained univariate normal distribution, the second and third complete conditionals are inverse gamma distributions (Gelfand, Smith, and Lee, 1992). Equation (6.9) is estimated by

$$\widehat{m}\left(y\,|g\right)=L/\sum_{\ell=1}^{L}1/f\left(y\left|\underline{\mu}_g^{(\ell)},\underline{\tau}_g^{2(\ell)},\gamma_g^{2(\ell)}\right.\right). \tag{6.10}$$

Finally, using (6.10),

$$\widehat{\pi}\left(g=g_0|y\right)=\frac{\widehat{m}\left(y\,|g=g_0\right)}{\sum_{g'=1}^{64}\widehat{m}\left(y\,|g=g'\right)}. \tag{6.11}$$

Because g is defined by t_1 and t_2, $\widehat{\pi}\left(g|y\right)$ can be represented by a 16×4 matrix whose margins are $\widehat{\pi}\left(t_1|y\right)$ and $\widehat{\pi}\left(t_2|y\right)$.

We omit the details of the MCMC, because the Gibbs sampler part is conditional on an order restriction, and is described in Gelfand, Smith, and Lee (1992). Markov chains of order-restricted variables often have high autocorrelation, which can be reduced by appropriate thinning.

An alternative to (6.9) is to estimate

$$m(y\,|g) = \int_0^\infty \int_{\Omega_1} \int_{\Omega_g} f\left(y|\underline{\mu}, \underline{\tau}^2, \gamma^2, g\right) p\left(\underline{\mu}, \underline{\tau}^2, \gamma^2\,|g\right) d\underline{\mu} d\underline{\tau}^2 d\gamma^2$$

with

$$\widehat{m}(y\,|g) = \sum_{\ell=1}^{L} f\left(y\left|\underline{\mu}_g^{(\ell)}, \underline{\tau}_g^{2(\ell)}, \gamma_g^{2(\ell)}\right.\right) \Big/ L,$$

where $\underline{\mu}_g^{(\ell)}$, $\underline{\tau}_g^{2(\ell)}$, and $\gamma_g^{2(\ell)}$, $\ell = 1, \ldots, L$ are Markov chains from $p\left(\underline{\mu}, \underline{\tau}^2, \gamma^2\,|g\right)$. This method is unsatisfactory because when the data are not congruent to an order restriction, the $\underline{\mu}_g^{(\ell)}$ are in the extreme tails of f.

5 Results

The posterior distribution $\widehat{\pi}\,(g|y)$ is given in Table 6.1. Table 6.1 has 4 columns corresponding to t_1 and 16 rows corresponding to t_2. The entries in Table 6.1 are the estimated probabilities that the rate of bone growth obeys the order restriction given by the column (t_1) and the row (t_2).

The column margin of Table 6.1 is $\widehat{\pi}\,(t_1|y)$ and is given in Table 6.2. The first row of Table 6.2 corresponds to age; the second row is the corresponding probability. More than 99% of the probability occurs at ages 12 and 13. Using Table 6.2, $\widehat{E}\,(t_1\,|y) = 12.667$ and $\widehat{SD}\,(t_1\,|y) = 0.386$.

The row margin of Table 6.1 is $\widehat{\pi}\,(t_2|y)$ and is given in Table 6.3. The first row of Table 6.3 corresponds to age, with the second row the corresponding probability. Using Table 6.3, $\widehat{E}\,(t_2\,|y) = 49.971$ and $\widehat{SD}\,(t_2\,|y) = 10.452$. A 95.7% credible region ranges from age 47 to age 51, which is a much smaller interval than $49.971 \pm 2\times\, 10.452$.

Finally, it is important to demonstrate that the $\mu\,(t)$ are consistent with the data, otherwise results about t_1 and t_2 are suspect. Figure 6.2 is a plot of the posterior expected age-specific means and observed rates against age. A posterior expected age-specific mean is defined as

$$E\left(\mu\,(t)\,|y\right) = \sum_{g=1}^{64} E(\mu\,(t)\,|g, y)\pi\,(g|y),$$

and is estimated by first averaging $\underline{\mu}_g^{(\ell)}$ over ℓ, conditional on g, $g = 1, \ldots, 64$, and then using (6.11) to average over the posterior distribution

$\pi(g \mid y)$

	11	12	13	14
45	8.046090e-011	1.233000e-008	1.177812e-008	1.032013e 022
46	3.522866e-008	5.427628e-006	7.178704e-006	1.663015e-020
47	3.293376e-006	1.580283e-003	5.276917e-002	4.175369e-017
48	2.788363e-005	6.172874e-003	1.450001e-002	9.480472e-018
49	4.744564e-005	1.467532e-002	4.209160e-003	4.938185e-014
50	6.177914e-004	2.806706e-001	5.240447e-001	1.299858e-014
51	1.069453e-004	1.685934e-002	4.080279e-002	9.735811e-016
52	5.825998e-005	2.190252e-003	1.275667e-002	2.217424e-017
53	7.441638e-006	5.327444e-003	4.135811e-003	5.434600e-016
54	1.107125e-005	1.924994e-003	1.402557e-002	3.694925e-017
55	9.979654e-007	9.882032e-004	4.170352e-004	2.627370e-017
56	1.107693e-005	1.678912e-004	4.424153e-004	2.888002e-019
57	2.026837e-007	3.418318e-005	7.434708e-005	6.081786e-016
58	8.885129e-007	4.495971e-005	3.690953e-005	3.369168e-018
59	2.129466e-007	1.330470e-004	5.469809e-005	3.121975e-018
60	1.111111e-007	2.948144e-005	2.559561e-005	7.753617e-018

TABLE 6.1. Posterior Distribution of the Order Restrictions. The columns represent t_1 and the rows represent t_2. The largest probabilities occur in the row corresponding to age 50.

$\pi(t_1 \mid y)$

age	11	12	13	14
prob	0.001	0.331	0.668	0

TABLE 6.2. Posterior Distribution of t_1. This distribution is the column margin from Table 6.1. The mode occurs at age 13.

$\pi(t_2 \mid y)$

age	45	46	47	48	49	50	51	52
prob	0	0	0.054	0.027	0.01902	0.805	0.0587	0.015

age	53	54	55	56	57	58	59	60
prob	0.009	0.015	0.001	0	0	0	0	0

TABLE 6.3. Posterior Distribution of t_2. This distribution is the row margin from Table 6.1. The mode occurs at age 50.

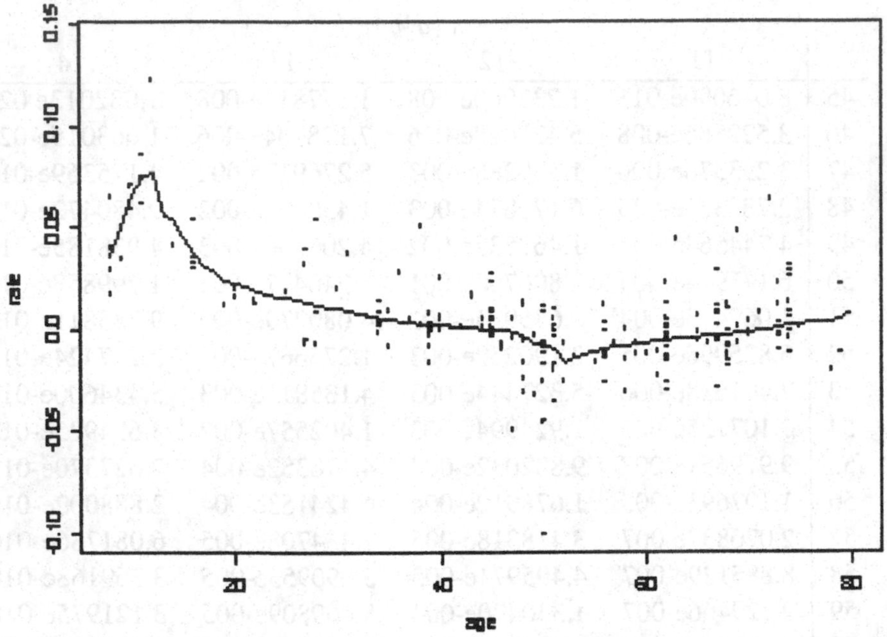

FIGURE 6.2. Posterior Expected Age-Specific Means and Rates by Age. The line represents the $E(\mu(t)\,|y)$, $t = 8, \ldots, 80$, and the dots (.) are the individual rates of bone growth.

of g. The posterior expected age-specific means in Figure 6.2 are consistent with the data. The cusp at age 50 is a result of $\widehat{\pi}(t_2 = 50\,|y) = 0.805$. It is possible to smooth the cusp using a prior distribution on τ_2^{-2} that has a larger expected value than the one in formula (6.7).

6 Comments

This paper provides a method for modeling rates of bone growth and loss using order-restricted inference. Although the science question is answered satisfactorily, several statistical problems remain. First, the prior distributions of the order-restricted parameters and the hyperparameters have considerable influence on the posterior distribution of the order-restricted parameters. A general assessment of their effect would help with the practical implementation of such order-restricted parameter models. There are several other questions of interest. For example, when is a proper noninformative prior distribution acceptable? Why does the pairwise difference prior distribution work when the noninformative prior does not? How does the number of the order-restricted parameters affect the inference?

Second, using a Gibbs sampler for every order restriction is inefficient because even though some of the order constraints have very low probability, a long sequence of parameters is sampled. If there existed a efficient technique to sample the space of order-restrictions, and to move over local maxima, then fewer of the variates with low posterior probability would be sampled.

7 Appendix: Use of locally uniform prior distributions together with order restrictions

We motivate the use of a prior distribution such as (6.5) using two simple examples that demonstrate the observations of Gelman (1996) about the influence of a seemingly noninformative prior distribution on the posterior distribution in constrained parameter problems. Let

$$y_i \,|\mu_i \overset{indep}{\sim} N(\mu_i, 1)\,,\; i = 1, \ldots, 70, \tag{6.1}$$

$$\mu_i \overset{iid}{\sim} N(0, 10000)\,,\;\; i = 1, \ldots, 70. \tag{6.2}$$

We use the notation μ_i instead of $\mu(t)$ to distinguish these simple examples from the bone problem. There is one observation for each value of the index, so by using (6.1) and (6.2), the unconstrained posterior distribution of the μ_i is closely approximated by

$$\pi\left(\underline{\mu}\,|data\right) \propto \prod_{i=1}^{70} e^{-\frac{1}{2}(\mu_i - y_i)^2}, \tag{6.3}$$

where $\underline{\mu} = (\mu_1, \ldots, \mu_{70})$. Note that $E(\mu_i|data) = y_i, i = 1, \ldots, 70$.

Next, assume the order restriction $\mu_i < \mu_{i+1}, i = 1, \ldots, 69$. Then the constrained posterior distribution of the μ_i denoted $\pi_c\left(\underline{\mu}\,|data\right)$, is proportional to (6.3), with the domain restricted to $\mu_i < \mu_{i+1}, i = 1, \ldots, 70$.

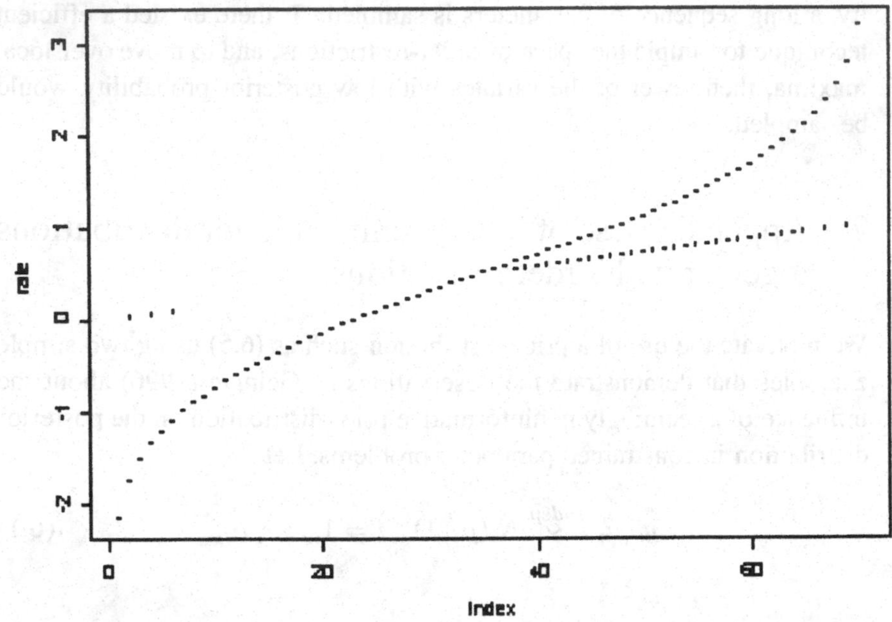

FIGURE 6.3. Data set D_1 and $E_c(\mu_i|D_1)$ by Index. The dots (.) represent the data and the dashes (-) represent the posterior expectations. For all index values the two points overlap.

Example 1

Suppose that we observe the data $D_1 = \{0, 2, 4, \ldots, 138\}$, corresponding to $y_1, \ldots, y_{70}$, and then use (6.3) and MCMC methods to calculate $E_c(\mu_i|D_1)$, the posterior expected value when the order restriction holds. Because the data are congruent to the order restriction (i.e., increasing), and because (6.2) is noninformative, we would expect that $E_c(\mu_i|D_1) = y_i$. Also, from the frequentist perspective, using D_1, (6.1), and $\mu_i < \mu_{i+1}, i = 1, \ldots, 69$, the maximum likelihood estimator of μ_i is y_i, $i = 1, \ldots, 70$. Confirming our intuition, Figure 6.3 (a plot of the data set D_1 and the $E_c(\mu_i|D_1)$ against the index $i = 1, \ldots, 70$) shows that the plot of D_1 and the plot of $E_c(\mu_i|D_1)$ overlap. That is, the $E_c(\mu_i|D_1)$ agree with the data.

Example 2

The second example is similar to the first example, but uses a different data set. As in the first example, we use (6.3) and the data $D_2 = \{0, 1/69, 2/69, \ldots, 68/69, 1\}$, corresponding to $y_1, \ldots, y_{70}$, to estimate $E_c(\mu_i|D_2)$. Like data set D_1, data set D_2 is congruent with the order restriction, so we would again expect that $E_c(\mu_i|D_2) = y_i$. Figure 6.4 is a plot of the data set D_2 and the $E_c(\mu_i|D_2)$ against the index $i = 1, \ldots, 70$. In this example, the $E_c(\mu_i|D_2)$ near the ends of the sequence are more than two posterior standard deviations from the y_i. Near the middle of the sequence, the $E_c(\mu_i|D_2)$ appear more consistent with the data.

The only difference in the two examples is the data, but in the second example the data seem to influence the posterior distribution less than in the first example. It is difficult to determine analytically why the $E_c(\mu_i|D_2)$ behave pathologically in Figure 6.4. Under the order restrictions closed forms of the integrals of the μ_j,needed to obtain the marginal posterior $\pi_c(\mu_i|D_2)$, $i = 1, \ldots, 70$, do not exist except for a few special cases. With analytic results, we might have found a general prior distribution that would correct for the problem observed in the second example (Figure 6.4). Instead, we use a prior distribution that fixes the "symptoms" of the problem, that is, one that, in conjunction with the constraints, will not overwhelm the data.

Two observations may aid in determining a prior distribution for the μ_i. First, in example 1, the data were spread so that each observation is 2 standard deviations from the other ($D_1 = \{0, 2, 4, \ldots, 138\}$). Then the $E_c(\mu_i|D_1)$ agree with the data, and we infer that (6.1) and (6.2) is a reasonable model. Second, the data D_2 seem close together relative to the variance in (6.1); that is, the data are correlated. Because (6.1) and (6.2) do not model the correlation in D_2 and that is the case for which the model fails, it is important to model the correlation.

There are many ways to model correlation in data. Markov random field prior distributions on location parameters are a flexible class of distributions because it is easy (by adjusting the smoothing parameters) to account for changes in the correlation of the data over time.

These two simple examples have ramifications for the bone data. Although t_1 and t_2 are our parameters of primary interest, if the $E_c(\mu(t)|y)$ are not consistent with the data then inference for t_1 and t_2 is suspect. We observe the same pathological behavior that occurred in the second example (Figure 6.4) when a proper noninformative prior distribution is used for the $\mu(t)$.

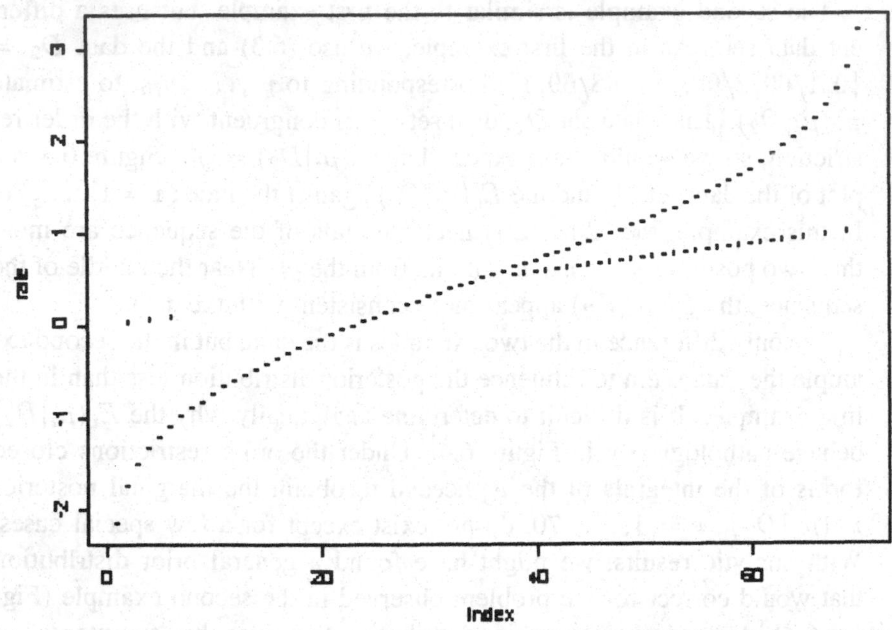

FIGURE 6.4. Data set D_2 and $E_c(\mu_i|D_2)$ by Index. The dots (.) represent the data and the dashes (-) represent the posterior expectations. Expectations near the ends of the sequence are far from the data.

Finally, in the course of our research, we found that the pathological behavior in the second example (Figure 6.4) also occurs with posterior distributions that have finite support. In addition, the pathological behavior in the second example (Figure 6.4) is less pronounced if there are few order-restricted quantities.

References

Bernardinelli, L., Pascutto, C., Best, N.G., and Gilks, W. R. (1997), "Disease Mapping with Errors in Covariates," *Statistics in Medicine*, 16, 741-752.

Besag, J., York, J., and Mollie, A. (1991), "Bayesian Image Restoration, with Two Applications in Spatial Statistics," *Annals of the Institute of Statistical Mathematics*, 43, 1-59.

Gelfand, A. E., Smith, A. F. M., and Lee, T. (1992), "Bayesian Analysis of Constrained Parameter and Truncated Data Problems Using Gibbs Sampling," *Journal of the American Statistical Association*, 87, 523-532.

Gelman, A. (1996), "Bayesian Model Building by Pure Thought: Some Principles and Examples," *Statistica Sinica*, 6, 215-232.

Hui, S. L., and Berger, J. O. (1983), "Empirical Bayes Estimation of Rates in Longitudinal Studies," *Journal of the American Statistical Association*, 78, 753-760.

Hui, S. L., Slemenda, C. W., and Johnston, C. C. Jr. (1990), "The Contribution of Bone Loss to Postmenopausal Osteoporosis," *Osteoporosis International*, 1, 30-34.

Hui, S. L., Zhou, L., Evans, R., Slemenda, C. W., Peacock, M., Weaver, C. M., McClintock, C., and Johnston, C. C. Jr. (1999), "Rates of Growth and Loss of Bone Mineral in the Spine and Femoral Neck in White Females," *Osteoporosis International*, 9, 200-205.

Matkovic, V., Jelic, T., Wardlaw, G. M., Ilich, J. Z., Goel, P. K., Wright, J. K., Andon, M. B., Smith, K. T., and Heaney, R. P. (1994), "Timing of Peak Bone Mass in Caucasian Females and its Implication for the Prevention of Osteoporosis," *Journal of Clinical Investigations*, 93(2), 799-808.

Nandram, B., Sedransk, J., and Smith, S. J. (1997), "Order-restricted Bayesian Estimation of the Age Composition of a Population of Atlantic Cod," *Journal of the American Statistical Association*, 92, 33-40.

Estimating Genotype Probabilities in Complex Pedigrees

Soledad A. Fernández
Rohan L. Fernando
Alicia L. Carriquiry
Bernt Guldbrandtsen

ABSTRACT Probability functions such as likelihoods and genotype probabilities play an important role in the analysis of genetic data. When genotype data are incomplete Markov chain Monte Carlo (MCMC) methods, such as the Gibbs sampler, can be used to sample genotypes at the marker and trait loci. The Markov chain that corresponds to the scalar Gibbs sampler may not work due to slow mixing. Further, the Gibbs chain may not be irreducible when sampling genotypes at marker loci with more than two alleles. These problems do not arise if the genotypes are sampled jointly from the entire pedigree. When the pedigree does not have loops, a joint sample of the genotypes can be obtained efficiently *via* modification of the Elston-Stewart algorithm. When the pedigree has many loops, obtaining a joint sample can be time consuming. We propose a method for sampling genotypes from a pedigree so modified as to make joint sampling efficient. These samples, obtained from the modified pedigree, are used as candidate draws in the Metropolis-Hastings algorithm.

1 Introduction

Determining genotype probabilities is important in genetic counseling, linkage analysis and in genetic evaluation programs. When performing genetic counseling, for example, it is important to know which individuals in a population are probable carriers of a deleterious recessive allele. The first methods for determining genotype probabilities were developed in human genetics by Elston and Stewart (1971) and were reviewed by Elston and Rao (1978) and by Elston (1987). Several human genetics computer packages are available to compute genotype probabilities. In livestock, pedigrees are usually much larger than in humans because animals, specially males, have multiple mates. Thus, the application of computer intensive methods developed for humans will often be difficult or inappropriate for livestock data.

In this paper, we consider the problem of estimating genotype probabilities in a dog pedigree. The pedigree consists of 3,052 dogs (Labrador Retrievers) from "The Seeing Eye, Inc". The trait of interest is a disease called progressive retinal atrophy (PRA). PRA is a genetic disorder of the eye and it is inherited as a simple autosomal recessive. Thus, the dog is affected when it has the homozygous recessive genotype for the disease locus. If the dog has the heterozygous genotype (Aa or aA) then it is a carrier. This disease can only be diagnosed by either an opthalmic exam after the dog is five years old or by a very expensive electro-retinal-gram (ERG) after the dog is 18 months of age. Therefore, it is important to know which dogs are at highest risk of transmitting the PRA gene to their offspring and which dogs are at lower risk both of transmitting the gene and of being PRA affected. There are 33 affected dogs in this pedigree. That is, we only know the genotype and phenotype for 33 dogs. The rest of the dogs could have the genotype aa but not yet detected, could be carriers (Aa) or could be non-carriers (AA).

To obtain estimates of the genotype probabilities, the likelihood of the pedigree is needed. The likelihood L is obtained as

$$L \propto \sum_{\mathbf{g}} f(\mathbf{y}|\mathbf{g})P(\mathbf{g}), \qquad (7.1)$$

where $\mathbf{y}$ is the vector of phenotypes and $\mathbf{g}$ is the vector of genotypes, $f(\mathbf{y}|\mathbf{g})$ are the conditional probabilities of $\mathbf{y}$ given $\mathbf{g}$, $P(\mathbf{g})$ are the genotype probabilities. The summation is over all possible genotypes for all individuals in the pedigree. The computations involved in the likelihood are not feasible except in trivial examples. For example, assume that there are two possible alleles for a locus, resulting in three possible genotypes for every individual in the pedigree (AA, Aa or aA and aa). If the pedigree consists of 100 individuals then 3^{100} summations need to be performed in order to compute the likelihood as in (7.1).

For pedigrees without loops, the likelihood can be computed efficiently using the Elston-Stewart algorithm (Elston and Stewart, 1971), which is also called *peeling*. For small pedigrees (about 100 members) with loops, extensions of the Elston-Stewart algorithm have been developed for evaluating the likelihood (Lange and Elston, 1975; Cannings et al., 1978: Lange and Boehnke, 1983; Thomas, 1986a,b). For large pedigrees with complex loops exact calculations are not possible and approximations are used (Van Arendonk et al., 1989; Janss et al., 1992; Stricker et al., 1995; Wang et al., 1996).

Van Arendonk et al. (1989) presented an iterative algorithm to calculate genotype probabilities of all members in an animal pedigree. Some limitations in their algorithm were removed by Janss et al. (1992). Their method can be used to approximate the likelihood for large and complex pedigrees with loops. Stricker et al. (1995) also proposed a method to approximate the likelihood in pedigrees with loops. Their method is based on an algorithm that cuts the loops. If the pedigree does not have loops, the computation of the likelihood is exact. In 1996, Wang et al. proposed a new approximation to the likelihood of a pedigree with loops by cutting all loops and extending the pedigree at the cuts. This method makes use of iterative peeling. They showed that the likelihood computed by iterative peeling is equivalent to the likelihood computed from a cut and extended pedigree.

When a mixed model of inheritance is used, the likelihood is not easy to obtain. Under this model, the phenotypic values of individuals in the pedigree cannot be assumed to be conditionally independent given the pedigree members because they are also influenced by the polygenic loci. Markov chain Monte Carlo (MCMC) methods have been proposed to overcome these problems. These MCMC methods give exact estimates to any desired level of accuracy. As Jensen and Sheehan (1998) observed, the genotypes in a pedigree can be sampled according to a Markovian process because a neighborhood system can be defined on a pedigree such that the genotype of an individual, conditional on the neighbors or close relatives, is independent of the remaining pedigree members. This local dependency makes MCMC methods, such as the Gibbs sampler, very easy to implement and provides a strategy to estimate posterior genotype probabilities that are difficult to calculate otherwise.

When using the Gibbs sampler, however, mixing can be very slow due to the dependence between genotypes of parents and progeny (Janss et al., 1995). The larger the progeny groups, the stronger the dependence and thus the Gibbs chains do not move. When this happens it is said that the chains are reducible "in practice". Mixing can be improved by applying Gibbs sampling to blocks (Jensen et al., 1995; Janss et al., 1995). This procedure is called "blocking Gibbs Sampling" and consists of sampling a block of genotypes jointly. In this approach, the blocks are typically formed by sub-families in the pedigree. Blocking Gibbs can be applied to very large pedigrees provided there are no complex loops. The efficiency of blocking depends on the pedigree structure and on the way that those blocks are built.

In this paper we propose a method to sample genotypes from large and complex pedigrees, and apply the proposed algorithm to estimate the genotype probabilities in the Labrador Retriever pedigree under study. We use the Metropolis-Hastings algorithm (Hastings, 1970; Metropolis et al., 1953) and sample genotypes (candidates) jointly from a proposal distribution that is close to the true posterior distribution of interest. When there are no loops, our proposal is the "true" posterior distribution and the Metropolis-Hastings algorithm reduces to direct sampling. When there are loops in the pedigree, we construct our proposal distribution by first sampling genotypes using iterative peeling. A variation of this approach consists in peeling the pedigree exactly up to the point where the complexity of loops makes it difficult to continue, and then using Metropolis-Hastings to sample from a "partially" peeled pedigree.

The paper is organized as follows. In Section 2, we review the method of peeling and show how it is used to sample genotypes. The approach we propose for sampling genotypes is discussed in Section 3.

In Section 4, we return to the case study, and apply our methods to the Labrador Retriever pedigree provided by "The Seeing Eye, Inc". We present the results of the analysis and briefly discuss the problem of assessing the performance of our method.

Finally, a brief conclusion and summary remarks are given in Section 5.

2 The peeling approach for sampling genotypes

Before describing iterative peeling we discuss the Elston-Stewart algorithm known as *peeling*.

Consider the simple pedigree shown in Figure 7.1. To introduce the principles involved in peeling we show how to sample genotypes from $f(\mathbf{g}|\mathbf{y})$ for a monogenic trait in pedigrees without loops. To obtain a random sam-

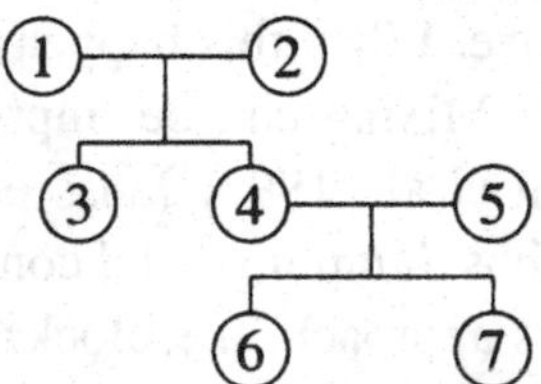

FIGURE 7.1. Simple two-generational pedigree.

ple from $f(\mathbf{g}|\mathbf{y})$, we can use a rejection sampler based on $f(\mathbf{g}|\mathbf{y})$, but this may be very inefficient.

Instead, we sample individuals sequentially. To obtain a sample from $f(g_1, g_2, g_3, g_4, g_5, g_6, g_7|\mathbf{y})$ in Figure 7.1, we first sample the genotype for individual 1 from $f(g_1|\mathbf{y})$. Next we sample g_2 from $f(g_2|g_1, \mathbf{y})$, g_3 from $f(g_3|g_1, g_2, \mathbf{y})$, and so on. To compute $f(g_1|\mathbf{y})$ we use peeling (Elston and Stewart, 1971; Cannings et al., 1978). The first step in computing $f(g_1|\mathbf{y})$ is to compute the likelihood.

The likelihood for the pedigree in Figure 7.1 can be written as

$$L \propto \sum_{g_1}\sum_{g_2}\cdots\sum_{g_7} h(g_1)h(g_2)h(g_1, g_2, g_3)h(g_1, g_2, g_4)h(g_5) \times h(g_4, g_5, g_6)h(g_4, g_5, g_7), \quad (7.2)$$

where

$$h(g_j) = P(g_j)f(y_j|g_j) \quad (7.3)$$

if j is a founder, that is if the parents of j are not in the pedigree. The penetrance function, $f(y_j|g_j)$, represents the probability that an individual with genotype g_j has phenotype y_j. The founder probability denoted $P(g_j)$, represents the prior probability that an individual has genotype g_j. If individual j is not a founder then

$$h(g_m, g_f, g_j) = P(g_j|g_m, g_f)f(y_j|g_j), \quad (7.4)$$

where g_m and g_f are the genotypes for the mother and father of individual j. The transition probability $P(g_j|g_m, g_f)$, is the probability that an individual has genotype g_j given parental genotypes g_m and g_f.

Suppose each g_j can take on one of three values (AA, Aa and aa). Then L as given by (7.2) is the sum of 3^7 terms. The number of computations is exponential in the number of genotypes in the expression.

Computing the likelihood as given by (7.2) is feasible only for small pedigrees. The Elston-Stewart algorithm provides an efficient reordering of the additions and multiplications in computing the likelihood. Thus, L in (7.2) is rearranged as

$$\sum_{g_1}\sum_{g_2} h(g_1)h(g_2)\sum_{g_3} h(g_1, g_2, g_3)\sum_{g_4} h(g_1, g_2, g_4) \times \sum_{g_5} h(g_5)\sum_{g_6} h(g_4, g_5, g_6)\sum_{g_7} h(g_4, g_5, g_7). \quad (7.5)$$

Note that (7.5) is identical in value to (7.2) but is computationally more efficient. For example, consider the summation over g_7. In (7.2) this summation is done over all combinations of values of g_1, g_2, g_3, g_4, g_5 and g_6.

However, the only function involving g_7, is $h(g_4, g_5, g_7)$, which depends only on two other individual genotypes (g_4 and g_5). In (7.5), the summation over g_7 is done only for all combinations of values of g_4 and g_5. An expression involving g_5 and g_4 is obtained after summing out g_7. These expressions are called *cutsets* and must be stored to be used in the final computation of the likelihood.

For example, after peeling g_7 we obtain a cutset of size 2, $c_7(g_4, g_5) = \sum_{g_7} h(g_4, g_5, g_7)$.

To compute L efficiently, the order of peeling is critical. For example, consider peeling g_1 as the first step, so the likelihood can be written as

$$L \propto \sum_{g_2}\sum_{g_3}\cdots\sum_{g_7} h(g_2)h(g_5)h(g_4, g_5, g_6)h(g_4, g_5, g_7)c_1(g_2, g_3, g_4),$$

where $c_1(g_2, g_3, g_4) = \sum_{g_1} h(g_1)h(g_1, g_2, g_3)h(g_1, g_2, g_4)$.

The result, $c_1(g_2, g_3, g_4)$, from peeling g_1 is a cutset of size three, and its computation involves summing over g_1 for all genotype combinations of g_2, g_3 and g_4. Computing $c_7(g_4, g_5)$ has lower storage and computational requirements than computing $c_1(g_2, g_3, g_4)$. The storage and computational requirements would be similar for peeling g_3 and g_6 in the first step. Peeling g_4 in the first step would be even more costly than peeling g_1, g_2 or g_5.

Thus, to evaluate the likelihood for this pedigree we first need to define the peeling order. The peeling order is determined by the greedy algorithm. The greedy algorithm picks the locally best choice at each step, without concern for the impact on future choices. Thus, the peeling algorithm is the following.

1. List all the individuals in the pedigree that need to be peeled.

2. For each individual determine the size of the resulting cutset after peeling the individual.

3. Peel the individual with the smallest cutset.

4. Repeat steps 2 and 3 until all the individuals are peeled.

In this case, the peeling order could be: 3, 1, 2, 7, 6, 4 and 5. Once all individuals have been peeled we sample individual's genotypes in the reverse order to which they were peeled (reverse peeling, Heath (1998)). In this example, after peeling individual 5 we compute the marginal probability for 5 as

$$f(g_5|\mathbf{y}) = \frac{P(g_5)f(y_5|g_5)c_5(g_5)}{L}.$$

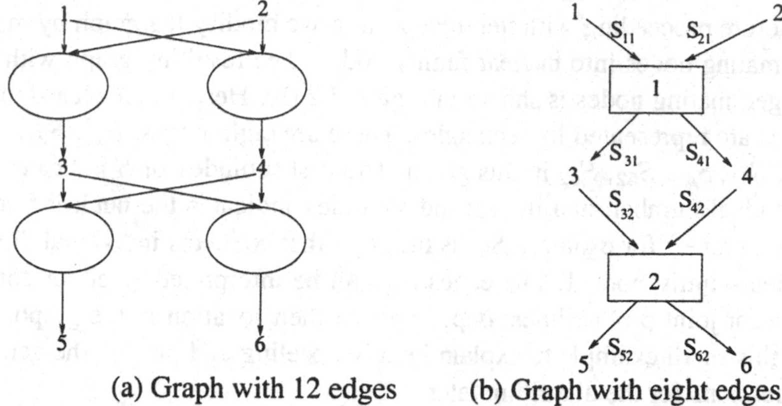

(a) Graph with 12 edges (b) Graph with eight edges

FIGURE 7.2. Graph representation of a two-generational pedigree with loops.

Once $f(g_5|\mathbf{y})$ has been obtained, we sample g_5 using the inverse cumulative function. Next, we compute

$$f(g_4|g_5, \mathbf{y}) = \frac{f(y_4|g_4)c_2(g_4)c_7(g_4, g_5)c_6(g_4, g_5)}{\sum_{g_4} f(y_4|g_4)c_2(g_4)c_7(g_4, g_5)c_6(g_4, g_5)},$$

and then we sample g_4 from $f(g_4|g_5, \mathbf{y})$. Repeatedly applying this procedure, we eventually generate a sample from the joint distribution of all genotypes for the entire pedigree. The sampling sequence in this case is: sample g_5 from $f(g_5|\mathbf{y})$, sample g_4 from $f(g_4|\mathbf{y}, g_5)$, and so on.

In pedigrees with complex loops, peeling methods as described above are not feasible. The reason is that the cutsets generated after peeling some individuals are larger when there are loops in the pedigree.

3 Iterative Peeling and Metropolis-Hastings algorithm to sample genotypes

Peeling methods cannot be applied when pedigrees are large and have complex loops. Iterative peeling (Van Arendonk et al., 1989; Janss et al., 1992; Wang et al., 1996), however can be used to get approximate results. To describe iterative peeling, it is convenient to present the pedigree as a directed graph (Figure 7.2 (a)).

Before peeling, the graph contains individual nodes and mating nodes. Each individual node is indicated by the individual identification number; they correspond to the penetrance functions, and in the case of founders, also include the founder probability function. Each mating node is indicated by an oval, which corresponds to the transition probability function. The edges in the graph connect the mating nodes with the parents and with the offspring.

Before proceeding with iterative peeling we modify the graph by merging mating nodes into nuclear family nodes. The resulting graph with the merged mating nodes is shown in Figure 7.2 (b). Here, the nuclear-family nodes are represented by rectangles. There are eight edges: S_{11}, S_{21}, S_{31}, S_{32}, S_{41}, S_{42}, S_{52}, S_{62} in this graph. The first subindex of S indicates the individual number, and the second subindex indicates the nuclear-family node number; for example S_{31} is the edge that connects individual 3 with nuclear-family node 1. The edges S_{ij} can be interpreted as either conditional or joint probabilities, depending on their location in the graph. We use this small example to explain iterative peeling and present the general expressions for the algorithm later.

Suppose we want to sample the genotype for individual 1 from $f(g_1|\mathbf{y})$. We first obtain an estimate for the edge probability S_{11}, connecting individual 1 to the rest of the pedigree through nuclear family 1. Once S_{11} is computed, the genotype probabilities can be obtained from the normalized values of $f(y_1|g_1)P(g_1)S_{11}$. Below we describe how to iteratively compute S_{11}.

We first initialize all the edge probabilities. In general, all edge probabilities are initialized to 1. For this example, however it is convenient to set S_{41} to be equal to the founder genotype probabilities. Once the edges are initialized we iteratively update edge probabilities using the phenotypes and the current values of the appropriate edges (explained below) of all the individuals in the corresponding nuclear family. Thus, we update S_{11} as

$$S_{11} = \sum_{g_2}\sum_{g_3}\sum_{g_4} f(y_2|g_2)P(g_2)f(y_3|g_3)P(g_3|g_1,g_2) \times f(y_4|g_4)P(g_4|g_1,g_2)S_{32}S_{42}.$$

At this stage, S_{11} is the conditional probability $f(y_2, y_3, y_4|g_1)$. Note that the edges that contributed to updating S_{11} are those that connect the members of nuclear family 1 to other nuclear families.

Similarly S_{21} is updated as

$$S_{21} = \sum_{g_1}\sum_{g_3}\sum_{g_4} f(y_1|g_1)P(g_1)f(y_3|g_3)P(g_3|g_1,g_2) \times f(y_4|g_4)P(g_4|g_1,g_2)S_{32}S_{42},$$

and is the conditional probability $f(y_1, y_3, y_4|g_2)$. Next, we update S_{31} as

$$S_{31} = \sum_{g_1}\sum_{g_2}\sum_{g_4} f(y_1|g_1)P(g_1)f(y_2|g_2)P(g_2)P(g_3|g_1,g_2) \times f(y_4|g_4)P(g_4|g_1,g_2)S_{42},$$

FIGURE 7.3. Cut and extended graphs

which is the joint probability $f(y_1, y_2, y_4, g_3)$. Next, we update S_{32} as,

$$S_{32} = \sum_{g_4}\sum_{g_5}\sum_{g_6} f(y_5|g_5)P(g_5|g_3, g_4)f(y_6|g_6)P(g_6|g_3, g_4)f(y_4|g_4)\underbrace{S_{41}}_{P(g_4)},$$

which is the conditional probability $f(y_4, y_5, y_6|g_3)$. Note that in these three cases, when we multiplied by an edge probability we used the initial values.

Next, we update S_{41} as

$$S_{41} = \sum_{g_1}\sum_{g_2}\sum_{g_3} f(y_1|g_1)P(g_1)f(y_2|g_2)P(g_2)f(y_3|g_3) \times$$
$$P(g_3|g_1, g_2)P(g_4|g_1, g_2)\underbrace{S_{32}}_{f(y_4,y_5,y_6|g_3)}.$$

In this case, the edge probability S_{32} was already updated once. Thus, the value of $S_{41} = f(y_1, y_2, y_3, y_{4^*}, y_5, y_6, g_4)$ is the joint probability of the genotype of individual 4 and of all the phenotypic values connected to 4 through nuclear family 1 in the cut-extended pedigree shown in Figure 7.3 (a).

Next, we update S_{42} as

$$S_{42} = \sum_{g_3}\sum_{g_4}\sum_{g_5} f(y_5|g_5)P(g_5|g_3, g_4)f(y_6|g_6)P(g_6|g_3, g_4)f(y_3|g_3) \times$$
$$P(g_3|g_1, g_2)\underbrace{S_{31}}_{f(y_1,y_2,y_4,g_3)}.$$

Again, in this case we use an edge probability that was already updated, and thus $S_{42} = f(y_1, y_2, y_3, y_{4^*}, y_5, y_6|g_3)$, which is the conditional probability of all the phenotypic values connected to 4 through nuclear family 2 in the cut-extended pedigree shown in Figure 7.3 (b), given the genotype of individual 4.

Each subsequent iteration results in further extensions to a cut pedigree. After a sufficient number of iterations we sample genotypes as follows from the iteratively peeled pedigree. First we sample the genotype of individual 1 from $f(g_1|\mathbf{y})$, which is computed using S_{11} as described above. Next, to sample the genotype of 2 we update S_{21} to reflect the sampled value for the genotype of 1 as

$$S_{21} = \sum_{g_3}\sum_{g_4} f(y_1|g_1)P(g_1)f(y_3|g_3)P(g_3|g_1,g_2)f(y_4|g_4)P(g_4|g_1,g_2)S_{32}S_{42},$$

where g_1 is the sampled value for the genotype of 1. Using this updated value for S_{21}, $f(g_2|\mathbf{y},g_1)$ is computed as

$$f(g_2|\mathbf{y},g_1) = \frac{f(y_2|g_2)P(g_2)S_{21}}{\sum_{g_2} f(y_2|g_2)P(g_2)S_{21}}.$$

This process is continued until all individuals are sampled. We propose to use these sampled genotypes as the proposal distribution in a Metropolis-Hastings step to accept or reject the candidate draws.

We now provide the general expressions for updating edge probabilities in iterative peeling. Let S_{js} be an edge between individual j and nuclear-family node s. If j is a parent in the nuclear family, S_{js} is computed iteratively as,

$$S_{js} = \sum_{g_p} R_{sp} \prod_{k\in C_s} [\sum_{g_k} \Pr(g_k|g_j,g_p)R_{sk}], \tag{7.6}$$

where p is the other parent in the nuclear family, C_s is the set of children in nuclear family s,

$$R_{sl} = f(y_l|g_l)P(g_l)\Big(\prod_{\substack{e\in E_l\\ e\neq s}} S_{le}\Big) \tag{7.7}$$

for $l = k,p$, E_l is the set of edges for individual l, $f(y_l|g_l)$ is the penetrance function and $P(g_l)$ is the founder probability, if individual l is not a founder then $P(g_l) = 1$. If j is a child in the nuclear family, S_{js} is computed iteratively as

$$S_{js} = \Pr(g_j|g_m,g_f)\sum_{g_m,g_f} R_{sm}R_{sf} \prod_{\substack{k\in C_s\\ k\neq j}} [\sum_{g_k} \Pr(g_k|g_m,g_f)R_{sk}], \tag{7.8}$$

where m and f are the parents in the nuclear-family node.

All edge probabilities are iteratively updated using (7.6) and (7.8). After a sufficient number of iterations, we sample genotypes for all individuals in the pedigree. We start from an arbitrary individual, and sample its genotype using the marginal probability function

$$f(g_j|y_j) = \frac{f(y_j|g_j)P(g_j)\prod_s S_{js}}{\sum_{g_j} f(y_j|g_j)P(g_j)\prod_s S_{js}}, \tag{7.9}$$

where the product of S_{js} is over all edges for j. Then we sample a neighbor conditional on the sampled genotypes as follows. A neighbor is defined as any individual who is also a member of those nuclear-family nodes to which the sampled individual belongs to. First, all edges of the individual to be sampled are updated to reflect the already sampled genotypes. To update edges, we use (7.6) and (7.8), but the summations are only over the unsampled genotypes. Now to sample the genotype conditional on the already sampled genotypes we use (7.9) with the edges that were updated for the sampled genotypes.

3.1 Metropolis-Hastings algorithm

The Metropolis-Hastings acceptance probability is

$$\eta = min\left(1, \frac{\pi(g_c)q(g_{prev}|g_c)}{\pi(g_{prev})q(g_c|g_{prev})}\right), \tag{7.10}$$

where π is the target distribution and q is the proposal distribution, g_{prev} is the accepted draw from the previous round and g_c is the sampled candidate from the present round. The chain moves from g_{prev} to g_c with probability η, and it stays at g_{prev} with probability $1-\eta$; in this case the draw is rejected. The key step is to choose a good proposal distribution so that the rejection rate is minimized. We consider the special case of *independence sampling*: instead of $q(g_c|g_{prev})$ and $q(g_{prev}|g_c)$, we use $q(g_{prev})$ and $q(g_c)$. Thus

$$\eta = min\left(1, \frac{\pi(g_c)q(g_{prev})}{\pi(g_{prev})q(g_c)}\right). \tag{7.11}$$

We sample genotypes from the iteratively peeled pedigree and use them in the Metropolis-Hastings step to be rejected or accepted. We use the following expression to obtain $\pi(.)$ on the true pedigree,

$$\pi(\mathbf{g}) \propto \prod_{j=1}^{n_1} P(g_j) \prod_{j=n_1+1}^{n} P(g_j|g_{f_j}, g_{m_j}), \tag{7.12}$$

where g_{f_j}, g_{m_j} are the genotypes of the parents of individual j and n_1 is the number of founders. In this example, $\pi(\mathbf{g})$ is

$$\pi(\mathbf{g}) \propto P(g_1)P(g_2)P(g_3|g_1,g_2)P(g_4|g_1,g_2)P(g_5|g_3,g_4)P(g_6|g_3,g_4).$$

To compute $q(.)$ we multiply the probabilities that were used in the sampling process described above. For example, for this pedigree $q(\mathbf{g})$ is

$$q(\mathbf{g}) = f(g_1|\mathbf{y})f(g_2|\mathbf{y},g_1)\cdots f(g_6|g_1,g_2,g_3,g_4,g_5,\mathbf{y}).$$

3.2 *Improving efficiency of sampler*

The efficiency of the sampler can be improved by combining exact peeling with iterative peeling. Exact peeling (Section 2) is used until the size of cutsets gets large enough to make computations infeasible. Then, iterative peeling is used, which as discussed above is equivalent to cutting and extending the remaining loops.

After this combination of peeling and iterative peeling is done, the graph contains two types of edges: edges between individuals and nuclear-family and between individuals and cutset nodes. Both types of edges need to be updated. The edges between individuals and nuclear-family nodes are updated as described in Section 3. The edges between individuals and cutset nodes are updated as

$$S_{js} = \sum c_s(g_{s_1},\cdots,g_{s_n}) \prod_{\substack{l \in c_s \\ l \neq j}} R_{sl}, \tag{7.13}$$

where the summation is over the genotypes of the individuals included in cutset s, except for the genotype of individual j, and $s_1,\cdots,s_n$ are the individuals in cutset s.

After iteratively updating all the edge probabilities, we sample genotypes for all individuals in the pedigree. First we sample genotypes for the individuals that were not peeled out. We start from an arbitrary "unpeeled" individual, and we sample its genotype using the marginal probability function $f(g_j|\mathbf{y})$. Then we sample all its unsampled neighbors using conditional probabilities (as we did in the small example for individual 2). Before obtaining a new sample, edges are updated to reflect the already sampled genotypes. A neighbor is defined as any individual who is also a member of those nuclear-family or cutset nodes, to which the sampled individual belongs to. Once all remaining individuals are sampled, we sample genotypes of the "peeled" individuals in the inverse order of peeling (see Section 2).

4 Estimation of genotype probabilities in the Labrador Retriever pedigree

The genotype probabilities were obtained by direct calculation and by sampling. Direct calculation includes exact peeling (only possible for small pedigrees) and approximation (iterative peeling without the Metropolis-Hastings step). The sampling method is the method proposed in this paper.

The "direct calculation" approaches will be compared to the proposed method. The proposed sampling method can be used to estimate the genotype probabilities by sampling from a proposal distribution generated by iterative peeling for the entire pedigree or by peeling exactly up to a certain point and then perform iterative peeling.

4.1 Application of the methods in a real pedigree

A real dog pedigree was used to test the performance of the sampling algorithm that we just described. The trait of interest in this pedigree is a disease called progressive retinal atrophy (PRA). This disease is transmitted by a recessive allele and the dog is affected when it has the recessive homozygous genotype (*aa*). The pedigree consists of 3,052 dogs (Labrador Retrievers) from "The Seeing Eye, Inc", and 33 of them are known to have the disease. That is for these 33 dogs we know the genotype and phenotype. For the rest of the dogs we are interested in obtaining estimates of the genotype probabilities to determine which dogs are at highest risk of transmitting the PRA gene to their offspring and which dogs are at lower risk of both transmitting the gene and of being PRA affected.

Exact peeling methods cannot be used in this pedigree because there are 679 loops that need to be cut. Thus, we can approximate the genotype probabilities by iterative peeling (without sampling) or we can sample genotypes and then compute the genotype probabilities by using the proposed method. We used two different variations of the proposed method to compute genotype probabilities. In the first variation we peeled exactly the pedigree up to cutsets of size 4 (C_4). We then iteratively peeled the remaining core pedigree and sampled genotypes according to Metropolis-Hastings algorithm as described above. In the second variation, we exactly peeled the pedigree up to cutsets of size 5 (C_5). The chain length in both cases (C_4 and C_5) was 20,000 (we discarded the first 10,000). To compare the results we computed the absolute difference between the genotype probabilities for each animal under the two variations of the proposed method. Those results are shown in Table 7.1. We observe that there are no large differences among variations of the proposed method. We, also

	Range	Mean	Std. Dev
$P(AA)$	0 to 0.1201	1.6×10^{-2}	1.7×10^{-2}
$P(Aa)$	0 to 0.1431	2.9×10^{-2}	2.3×10^{-2}
$P(aa)$	0 to 0.1431	2.7×10^{-2}	2.1×10^{-2}

TABLE 7.1. Range, mean and standard deviations of the absolute differences between probabilities for the two variations of the proposed method.

computed the absolute difference between the genotype probabilities estimated by applying the two versions of the proposed method (C_4 and C_5) and the approximate method (iterative peeling without sampling). These results are shown in Table 7.2. The rejection rate for the C_4 and C_5 ap-

	Range	Mean	Std. Dev
C_4			
$P(AA)$	0 to 0.074686	5.6×10^{-3}	7.0×10^{-3}
$P(Aa)$	0 to 0.115085	1.4×10^{-2}	1.6×10^{-2}
$P(aa)$	0 to 0.129718	1.2×10^{-2}	1.4×10^{-2}
C_5			
$P(AA)$	0 to 0.10326	1.6×10^{-2}	1.7×10^{-2}
$P(Aa)$	0 to 0.17068	3.3×10^{-2}	2.6×10^{-2}
$P(aa)$	0 to 0.16536	3.0×10^{-2}	2.5×10^{-2}

TABLE 7.2. Range, mean and standard deviations of the absolute differences between probabilities computed by the two variations of the proposed method and the approximate method.

proaches were 53% and 83%, respectively. The computation times were about the same for both versions of the proposed method.

We also tried running the program for cutset size=0, that is using the proposal generated by performing iterative peeling for the entire pedigree (with no exact peeling). After three days the program was at iteration 568 and we decided to stop the execution.

4.2 *Results: proportion of dogs carrying the PRA allele*

The estimated numbers of affected and carrier animals are presented in Table 7.3. The numbers differ a little from one variation of the method to the other. These small differences may be due to, for example, those animals for which the probability of genotype aa is estimated to be very high. In one version of the approach some of those animals may be included in the $0.9 < P(aa) < 1$ category, while in the other they may be counted among the $P(aa) = 1$ category. The 33 dogs known to have the disease were always sampled as affected, thus they are all in the category of $P(aa) = 1$.

Genotype probability	No. of animals C_4	C_5
$0.5 \leq P(aa) < 0.6$	519	465
$0.6 \leq P(aa) < 0.7$	534	580
$0.7 \leq P(aa) < 0.8$	298	299
$0.8 \leq P(aa) < 0.9$	108	153
$0.9 \leq P(aa) < 1$	157	140
$P(aa) = 1$	41	56
$0.5 \leq P(Aa) < 0.6$	515	460
$0.6 \leq P(Aa) < 0.7$	201	195
$0.7 \leq P(Aa) < 0.8$	64	64

TABLE 7.3. Estimated number of affected and carrier animals.

4.3 *Assessing the performance of the algorithm*

To assess the performance of the algorithm we used a small pedigree with loops. We considered the inheritance at a single biallelic disease locus. This small pedigree consists of 77 individuals, and two of them are affected. There are four generations in this pedigree and large families (more than five offspring per family). This pedigree also has a few loops. We sampled genotypes for all the individuals and computed the genotype probabilities.

In this small pedigree we can perform exact calculations by exact peeling. These exact calculations were verified with the results obtained from a software package for pedigree analyses (Hasstedt, 1994, PAP). The probabilities obtained by PAP can be thought of as true results. The range,

	Range	Mean	Std. Dev
$P(AA)$	0 to 4.8×10^{-5}	2.5×10^{-5}	1.6×10^{-5}
$P(Aa)$	0 to 4.9×10^{-5}	2.3×10^{-5}	1.4×10^{-5}
$P(aa)$	0 to 4.9×10^{-5}	2.2×10^{-5}	1.4×10^{-5}

TABLE 7.4. Range, mean and standard deviations of the absolute differences between probabilities computed by PAP and exact peeling.

means and standard deviations of the absolute differences between genotype probabilities from our algorithm and PAP for the 77 individuals are shown in Table 7.4. In Table 7.4 we observe that the genotype probabilities computed by the two methods do not differ. The small differences are due to rounding errors.

We then compare the results from PAP with estimates from the proposed sampling method where no exact peeling was done (Table 7.5) and also with the estimates from the proposed method where exact peeling was done until the cutset size was 4 and then iterative peeling was done for the rest of the pedigree (Table 7.6). The length of the chain was 10,000 iterations and we discarded the first half, thus the genotype probabilities are obtained based on the second half of the chain. Tables 7.5 and 7.6 show the ranges, means and standard deviations for the absolute differences between PAP and the proposed method with no partial peeling and with partial peeling, respectively. We observe that the results are similar, indicating that for this small pedigree there is no advantage in partially peeling the pedigree prior to sampling.

	Range	Mean	Std. Dev
$P(AA)$	0 to 2.1×10^{-2}	5.6×10^{-3}	5.4×10^{-3}
$P(Aa)$	0 to 2.5×10^{-2}	6.9×10^{-3}	5.8×10^{-3}
$P(aa)$	0 to 2.2×10^{-2}	6.3×10^{-3}	5.3×10^{-3}

TABLE 7.5. Range, mean and standard deviations of the absolute differences between probabilities computed by PAP and proposed method with no exact peeling.

The results presented in Tables 7.5 and 7.6 show that the proposed method yields accurate estimates. If we increase the number of draws, the estimates improve even more.

	Range	Mean	Std. Dev
$P(AA)$	0 to 2.0×10^{-2}	3.9×10^{-3}	3.9×10^{-3}
$P(Aa)$	0 to 1.9×10^{-2}	5.6×10^{-3}	3.9×10^{-3}
$P(aa)$	0 to 1.6×10^{-2}	5.6×10^{-3}	3.8×10^{-3}

TABLE 7.6. Range, mean and standard deviations of the absolute differences between probabilities computed by PAP and proposed method with exact peeling (cutset size=4).

In Table 7.7 we present the absolute difference between the true probabilities and the approximate iterative peeling method without sampling (approximate method) for the three possible genotypes for the 77 individuals. Comparing results from Tables 7.5 and 7.6 with those in Table 7.7

	Range	Mean	Std. Dev
$P(AA)$	0 to 3.2×10^{-2}	1.0×10^{-2}	6.7×10^{-3}
$P(Aa)$	0 to 5.1×10^{-2}	2.3×10^{-2}	1.7×10^{-3}
$P(aa)$	0 to 4.2×10^{-2}	1.4×10^{-2}	1.3×10^{-3}

TABLE 7.7. Range, mean and standard deviations of the absolute differences between probabilities computed by PAP and approximate method.

reveal that our method yields estimates closer to the truth than the approximation.

The rejection rates were 29% and 5% for the proposed method with no exact peeling and with exact peeling up to cutset size=4, respectively.

In general, it seems that it is more efficient to peel exactly the pedigree as much as possible and then perform iterative peeling to the remaining core pedigree to obtain the proposal distribution to be used in the Metropolis-Hastings step. We cannot peel too deeply because then the cutsets become large, increasing the expense in computation time and memory. The optimal size of cutsets for efficient computation is hard to determine. We are currently investigating whether it is possible to decide, for a given pedigree, what is the best strategy in terms of cutset size.

5 Summary and Conclusions

Estimating the probability of genotypes at biallelic loci is non-trivial when pedigrees are large and contain loops. In this case, scalar Gibbs sampling approach cannot be used, as the chains are reducible in practice. We propose a more general Metropolis-Hastings approach to sampling genotypes jointly from complex pedigrees, in which candidate draws are obtained from modified pedigrees. These modified pedigrees are obtained by applying extensions of traditional peeling methods, and are used as candidate draws in the Metropolis-Hastings step. The resulting Markov chains satisfy the assumptions that are required for good performance of MCMC methods.

The method for sampling genotypes was developed to address the problem of estimating genotype probabilities of the animals transmitting progressive retinal atrophy (PRA) in Labrador Retrievers. The pedigree data of interest, collected by veterinarians at "The Seeing Eye, Inc." included over 3,000 animals, and had over 600 closed loops created by inbreeding and multiple matings. A summary of the results of this pedigree analysis is given in Table (7.3).

Acknowledgments

The authors are grateful to Dr. Eldin A. Leighton, Director of Canine Genetics of "The Seeing Eye, Inc.", who provided the pedigree data used in this analysis.

References

Cannings, C., Thompson, E. A., and Skolnick, E. (1978). Probability functions on complex pedigrees. *Adv. Appl. Prod.*, 10:26–61.

Elston, R. C. (1987). Human quantitative genetics. In G., W. B. E. E. G. M. N., editor, *Proc. 2nd. Int. Conf. Quant. Genet.*, pages 281–282, Sinauer, Sunderland.

Elston, R. C. and Rao, D. C. (1978). Statistical modeling and analysis in human genetics. *Annu Rev Biophys Bioeng*, (7):253–286.

Elston, R. C. and Stewart, J. (1971). A general model for the genetic analysis of pedigree data. *Hum Hered*, 21:523–542.

Hasstedt, S. J. (1994). *Pedigree Analysis Package*. Department of Human Genetics, University of Utah, Salt Lake City, revision 4.0 edition.

Hastings, W. K. (1970). Monte Carlo sampling methods using Markov chains and their applications. *Biometrika*, 57:97–109.

Heath, S. C. (1998). Generating consistent genotypic configurations for multi-allelic loci and large complex pedigrees. *Human Heredity*, 48:1–11.

Janss, L. L. G., der Werf J. H. J., V., and van Arendonk J. A. M. (1992). Detection of a major gene using segregation analysis in data from generations. In *Proc. Eur. Assoc. Anim. Prod.*

Janss, L. L. G., Thompson, R., and Van Arendonk, J. A. M. (1995). Application of Gibbs sampling for inference in a mixed major gene-polygenic inheritance model in animal populations. *Theor. Appl. Genet.*, 91:1137–1147.

Jensen, C. S., Kong, A., and Kjærulff, U. (1995). Blocking Gibbs Sampling in very large probabilistic expert systems. *International-Journal of Human Computer Studies*, 42:647–66.

Jensen, C. S. and Sheehan, N. (1998). Problems with determination of non-communicating classes for Monte Carlo Markov chain applications in pedigree analysis. *Biometrics*, 54:416–425.

Lange, K. and Boehnke, M. (1983). Extensions to pedigree analysis. V. Optimal calculation of Mendelian likelihoods. *Hum. Hered.*, 33:291–301.

Lange, K. and Elston, R. C. (1975). Extensions to pedigree analysis. i. Likelihood calculations for simple and complex pedigrees. *Hum. Hered.*, 25:95–105.

Metropolis, N., Rosenbluth, A. W., Rosenbluth, M. N., Teller, A. H., and Teller, A. H. (1953). Equation of state calculation by fast computing machines. *Journal of Chemical Physics*, 21:1087–1092.

Stricker, C., Fernando, R. L., and Elston, R. C. (1995). An algorithm to approximate the likelihood for pedigree data with loops by cutting. *Theor. Appl. Genet.*, 91:1054–1063.

Thomas, A. (1986a). Approximate computations of probability functions for pedigree analysis. *IMA J Math Appl Med Biol*, 3:157–166.

Thomas, A. (1986b). Optimal computations of probability functions for pedigree analysis. *IMJ J Math Appl Med Biol*, 3:167–178.

Van Arendonk, J. A. M., Smith, C., and Kennedy, B. W. (1989). Method to estimate genotype probabilities at individual loci in farm livestock. *Theor. Appl. Genet.*, 78:735–740.

Wang, T., Fernando, R. L., Stricker, C., and Elston, R. C. (1996). An approximation to the likelihood for a pedigree with loops. *Theor. Appl. Genet.*, 93:1299–1309.

Identifying Carriers of a Genetic Modifier Using Nonparametric Bayesian Methods

Peter D. Hoff
Richard B. Halberg
Alexandra Shedlovsky
William F. Dove
Michael A. Newton

ABSTRACT Animals in a certain population of mice each carry a mutant allele called $\star$ with probability one-half. This population is bred to a strain of mice which carry the *Min* allele of the *APC* gene, an allele which results in the development of intestinal tumors. Offspring from this cross are genotyped for *Min* and tumor counts are recorded. It is assumed that offspring carrying only *Min* have tumor counts distributed according to P_1, while offspring having both *Min* and the $\star$ allele have tumor counts distributed according to P_2, a probability measure assumed to be stochastically smaller than P_1. Presence of the *Min* allele is observable, but presence of the $\star$ allele is not. Given the tumor count data and assuming the stochastic ordering constraint, our goal is to estimate P_1, P_2 and the the unobserved genotype information. This is done by putting a nonparametric prior on the space of all pairs of stochastically ordered tumor count distributions, and computing posterior quantities of interest using MCMC.

1 Introduction

People with familial adenomatous polyposis (FAP) develop hundreds to thousands of benign tumors of the colon, which if untreated eventually progress to become carcinomas. The disease results from an inherited mutation in the adenomatous polyposis coli (*APC*) gene. The *Min* mutation in the mouse homologue of *APC* results in a phenotype very similar to human FAP. Mice with the *Min* mutation thus provide a model for studying this type of inherited colon cancer (Dietrich et al., 1993).

In a mutagenesis experiment, a mouse is obtained which shows signs of carrying a mutant allele at a modifier gene, suppressing the tumor-causing effects of *Min*. In order to genetically map the location of the modifier gene, it is necessary to breed and identify a group of animals carrying the modifier allele. Although inheritance of the modifier is not directly observable, animals resulting from a breeding experiment carry the modifier with probabilities determined by the rules of Mendelian inheritance. Conditional upon the unobserved pattern of inheritance, each animal is modeled as having a tumor count sampled from either a carrier or a non-carrier probability distribution. Our goal is to estimate the two probability distributions and identify likely carriers and non-carriers of the modifier, assuming only that the tumor count distributions are stochastically ordered.

We take a nonparametric Bayesian approach, putting a prior on the space of pairs of stochastically ordered distributions. Such a prior can be constructed indirectly by putting a Dirichlet prior on the set of bivariate distributions of latent observations, members of this set having support only on ordered pairs of points. The marginals of such distributions will follow the stochastic ordering, and thus the Dirichlet prior on distributions of latent observations induces a prior on pairs of stochastically ordered distributions. This technique of modeling a collection of constrained distributions via an unconstrained latent distribution is discussed by Hoff (2000) in the context of maximum likelihood estimation.

Although construction of our prior is straightforward, computation of posterior quantities is quite difficult. We construct a Markov chain to generate approximate samples from the posterior. In order to achieve sufficient mixing in our sequence of posterior samples, our chain uses a combination of Gibbs and Hastings updates, based on full and partial conditioning (Besag, Green, Higdon and Mengersen, 1993).

2 Breeding Scheme

A kindred founder mouse is suspected of carrying an allele, referred to hereafter as $\star$, which suppresses the tumor-causing effects of *Min*. This kindred founder is bred to the *BTBR* strain of mice to produce a new population, members of which carry the $\star$ allele independently with probability one-half. Animals in this population are referred to as subkindred founders, for which presence or absence of $\star$ is unobserved.

Subkindred founders are bred to the *B6 Min/+* strain of mice, members of which carry one copy of the *Min* allele, causing intestinal tumor growth. From this cross, the resulting offspring carrying *Min* are identified by genotyping and their tumor counts are recorded. This group of mice is referred to as the NF population. Note that if a subkindred founder carries the $\star$ allele, then so will roughly half of its NF offspring.

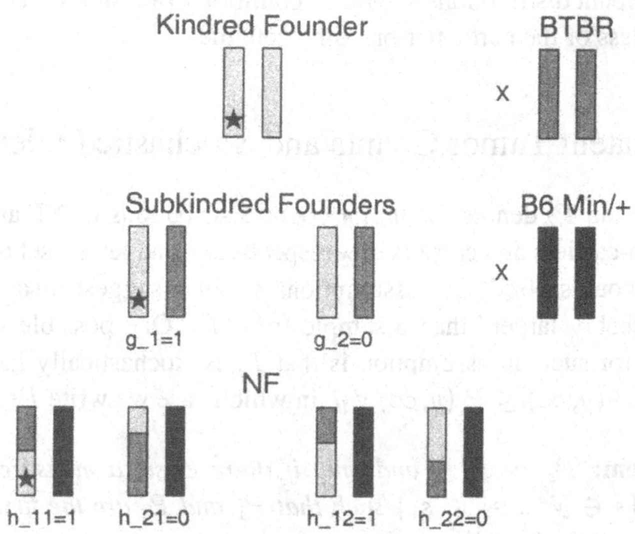

FIGURE 8.1. Basic Breeding Scheme

This breeding scheme and model of inheritance are outlined in Figure 8.1. We let g_j be the random variable indicating presence or absence of the $\star$ allele in the jth subkindred founder, $j = 1, \ldots, m$, and note the Mendelian model of inheritance implies $g_1, \ldots, g_m$ are i.i.d. Bernoulli(1/2) random variables. One goal of this paper is to estimate the g_j's from the NF tumor count data. These estimates of carrier status will be used in future work to genetically map the location of the modifier gene.

Each NF animal inherits one set of chromosomes from its subkindred parent and one set from its *B6 Min/+* parent. Because of the possibility of chromosomal crossing-over, each chromosome inherited by an NF animal from its subkindred parent may be a mixture of chromosomal material from the kindred founder and the *BTBR* strain. In the region of the genome where $\star$ resides, the probability that a particular NF mouse has chromosomal material from the kindred founder is one-half. We denote the indicator of this event for the ith mouse in subkindred j as $h_{(i,j)}$. The $h_{(i,j)}$'s are i.i.d. Bernoulli(1/2) random variables, and are independent of the g_j's. The indicator of the event that mouse (i, j) has the $\star$ allele can be written as $x_{(i,j)} = g_j h_{(i,j)}$.

We note that some of the data analyzed in this paper were generated using slightly different breeding schemes. However, the basic structure of the above model is applicable to all of them, and the carrier and non-carrier tumor count distributions should be common to all subkindred populations, regardless of the particular breeding scheme.

3 Latent Tumor Counts and Stochastic Ordering

Let P_1 and P_2 denote the tumor count distributions of NF animals which are non-carriers and carriers of $\star$ respectively, and let the set of all possible tumor counts be $\mathcal{Y}$. Our assumptions about $\star$ suggest a sample from P_1 is "probably larger" than a sample from P_2. One possible mathematical model for such an assumption is that P_1 is stochastically larger than P_2, that is $P_2(y,\infty) \leq P_1(y,\infty)\ \forall\, y$, in which case we write $P_2 \preceq P_1$.

Theorem: $P_2 \preceq P_1$ *if and only if there exists a measure* P *on* $\mathcal{S} = \{s \in \mathcal{Y}^2 : s_2 \leq s_1\}$ *such that* P_1 *and* P_2 *are the first and second marginals of* P.

The above result can be proven directly as by Lehmann (1986, Section 3.3), or can be seen as an application of a Choquet-type theorem, as described by Hoff (2000). Using this parametrization, an observation y distributed according to P_k, $k = 1, 2$ can be modeled as follows:

- Sample $s \sim P$;
- observe $y = s_k$.

We can think of s as being partially observed latent data, and y as the observed data. Estimating P_1 and P_2 subject to the stochastic ordering constraint can be done via unconstrained estimation of the measure P. In this way, a constrained estimation problem can be rewritten as an unconstrained missing-data problem, which is often easier to solve.

This parametrization provides a natural interpretation of the stochastic ordering constraint: We assume the tumor count y of each animal in our experiment is a deterministic function of $x \in \{0, 1\}$, the indicator of the presence of $\star$, and other unrecorded information $\omega \in \Omega$, so $y = y(\omega, x)$. If we assume the presence of $\star$ reduces tumor count, then it is natural to suppose $y(\omega, 1) \leq y(\omega, 0)\ \forall\, \omega$, i.e. all else being equal, the presence of $\star$ will not increase tumor count. Now define $s(\omega) = \{s_1(\omega), s_2(\omega)\} = \{y(\omega, 0), y(\omega, 1)\}$ as the vector of latent tumor counts. Any probability measure on Ω induces a canonical measure P on s so that $s_2 \leq s_1$ a.s. P. Furthermore, the marginals of P will satisfy the ordering $P_2 \preceq P_1$.

4 A Hierarchical Model

Our goal is to estimate P_1, P_2, and the missing subkindred genotype information $g_1, \ldots, g_m$ from the observed tumor count data. A nonparametric Bayesian approach involves a prior for (P_1, P_2) having support on all pairs of stochastically ordered measures on $\mathcal{Y}$. Such a prior is induced by the construction of a prior for the latent tumor count distribution P: If the support of the prior for P includes all possible distributions of ordered latent tumor counts in $\mathcal{Y}^2$, then by virtue of the theorem, the induced prior on the marginals has support on all pairs (P_1, P_2) such that $P_2 \preceq P_1$.

In this paper, our prior for P is based upon a simple parametric family of probability measures for latent tumor counts. Our uncertainty about the adequacy of the parametric family is quantified by assuming P is a sample from a Dirichlet process, centered around a base measure which is a member of the parametric family. A parametric prior on the base measure results in a nonparametric hierarchical prior for P.

4.1 A Parametric Model For Latent Tumor Counts

Suppose a set of cells in an organism have a certain probability of developing into tumors independently of one another. A model for total tumor count would then be a binomial distribution. Since the probability of tumorigenesis is typically quite small, and the number of cells in question is quite large, the binomial model of tumor counts can be well approximated by a Poisson model. Now suppose we are looking at a population of tumor counts, obtained from a population of organisms, each of whom have potentially different rates of tumorigenesis. Assuming a gamma prior for the population of rates, the resulting distribution of tumor counts follows a negative binomial model, a two parameter family of distributions with support on the nonnegative integers, with a density given by

$$p_{negbin}(s|\theta,\gamma) = \frac{\Gamma(s+\gamma)}{\Gamma(s+1)\Gamma(\gamma)}\left(\frac{\gamma}{\gamma+\theta}\right)^{\gamma}\left(\frac{\theta}{\gamma+\theta}\right)^{s}.$$

With this parametrization, $\mathrm{E}(s|\theta,\gamma) = \theta$, and $\mathrm{Var}(s|\theta,\gamma) = \theta(1+\theta/\gamma)$. Modeling tumor counts using the negative binomial distribution has been discussed before, for example in Drinkwater and Klotz (1981).

Our parametric model for latent tumor counts is as follows: We assume the tumor count s_1 of each non-carrier of $\star$ follows a negative binomial(θ,γ) distribution. We model tumor suppression by assuming each tumor that would have developed without $\star$ develops with probability p in the presence of $\star$, independently of the other tumors. This implies the conditional distribution of the suppressed tumor count s_2 given s_1 is binomial(s_1, p), and so the resulting joint distribution of (s_1, s_2) has support on $s_2 \leq s_1$. It is interesting to note that, unconditional on s_1, s_2 is distributed according to a negative binomial$(p\theta, \gamma)$ distribution, and so p can be interpreted as the multiplicative effect of $\star$ on mean tumor count.

4.2 *Nonparametric Extension*

Our knowledge of tumorigenesis suggests the above model is reasonable, but we would like to relax the strict parametric assumptions. This can be done by using the Dirichlet prior: A Dirichlet prior $\mathcal{D}(\alpha P_0)$ is a probability measure on a space of distributions parametrized by a positive weight parameter α and a base measure P_0. A probability measure P sampled from $\mathcal{D}(\alpha P_0)$ is "centered" around P_0 in the sense that if $x_1, \ldots, x_n|P$ are i.i.d. observations from P, then marginally $\mathrm{E}[\mathrm{E}_P(f(x_i))] = \mathrm{E}_{P_0}[f(x_i)]$. However, such observations are marginally correlated, and in particular

$$\mathrm{E}\left(\sum (x_i - \bar{x})^2/(n-1)\right) = \sigma_0^2 \frac{\alpha}{\alpha+1},$$

where σ_0^2 is the variance of a single observation under the base measure P_0. In fact, as $\alpha \to 0$, P converges to a point-mass measure with support on a random draw from P_0 (Sethuraman and Tiwari, 1981). See Ferguson (1973) or Blackwell and MacQueen (1973) for a more detailed account of the Dirichlet prior.

Our nonparametric model for the latent tumor count distribution P is a Dirichlet prior with a fixed α parameter and a random base measure. Our uncertainty about the base measure is quantified by a parametric prior π on a family of base measures, parametrized by

- the expectation of the non-suppressed tumor counts $\theta = E(s_1)$;
- the multiplicative effect p of $\star$, so that $p\theta = E(s_2)$;
- the expected sample variance of the non-suppressed tumor counts $\sigma^2 = E\left(\sum (s_{(i,j),1} - \bar{s}_1)^2/(n-1)\right)$.

More specifically, for a given value of $\phi = (\theta, p, \sigma^2)$, the base measure P_ϕ is given by

- $\gamma = \theta^2/(\sigma_0^2 - \theta)$, where $\sigma_0^2 = \sigma^2(\alpha+1)/\alpha$;
- $P_\phi(s_1) = p_{\text{negbin}}(s_1|\theta, \gamma)$;
- $P_\phi(s_2|s_1) = p_{\text{bin}}(s_2|s_1, p)$.

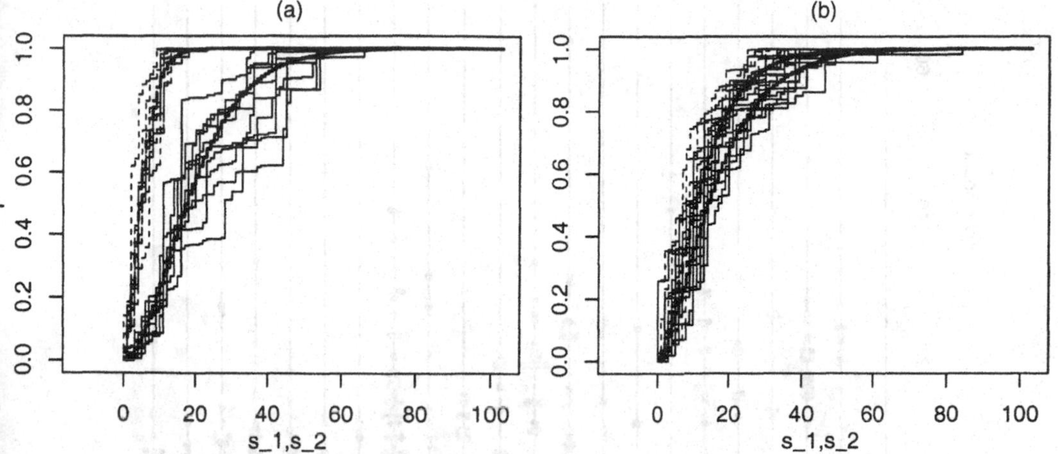

FIGURE 8.2. Marginal samples from $\mathcal{D}(\alpha P_\phi)$, with $\alpha = 10$ and (a) ϕ=(20,150,.25), (b) ϕ=(17,150,.75). Thick lines are marginals of P_ϕ, and solid and dashed lines represent the non-suppressed and suppressed groups respectively.

Given a prior π on ϕ, the complete hierarchical model is as follows:

- Hyperparameter: $\phi \sim \pi(\phi)$;
- Latent tumor count distribution: $P|\phi \sim \mathcal{D}(\alpha P_\phi)$;
- Latent observations: $s_{(1,1)}, \ldots, s_{(n_m,m)}|P \sim$ i.i.d. P;
- Genotype Information: $h_{(1,1)}, \ldots, h_{(n_m,m)}, g_1, \ldots, g_m \sim$ i.i.d. Bernoulli(1/2);
- Observed data: $y_{(i,j)} = \begin{cases} s_{(i,j),1} & \text{if } h_{(i,j)}g_j = 0; \\ s_{(i,j),2} & \text{if } h_{(i,j)}g_j = 1. \end{cases}$

Animals with the *Min* allele without $\star$ have been well studied, populations of such mice having average tumor counts of roughly 20 and a population variance of about 150. We therefore use a gamma(40,0.5) prior for θ and gamma(150,1) prior for σ^2 to reflect our uncertainty about these parameters. The effect of $\star$ is not known; this uncertainty is quantified by a uniform prior for p on the interval $(0, 1)$.

The α parameter determines, among other things, how close the tumor count measure P is to the base measure P_ϕ for a given ϕ. For each of two different P_ϕ we have drawn 10 samples from a $\mathcal{D}(\alpha P_\phi)$ distribution with $\alpha = 10$ and plotted the resulting marginals in the two panels of Figure 8.2.

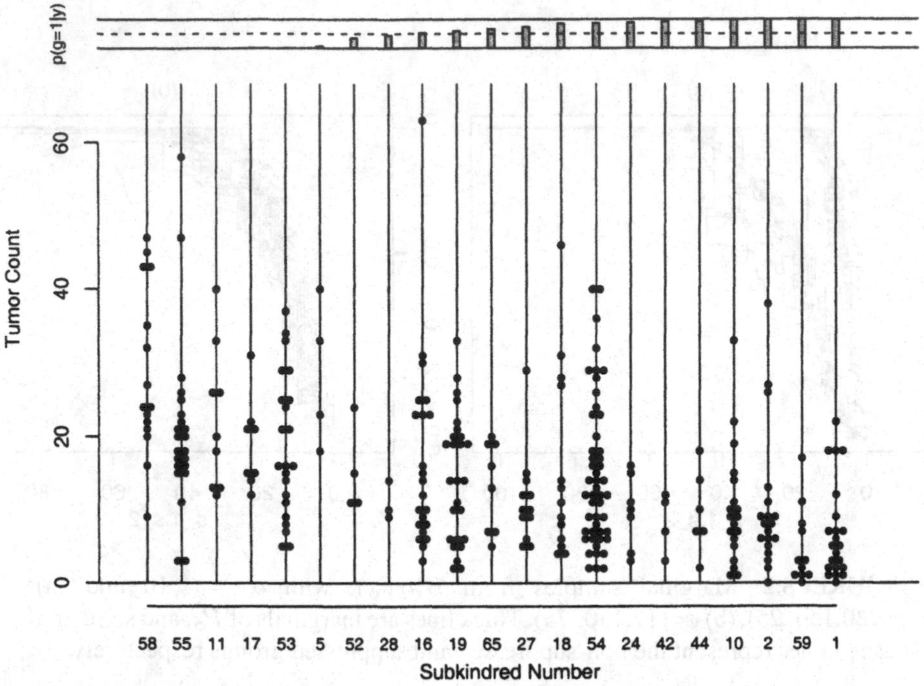

FIGURE 8.3. Some tumor count data.

Having studied other populations of such mice, we think this α-value of 10 roughly reflects our uncertainty about the fit of the parametric negative binomial/binomial model.

5 Data Analysis

Data were collected from 74 subkindreds, with tumor counts from 968 NF animals. Tumor counts ranged from zero to 79, with an average of 15.74 and a standard deviation of 11.51. Tumor count data from 21 subkindreds selected at random are shown in Figure 8.3. Each vertical line represents a subkindred, with dots plotted along a line representing the tumor counts of NF offspring from the corresponding subkindred founder.

Recall that a subkindred founder carrying $\star$ will pass the allele on to each of its NF offspring independently with probability one-half. We therefore expect tumor counts from such a subkindred to be an approximately equal mix of high and low values. Conversely, we expect mostly high tumor counts from animals in a subkindred lacking $\star$. With this in mind, we might categorize subkindred founders represented on the right-hand side of Figure 8.3 as likely carriers, and those on the left-hand side as likely non-carriers. One goal of this data analysis is to make our determination of carrier versus non-carrier status more precise.

5.1 Markov Chain Implementation

Given the observed tumor count data y, we wish to calculate posterior estimates of the tumor count distributions P_1 and P_2 (which are deterministic functions of P), and the subkindred genotype information $g = (g_1, \cdots, g_m)$. These posterior quantities of interest involve complicated integrals over high-dimensional spaces. Therefore, we approximate these integrals by empirical distributions of samples from a Markov chain whose stationary distribution is the desired posterior. To facilitate the sampling, we include the latent tumor counts $S = (s_{(1,1)}, \cdots, s_{(n_m,m)})$ and the parameter ϕ in the construction of our chain. Given current values (g^b, S^b, ϕ^b, P^b), one scan of the chain consists of

1. sampling $g^{b+1} \sim \pi(g|P^b, \phi^b, y) = \pi(g|P^b, y)$, a distribution of independent Bernoulli random variables;

2. sampling $S^{b+1} \sim \pi(S|g^{b+1}, P^b, \phi^b, y) = \prod_{i,j} \pi(s_{(i,j)}|g_j^{b+1}, P^b, y_{(i,j)})$, in which $s_{(i,j)}^{b+1}$ is distributed as P^b conditional on $s_{(i,j),1}^{b+1} = y_{(i,j)}$ if $g_j^{b+1} = 0$, and is distributed as P^b conditional on at least one component of $s_{(i,j)}^{b+1}$ being equal to $y_{(i,j)}$ if $g_j^{b+1} = 1$;

3. sampling ϕ^* from a symmetric random walk distribution, and accepting ϕ^* as ϕ^{b+1} with probability $\frac{\pi(\phi^*|g^{b+1}, S^{b+1}, y)}{\pi(\phi^b|g^{b+1}, S^{b+1}, y)} = \frac{\pi(S^{b+1}|\phi^*)\pi(\phi^*)}{\pi(S^{b+1}|\phi^b)\pi(\phi^b)}$;

4. sampling $P^{b+1} \sim \pi(P|S^{b+1}, g^{b+1}, \phi^{b+1}, y) = \mathcal{D}(\alpha P_{\phi^{b+1}} + n\hat{P}_{S^{b+1}})$, a Dirichlet distribution where $\hat{P}_{S^{b+1}}$ is the empirical distribution of the current state of S.

The updates (1) and (3) for g and ϕ are based on partial conditionals, that is, the conditional distributions given some, but not all, of the current values of the other components. Such partial conditioning is justified by noting that the above sampling scheme is equivalent to

Step 1: a Gibbs update for (g, S);

Step 2: a Gibbs update for S;

Step 3: a Hastings update for (ϕ, P);

Step 4: a Gibbs update for P.

The equivalence can be seen via the following general argument: Suppose we wish to estimate a generic joint distribution $\pi(x,y,z)$ by MCMC methods. In some cases, it may be more desirable to update x based on $\pi(x|z)$ rather than the full conditional $\pi(x|y,z)$. This can be done by sampling an x^* from a desired proposal distribution $J_1(x^*|x,z)$, then "pretending" to sample y^* from $\pi(y|x^*,z)$. The proposal (x^*,y^*) is then accepted with probability

$$\begin{aligned}\frac{\pi(x^*,y^*|z)}{\pi(x,y|z)}\frac{J(x,y|x^*,y^*,z)}{J(x^*,y^*|x,y,z)} &= \frac{\pi(x^*|z)\pi(y^*|x^*,z)}{\pi(x|z)\pi(y|x,z)}\frac{\pi(y|x,z)J_1(x|x^*,z)}{\pi(y^*|x^*,z)J_1(x^*|x,z)} \\ &= \frac{\pi(x^*|z)J_1(x|x^*,z)}{\pi(x|z)J_1(x^*|x,z)}.\end{aligned}$$

By using a full conditional for y^* and a proposal distribution for x^* that doesn't depend on y (for example a partial conditional as in Step 1 above, or a symmetric random walk as in Step 3), we have ensured our acceptance probability of (x^*,y^*) doesn't depend on y^*. Therefore, y^* doesn't actually need to be generated at this stage; instead, it can be updated at the next stage, using a potentially different proposal mechanism. For a more detailed discussion of partial conditioning, see Besag, Green, Higdon, and Mengersen (1995, Appendix 2). We base our Markov chain on partial conditionals for the reasons given below.

To improve mixing: Although the chain based on the full conditionals is irreducible, it doesn't mix very well. This is because the conditional distribution of g given S, P and y is often degenerate: Consider a single subkindred j whose founder has unknown genotype g_j. Note that $g_j = 0$ implies the event $E_j = \{s_{(i,j),1} = y_{(i,j)}, i = 1,\ldots,n_j\}$, i.e. if the subkindred founder does not carry $\star$, then the tumor counts of its offspring are non-suppressed. On the other hand, $E_j^c \Rightarrow \{g_j = 1\}$, so the full conditional of g_j is a point mass at one if E_j does not hold. Given $g_j^b = 1$, sampling an S^b to satisfy E_j is possible but extremely unlikely. This in turn makes the probability $\Pr(g_j^{b+1} = 0|g_j^b = 1)$ very small, and leads to poor mixing. This difficulty is avoided by sampling g^{b+1} conditional on P^b and y only.

To reduce calculations and numerical errors: In the case of the Hastings step for ϕ, one would typically base the acceptance probability of ϕ^* on $\pi(\phi^*|P,S,g,y)$, which reduces to $\pi(\phi^*|P)$. Since $P|\phi^*$ is distributed as a Dirichlet process, computing this acceptance probability involves calculating $\Gamma(\alpha P_{\phi^*}(s))$ for all possible latent tumor counts s. Many values of $P_{\phi^*}(s)$ will be extremely small, making the calculation of the Gamma function prone to numerical errors. On the other hand, the conditional distribution of ϕ^* given only S involves computing $\Gamma(\alpha P_{\phi^*}(s))$ only for those values of s occurring in the sampled set of the latent tumor counts S. As $P_{\phi^*}(s)$ is typically larger for sampled values of s, this modification of the sampling scheme tends to reduce not only the number, but also the magnitude of the errors made in computing the Gamma function.

Finally, we note the updates for ϕ can be done component by component. That is, proposals and acceptances can be made separately for θ,γ, and p, and so after one scan of the chain, ϕ^{b+1} could be the same as ϕ^b, or could differ at one, two, or three component values. This component by component method of updating was used to make the inference in the following section.

5.2 *Posterior Inference*

For the purpose of data analysis, all tumor count distributions were conditioned to lie on the integers from zero to ninety. The sampling scheme described above was coded in the C++ programming language, and was used to generate four chains of 200,000 scans each, recording output every 100th scan. The starting values of P^0 for the four chains were generated by sampling P^0 from $\mathcal{D}(\alpha P_{\phi^0})$, using four different values of ϕ^0, given in Table 8.1.

The output of the chain is very high-dimensional: For diagnostics we only report on sequences of mean tumor count for the non-suppressed and suppressed groups, $E(s_1|P)$ and $E(s_2|P)$. As can be seen in Figure 8.4, after about 20,000 scans the four separate sequences of $E(s_1|P)$ and $E(s_2|P)$ seem to have converged to similar distributions. We delete the first 50,000 scans from each chain to allow for burn in, and compute the sample acf values from the remaining 4×1500 scans (150,000 scans subsampled every 100th scan), given in Table 8.2.

The 4×1500 scans recorded after burn in were used to compute posterior quantities of interest, some of which are displayed in Figure 8.5. The first panel shows the posterior mean CDF's of the two stochastically ordered groups in heavy lines, with confidence bands in lighter lines. The confidence bands represent the range of the CDF's saved from the chain, that is every 100th sample after the first 50,000 scans. The second panel gives a contour plot of the joint posterior density of $E(s_1|P)$ and $E(s_2|P)$.

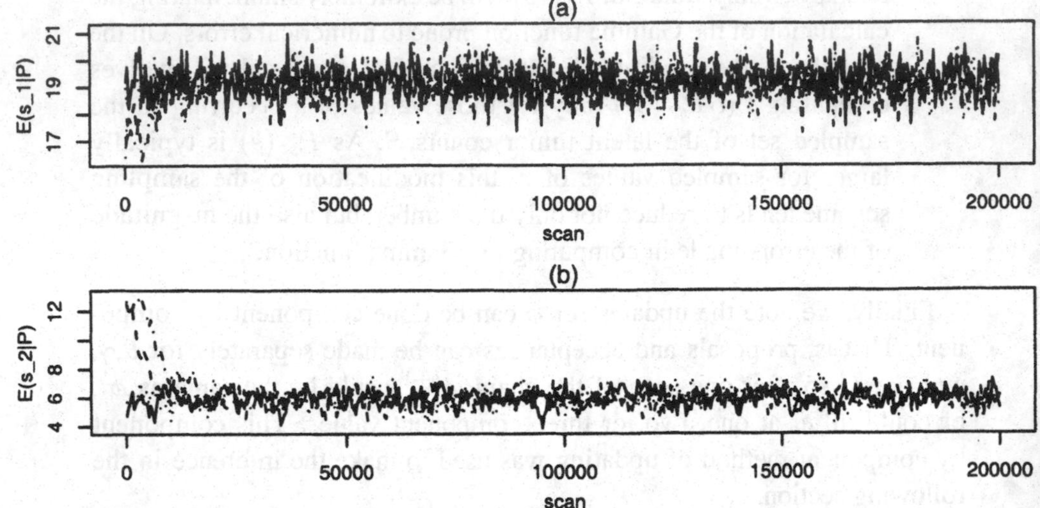

FIGURE 8.4. Sequences of (a) $E(s_1|P)$ and (b) $E(s_2|P)$ for four different chains.

Chain #	θ	σ^2	p
1	20	125	0.25
2	20	175	0.25
3	17	125	0.75
4	17	175	0.75

TABLE 8.1. Starting Values of $\phi^0 = (\theta, \sigma^2, p)^0$.

The contour lines represent highest posterior density regions of 20, 50, 80, 90, and 95 percent probability. The posterior means of these parameters are 19.11 and 5.89 respectively, with posterior standard deviations of 0.34 and 0.33 (based on weighted averages of within-chain and between-chain variances). The standard deviations of the tumor count distributions are estimated as 11.37 and 4.15. The third panel gives the marginal posterior distribution of the multiplicative effect p of $\star$ in the base model, which has a posterior mean and mode of .31, and a posterior standard deviation of .034. These three plots show the estimated effect of $\star$ to be quite large; giving about a 70% reduction in mean tumor count between the two populations. This is an important result, as an allele with such a large effect is of biological significance and warrants further study.

Another important piece of output from the Markov chain is the posterior distribution of the subkindred genotypes $g_1, \dots, g_m$ (the posterior

	Lag 100	Lag 500	Lag 1000	Lag 5000
acf($E(s_1\|P)$)	0.0590	0.0149	-0.0277	-0.0107
acf($E(s_2\|P)$)	0.7549	0.4637	0.2524	0.0102

TABLE 8.2. Sample Autocorrelation

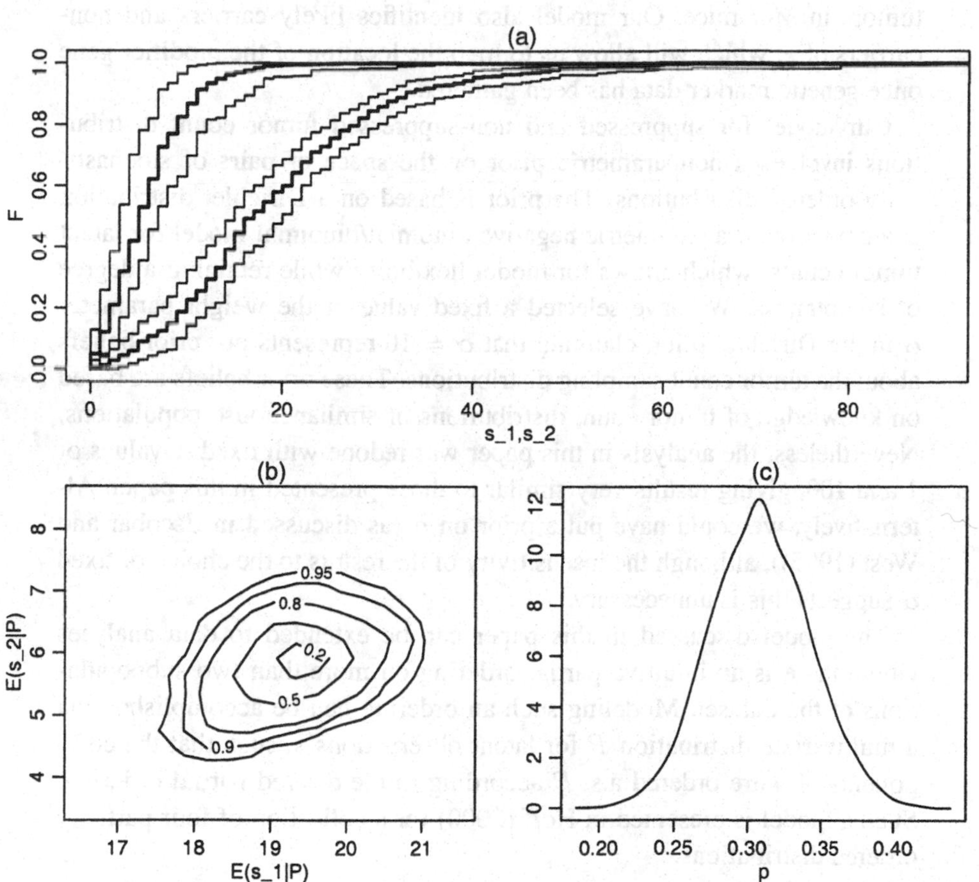

FIGURE 8.5. Posterior Quantities: (a) Bayes estimates of tumor count CDF's; (b) Posterior joint distribution of $E(s_1|P), E(s_2|P)$; (c) Posterior distribution of p.

expectation of these variables for some subkindreds are plotted along the top of Figure 8.3). This output allows us to identify likely carriers and non-carriers of $\star$, which in turn will aid us in the next stage of inquiry—mapping the location of the modifier gene in the mouse genome.

6 Discussion

We have developed a model which allows us to measure the effect of a modifier allele on intestinal tumor count. From our analysis, we estimate the effect of the $\star$ allele to be a 70% reduction in the number of intestinal tumors in *Min* mice. Our model also identifies likely carriers and non-carriers of $\star$, which will allow us to map the location of the modifier gene once genetic marker data has been gathered.

Our model for suppressed and non-suppressed tumor count distributions involves a nonparametric prior on the space of pairs of stochastically ordered distributions. The prior is based on a Dirichlet distribution centered around a parametric negative binomial/binomial model for latent tumor counts, which allows for model flexibility while retaining a degree of smoothness. We have selected a fixed value of the weight parameter α in our Dirichlet prior, claiming that $\alpha = 10$ represents our prior beliefs about the tumor count sampling distributions. These prior beliefs are based on knowledge of tumor count distributions of similar mouse populations. Nevertheless, the analysis in this paper was redone with fixed α-values of 1 and 100, giving results very similar to those presented in this paper. Alternatively, we could have put a prior on α (as discussed in Escobar and West (1995)), although the insensitivity of the results to the choice of fixed α suggests this is unnecessary.

The model discussed in this paper can be extended to data analyses where there is an intuitive partial ordering on more than two subpopulations of the dataset. Modeling such an ordering can be accomplished via a multivariate distribution P for latent observations s, such that the components of s are ordered a.s. P according to the desired partial ordering. Such a model is presented in Hoff (2000) for a collection of four partially ordered distributions.

Acknowledgments

Thanks to Linda Clipson of the McArdle Laboratory of Cancer Research, University of Wisconsin-Madison, for her help in maintaining these data. This research was supported in part by National Cancer Institute grants T32-CA09565-09 for the first author, F32-CA77946 for the second author, R37-CA63677 for the third and fourth authors, and R01-CA64364 for the fifth author.

References

Besag, J., Green, P., Higdon, D., and Mengersen, K. (1995). Bayesian Computation and Stochastic Systems (with discussion). *Statistical Science* **10** , 3-66.

Blackwell, D., and MacQueen, J.B. (1973). Ferguson Distributions via Pólya Urn Schemes. *The Annals of Statistics* **1**, 353-355.

Dietrich, W.F., Lander, E.S., Smith, J.S., Moser, A.R., Gould, K.A., Luongo, C., Borenstein, N., and Dove, W. (1993). Genetic Identification of *Mom-1*, a Major Modifier Locus Affecting *Min*-induced Intestinal Neoplasia in the Mouse. *Cell* **75**, 631-639.

Drinkwater, N.R., and Klotz, J.H. (1981). Statistical Methods for the Analysis of Tumor Multiplicity Data. *Cancer Research* **41**, 113-119.

Escobar, M.D., and West, M. (1995). Bayesian Density Estimation and Inference Using Mixtures. *Journal of the American Statistical Association* **90**, 577-588.

Ferguson, T.S. (1973). A Bayesian Analysis of some Nonparametric Problems. *The Annals of Statistics* **1**, 209-320.

Hoff, P.D. (2000). Constrained Nonparametric Maximum Likelihood via Mixtures. *The Journal of Computational and Graphical Statistics*, To appear.

Lehmann, E.L. (1986). *Testing Statistical Hypotheses*. Singapore: John Wiley and Sons.

Sethuraman, J., and Tiwari, R.C. (1982). Convergence of Dirichlet Measures and the Interpretation of Their Parameter. In *Statistical Decision Theory and Related Topics III*, eds. S.S. Gupta and J.O. Berger. New York: Academic Press, 305-315.

Bayesian Tools for EDA and Model Building: A Brainy Study

Steven N. MacEachern
Mario Peruggia

ABSTRACT We consider a strategy for Bayesian model building that begins by fitting a simple, default model to the data. Numerical and graphical exploratory tools, based on summary quantities from the default fit, are used to assess the adequacy of the initial model and to identify directions in which the fit can be refined. We apply this strategy to build a Bayesian regression model for a classic set of data on brain and body weights of mammalian species. We discover inadequacies in the traditional regression model through use of our exploratory tools. More sophisticated models point the way toward judging the adequacy of a theory on the relationship between body weight and brain weight, and also bear on the timeless question "do we have big brains?"

1 Introduction

The scientific literature is rich in allometric studies relating the body size, W, of various animal species to the size of a body part or to some other biological characteristic of interest, Y. A comprehensive overview of this field of research is given in Peters (1983). In the book, the author outlines the reasons why these empirical theories are developed, and explores their implications for ecology. Typically, the dependence of Y on W is summarized in terms of a power law of the form $Y = cW^b$. While Peters stresses the use of these types of models for prediction, he indicates that many ecologists postulate the existence of biological explanations for the empirical relations.

In particular, some values of the exponent b in the power relation seem to recur in many of the published studies, and various explanations have been proposed. For example, values of b near $3/4$ are often estimated from data relating the body size of an organism to its biological rates (metabolic, respiratory air flow, etc.). For responses other than rates, a "popular" magnet for estimates of the exponent is the value of $2/3$. The so-called "surface law" provides a justification for $b = 2/3$ that proceeds along the following lines. Assuming constant density, body size is a proxy for body volume and, therefore, (ignoring important geometrical differences among animal shapes) body surface should be proportional to $W^{2/3}$ (see Peters, 1983, and Calder, 1984, for supporting empirical evidence). Thus, if the response Y is dependent on surface area, it should also rise according to $W^{2/3}$.

An estimate of b close to $2/3$ has been calculated from several data sets of mammalian species in which an animal's brain mass is considered as the dependent variable Y (see Von Bonin, 1937; Jerison, 1955; and Appendix IVe in Peters, 1983). Gould (1977) justifies the validity of the surface law in this instance by noticing that "brain weight is not regulated by body weight, but primarily by the body surfaces that serve as end points for so many innervations." In summary, the surface law would support the following relation:

$$(\text{brain mass}) = c\,(\text{body mass})^{2/3}.$$

Very often, the estimation of the model parameters is done by least squares after transforming both W and Y to a logarithmic scale to achieve linearity and constant variance (Weisberg, 1985). For brain mass data, this implies the postulation of a simple linear regression (SLR) model of the form

$$\ln(\text{brain mass}) = a + b\ln(\text{body mass}),$$

and the surface law now states that the slope of the regression line should be equal to $2/3$.

The limitations associated with fitting SLR models have been pointed out by various authors who have noted how differences between taxonomic levels and similarities between closely related species may invalidate the modeling assumptions (see, for example, Jerison, 1955; and Bennett and Harvey, 1985.a and 1985.b). Among the alternative modeling strategies that have been proposed, one that accounts for the genetic proximity of different species has recently gained popularity. This methodology recasts the original problem of estimating b into an alternative estimation problem for phylogenetically independent contrasts (PIC). (The original reference is Felsenstein, 1985. See also Garland et al., 1992 for a more recent review and clarification of the methodology.)

In this paper we argue that Bayesian variance components (VC) models provide a natural means for describing the taxonomic and phylogenetic dependences present in allometric data. Unlike PIC, the Bayesian framework allows one to perform the estimation directly on a model for the observed data, as opposed to a model for derived independent contrasts. Thus, other meaningful parameters, beside the allometric scaling exponent, b, can be estimated and interpreted. We also make an argument for the need for EDA and diagnostic tools to aid Bayesian data analysis and model building.

We illustrate our modeling approach on a data set relating body mass to brain mass in mammalian species. First, we fit a SLR model. Next, we develop numerical and graphical diagnostic tools to validate the model, giving a brief and informal description of their main theoretical properties. The diagnostics indicate that the SLR fit is seriously inadequate and suggest directions for improvement. We then specify and fit a more sophisticated VC model that accounts for the taxonomic levels of the various species and indicates how, conceptually, phylogenetic information could also be incorporated. We conclude by noticing how the estimates of the allometric scaling parameter, b, from the SLR and VC models differ substantially. This has implications with regard to the validity of the surface law and to the question of whether or not the data point for humans is an outlier.

2 Simple Linear Regression

We considered the data on placental mammals reported in Sacher and Staffeldt (1974), retaining only those 100 species for which brain weights, Y, and body weights, W, were both available. A scatterplot of the data is presented in Figure 9.1.

In the first step of our analysis we used BUGS (Speigelhalter et al., 1995) to fit a standard Bayesian model regressing $Z_i = \ln(Y_i)$ on $X_i = \ln(W_i) - (1/100)\sum_{j=1}^{100}\ln(W_j)$. Specifically, assuming conditional independence at all hierarchical stages and following the notational conventions used in BUGS, we specified:

$$Z_i|a, b, \tau \sim N(a + bX_i, \tau), \quad i = 1, \ldots, 100,$$

$$a \sim N(0, 10^{-4}), \quad b \sim N(2/3, 20.25),$$

$$\tau \sim \Gamma(10^{-3}, 10^{-3}).$$

The recentering of the predictor W justifies the specification of independent priors for the regression coefficients a and b. Note that, even after

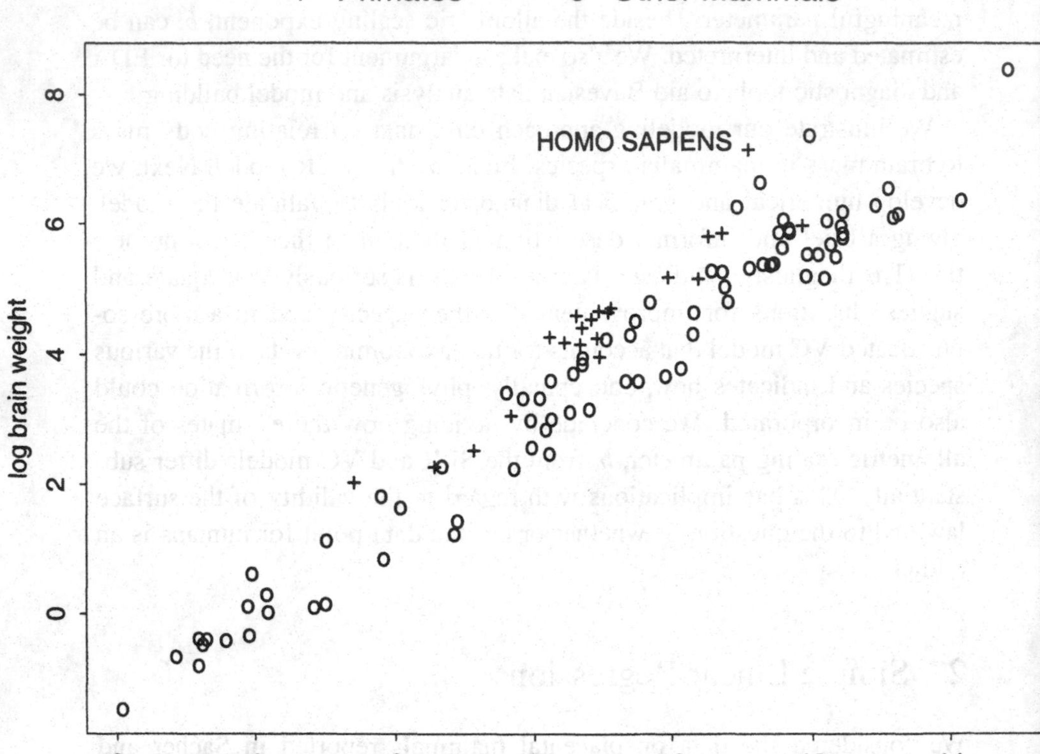

FIGURE 9.1. Scatterplot of Log Brain Weights vs. Log Body Weights for 100 Primates and Other Placental Mammals.

recentering, the slope b still represents the allometric scaling exponent in the original relation $Y = cW^b$. For this reason we centered the prior distribution of b at the value of 2/3, as suggested by the surface law. A prior precision of 20.25 corresponds to a standard deviation of $2/9$. Therefore the prior mean lies three standard deviations above zero and negative values of b are given a negligible probability of 0.0014. All other specifications were intended to reflect vague prior knowledge.

The estimated posterior mean of b based on a BUGS sample of size 1,000 was 0.724, with a 95% posterior probability interval given by (0.68, 0.76). The fact that the interval does not contain $2/3$ seemed to cast doubt on the surface law, but, before drawing hasty conclusions, we decided to assess the fit of the model. We did so by means of some novel EDA and diagnostic tools that are described in the following section.

3 Bayesian EDA and Diagnostic Tools

We describe the set-up for our diagnostics in some generality. We assume that a sample, $\{\theta_j\}_{j=1}^J$, from the full posterior distribution of the model parameters given the data is available. We associate a set of weights, $\{w_{\setminus i}(\theta_j)\}_{j=1}^J$, with the i-th observation in the data set by defining

$$w_{\setminus i}(\theta_j) = \frac{q_{\setminus i}(\theta_j)/q(\theta_j)}{\sum_{k=1}^J q_{\setminus i}(\theta_k)/q(\theta_k)},$$

where q and $q_{\setminus i}$ are, respectively, functions proportional to the full posterior density, p, and to the case-deleted posterior density, $p_{\setminus i}$, conditional on all observations except the i-th one. The weights are normalized to sum to unity for convenience.

The distribution of $v_{\setminus i}(\theta) = p_{\setminus i}(\theta)/p(\theta)$ (and consequently the empirical distribution of the $w_{\setminus i}(\theta_j)$) depends on the impact of the removed observation on the analysis. Similarly, the joint distribution of $v_{\setminus i}(\theta)$ and $v_{\setminus k}(\theta)$ depends on the simultaneous impact of observations i and k on the analysis. Our Bayesian diagnostics exploit these dependences.

For illustration, consider the case when, conditional on a realization of μ from a distribution $\pi(\mu)$, an i.i.d. sample $x_1, \ldots, x_n$ is drawn from a univariate normal distribution with mean μ and known variance σ^2. In this case, we have

$$\ln v_{\setminus i} = (1/2)\ln(2\pi\sigma^2) + (x_i - \mu)^2/(2\sigma^2),$$

and the joint distribution of $(\ln v_{\setminus i}, \ln v_{\setminus k})$ is supported on the parabola $S(\mu)$ defined parametrically, for $-\infty < \mu < \infty$, by the system of equations

$$\begin{cases} \ln v_{\setminus i} &= (1/2)\ln(2\pi\sigma^2) + (x_i - \mu)^2/(2\sigma^2) \\ \ln v_{\setminus k} &= (1/2)\ln(2\pi\sigma^2) + (x_k - \mu)^2/(2\sigma^2). \end{cases}$$

Owing to the simplicity of the example, the distribution of $(\ln v_{\setminus i}, \ln v_{\setminus k})$ is supported on a one-dimensional curve in the plane. This is the exception rather than the rule, but the essential features of this case are typical of much more general situations.

Suppose that, given the data, the posterior distribution of μ is normal with mean zero and standard deviation $1/3$. Then, the bulk of the posterior mass for μ will be concentrated on the interval $(-1, 1)$. Suppose also that two of the observations in the data set are $x_i = -0.7$ and $x_k = 0.7$. Because the two observations belong to opposite tails of the posterior for μ, the impact of dropping one or the other from the analysis will be opposite. In particular, if $x_i = -0.7$ is dropped, the posterior will shift to the right, while, if $x_k = 0.7$ is dropped, the posterior will shift to the left.

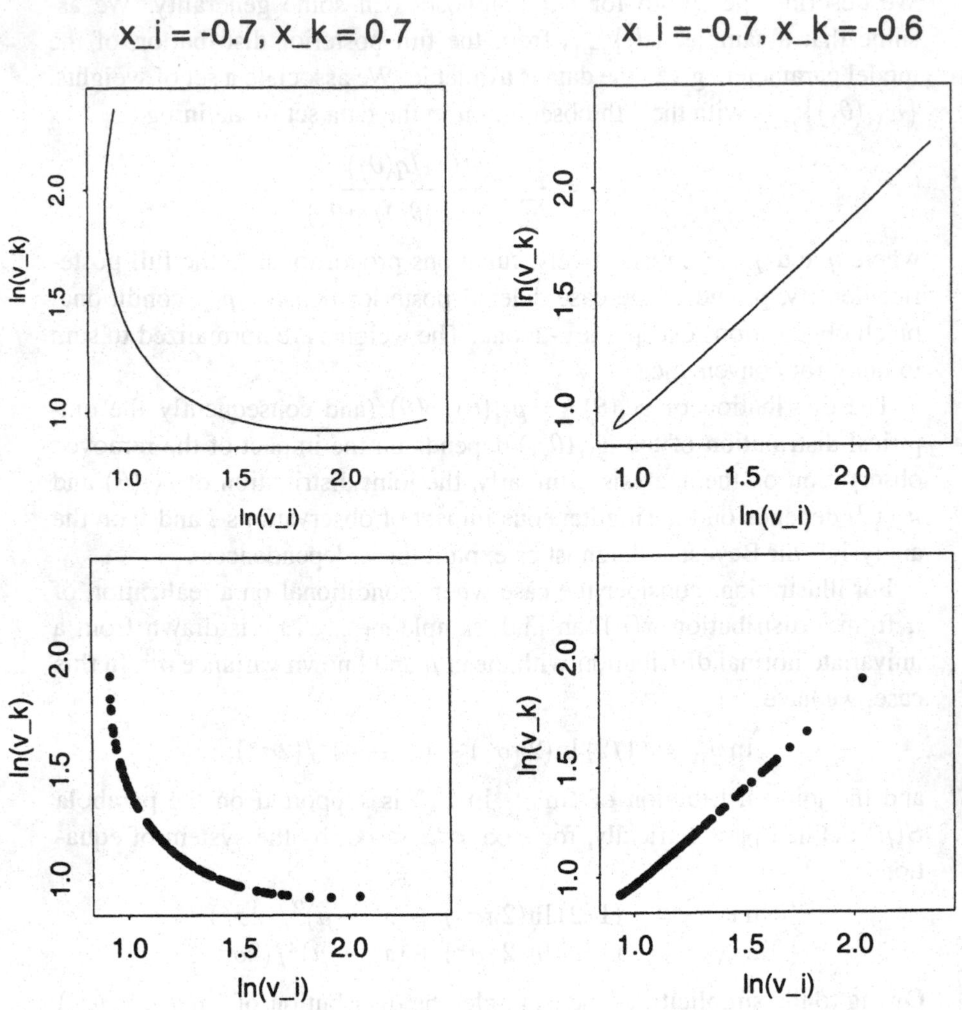

FIGURE 9.2. Support $S(\mu)$ of the Joint Distribution of $(\ln v_{\backslash i}, \ln v_{\backslash k})$ for Two Different Choices of Pairs of Observations (x_i, x_k) to be Dropped from the Analysis (top row) and Two Corresponding Realized Samples of Size 100 (bottom row). The displays of the supports are restricted to values of μ in the interval $(-1, 1)$.

Under these circumstances and assuming that $\sigma^2 = 1.0$, the support $S(\mu)$ is displayed in the top-left panel of Figure 9.2, for values of μ in the interval $(-1, 1)$.

The lower left panel of Figure 9.2 displays 100 realizations $\{(\ln v_{\backslash i}(\mu_j), \ln v_{\backslash k}(\mu_j))\}_{j=1}^{100}$, where the μ_j are 100 independent draws from the posterior of μ. These displays show that the realizations of $(\ln v_{\backslash i}(\mu), \ln v_{\backslash k}(\mu))$ tend to distribute themselves around the vertex of the parabola $S(\mu)$ (symmetrically, in this case, since x_i and x_k are equidistant from the posterior mean of μ).

Contrast this with the case depicted in the two right panels of Figure 9.2. All else is as before, but the observations dropped from the analysis are now $x_i = -0.7$ and $x_k = -0.6$. Since x_i and x_k are close to each other and both are smaller than the posterior mean of μ, the deletion of either of them will have a comparable impact on the posterior distribution of μ (mainly, a location shift to the right). As shown by the bottom-right panel of Figure 9.2, realizations of $(\ln v_{\backslash i}(\mu), \ln v_{\backslash k}(\mu))$ tend to distribute themselves along one of the branches of the support $S(\mu)$ and exhibit a large positive correlation. In the limit, letting x_i and x_k converge to a common value, the two branches of $S(\mu)$ would collapse into a halfline and samples from $(\ln v_{\backslash i}(\mu), \ln v_{\backslash k}(\mu))$ would be perfectly positively correlated.

Reverting to the notation introduced at the beginning of this section, the examples above illustrate rather general properties of the posterior joint distribution of $(\ln v_{\backslash i}(\theta), \ln v_{\backslash k}(\theta))$, and give a complementary explanation of the claim made in Bradlow and Zaslavsky (1997) that the sample covariance between $\{w_{\backslash i}(\theta_j)\}_{j=1}^{J}$ and $\{w_{\backslash k}(\theta_j)\}_{j=1}^{J}$ measures the "synergy" of two cases. The more observations i and j have a similar impact on the analysis, the more their realized case-deletion log-weights will tend to be highly correlated.

Our EDA and diagnostic tools take advantage of these properties to discover similarities and dissimilarities between observations that might not be adequately described by the model being fit. First, for all pairs of observations, we propose to examine the scatterplots of the log-weights constructed from a sample drawn (often by MCMC methods) from the full posterior. The displays should be arranged in a scatterplot matrix. Obviously, for large data sets, not all scatterplots can be displayed at once, and interesting subsets should be chosen on the basis of considerations pertaining directly to the situation being modeled. In particular, the examination of these scatterplots is useful to detect whether or not all observations in a given group (e.g., in a study of the health of patients in a cardiac rehabilitation program, all young MI patients, or all patients older than 60) have a similar influence on the analysis.

Our second diagnostic is based directly on summary measures of the log-weights corresponding to different observations. Specifically, let r_{ik} denote the sample correlation between $\{\ln w_{\setminus i}(\theta_j)\}_{j=1}^{J}$ and $\{\ln w_{\setminus k}(\theta_j)\}_{j=1}^{J}$, and define the distance d_{ik} between observations i and k by $d_{ik} = 1 - r_{ik}$. (Alternatively, d_{ik} could be defined in terms of the sample covariance or a non-Pearsonian correlation.) According to this distance, the more two observations have highly correlated case-deletion log-weights (and, hence, the more synergy they have), the closer they are. We use this distance below to perform a hierarchical clustering of the observations and determine if they group together according to recognizable patterns that might suggest directions for model improvement. The clustering can be implemented in a variety of ways, often leading to similar conclusions about the data.

4 Variance Components Regression

We now return to the analysis of the allometric data that we began presenting in Section 2. After processing the BUGS output of 1,000 MCMC draws from the full posterior distribution to compute estimates of the parameters in the SLR model, we used the same draws to construct the Bayesian diagnostics described in Section 3. Figure 9.3 displays a matrix scatterplot of case-deletion log-weights for selected pairs of observations.

The log-weights for the first three species, Pongo Pigmaeus, Pan Troglodytes, and Homo Sapiens (order primates, sub-order anthropoidea), exhibit high to very high correlations. The log-weights for Potos Flavus (order carnivora, sub-order fissipeda) exhibit moderate positive correlations with those of the three primates and strong negative correlation with those of Rattus Norvegicus (order rodentia, sub-order myomorpha). The log-weights for the latter appear to be uncorrelated or to have moderate negative correlations with those of the three primates.

The few scatterplots presented in Figure 9.3 embody many of the features common to the scatterplots of the case-deletion log-weights for the other pairs of observations and are indicative of the patterns that one might expect to observe in these types of displays. We notice that the log-weights of species belonging to the same taxonomic order tend to be positively correlated, thus indicating a synergistic impact on the analysis, while those of unrelated species tend to show no or negative correlations. The patterns that can be observed in the scatterplots are reminiscent of those displayed in the bottom panels of Figure 9.2. Now, however, the supports of the joint distributions of pairs of case-deletion log-weights are no longer supported on one-dimensional curves and the displays may exhibit substantial scatter.

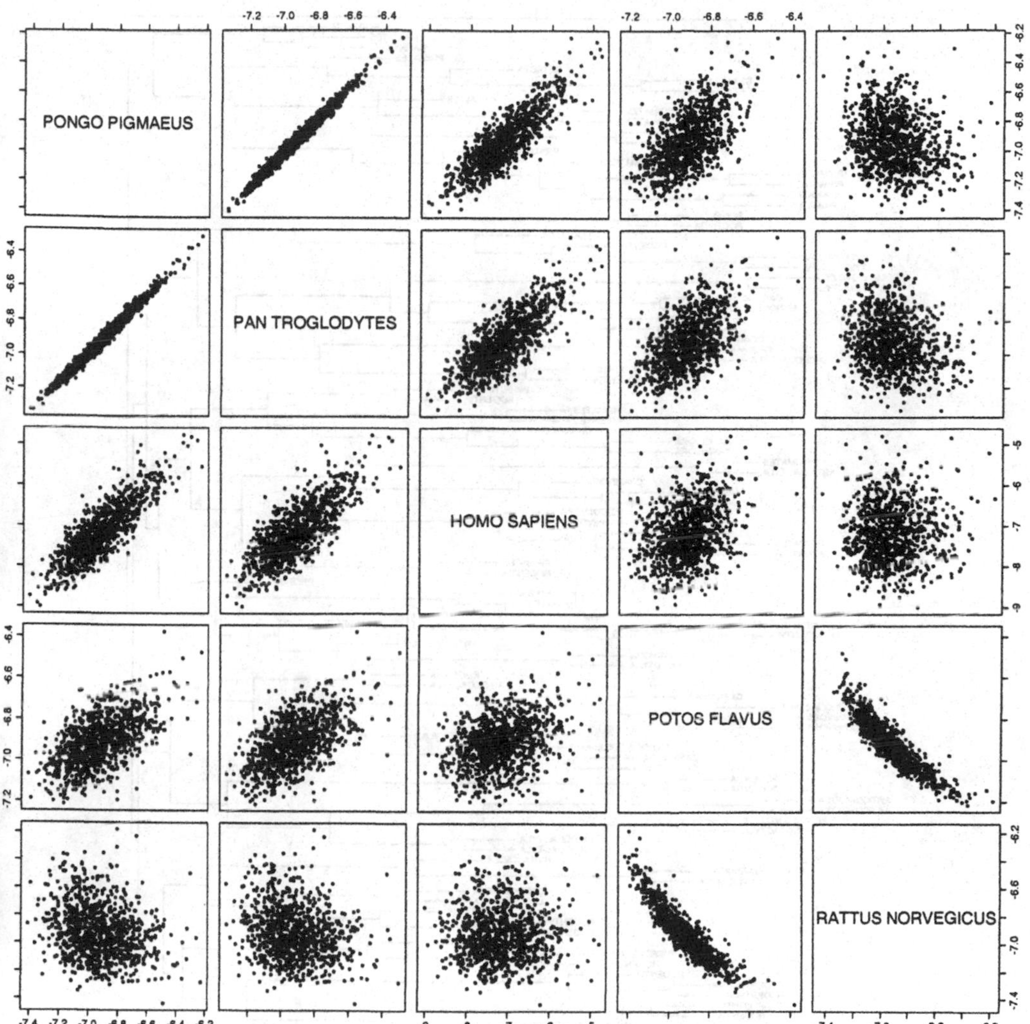

FIGURE 9.3. Matrix Scatterplot of Selected Sets of Case-Deletion Log-Weights.

The impression that species of the same order usually exert comparable influence on the analysis is confirmed by an examination of the clustering tree displayed in Figure 9.4.

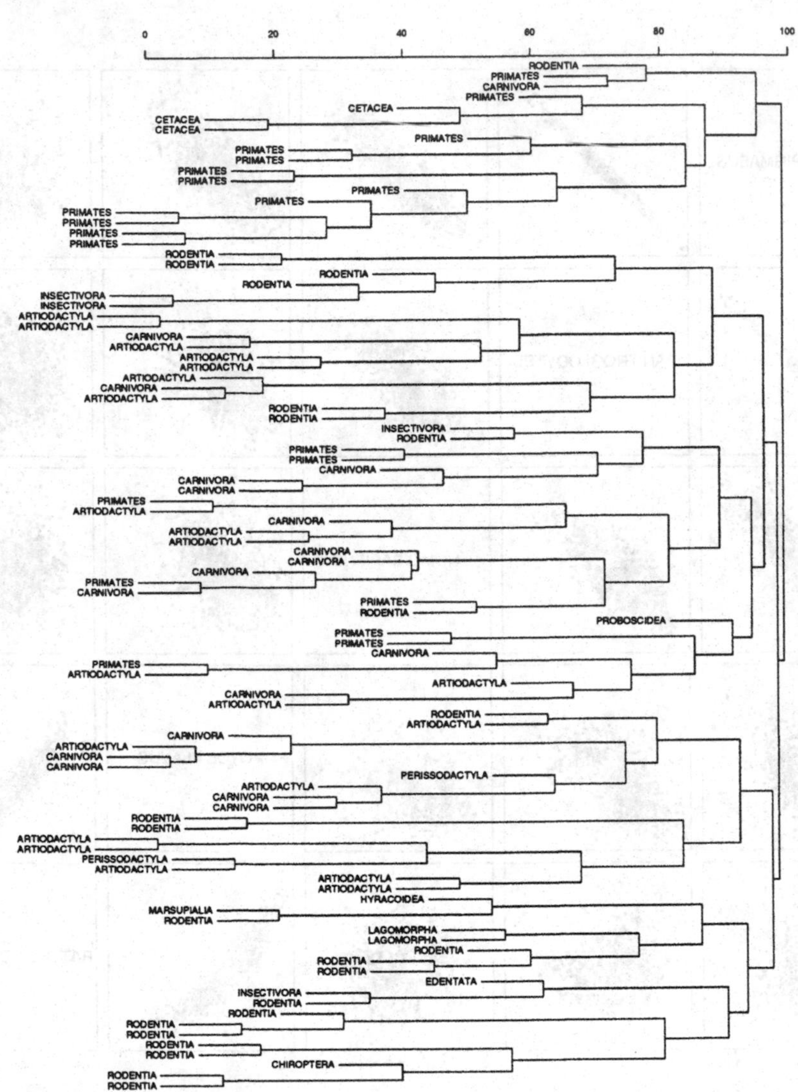

FIGURE 9.4. SLR Clustering Tree Labeled by Taxonomic Order. The tree illustrates how species of the same order tend to group together.

The tree was produced by the Splus function `hclust`, using a distance matrix with (i,k)-th entry given by d_{ik} (as defined on p. 352) and selecting the option `method = compact` (complete linkage clustering). Each leaf of the tree represents one of the species in the dataset, but, rather than labeling the leaves by the names of the corresponding species, the figure labels them by the taxonomic orders to which the species belong. Clustering of species belonging to the same order manifests itself in several branches of the tree. Labeling the tree by taxonomic sub-order (display not shown) provides evidence of additional sub-clustering.

The fact that the observations cluster according to recognizable and interpretable patterns sets off a warning signal. The exploratory diagnostic tools indicate that the SLR model fails to account for dependences in the error structure within taxonomic levels. A likely explanation of these dependences is that the taxonomy based on order and sub-order serves as a crude proxy for phylogenetic relatedness.

We decided to obviate to the deficiencies of the SLR model by fitting a VC model that incorporates random effect terms γ_ℓ and δ_m for the 13 orders and 19 sub-orders in the taxonomy. Specifically, let $\boldsymbol{\beta} = (a, b, \gamma_1, \ldots, \gamma_{13}, \delta_1, \ldots, \delta_{19})'$ and let $\boldsymbol{X}$ be the $100 \times (2 + 13 + 19 = 34)$ design matrix defined by adjoining a column of ones, the column X of recentered log body weights, 13 columns of 0-1 order indicators, and 19 columns of 0-1 sub-order indicators. Then, denoting by $\boldsymbol{X}'_i$ the i-th row of $\boldsymbol{X}$ and again assuming conditional independence at all stages of the hierarchy, we specified:

$$Z_i|\boldsymbol{\beta}, \tau \sim N(\boldsymbol{X}'_i\boldsymbol{\beta}, \tau), \quad i = 1, \ldots, 100,$$

$$a \sim N(0, 10^{-4}), \quad b \sim N(2/3, 20.25),$$

$$\gamma_\ell \sim N(0, \tau_\gamma), \quad \ell = 1, \ldots, 13, \quad \delta_m \sim N(0, \tau_\delta), \quad m = 1, \ldots, 19,$$

$$\tau \sim \Gamma(10^{-3}, 10^{-3}),$$

$$\tau_\gamma \sim \Gamma(10^{-3}, 10^{-3}), \quad \tau_\delta \sim \Gamma(10^{-3}, 10^{-3}).$$

The prior specifications for a, b, and τ are the same as those we made for the SLR model. The normal priors centered at zero for the random effects and the gamma priors for their precisions were meant to indicate vague prior knowledge. The priors for the precisions, however, should not be "too" vague, because the specification of the usual improper prior for this model would produce an improper posterior (Hobert and Casella, 1996).

Also in this case, we performed the estimation based on the output of 1,000 draws from the full posterior generated by BUGS. The estimated posterior mean of the allometric scaling parameter b was now 0.59 with a 95% posterior probability interval given by $(0.55, 0.64)$. Interestingly, once again, the interval did not contain the 2/3 value postulated by the surface law. However, while the SLR interval lay entirely above $2/3$, the VC interval lay entirely below it.

One question that has received much attention in the allometric literature on brain and body sizes of mammalian species is whether human brains are unusually large. The answer is often affirmative and is usually justified by a comparison of the residuals from the least squares fit of a SLR model. For example, we quote from Gould (1977) p. 183:

> To judge the size of our brain, we must compare it with the expected brain size for an average mammal of our body weight. On this criterion we are, as we had every right to expect, the brainiest mammal by far. No other species lies as far above the expected brain size for average mammals as we do.

To clarify, the "expected brain size" to which Gould alludes is the expected log brain size computed from the SLR least squares fit of log brain size on log body size for a data set similar to the one we analyzed.

If we were to interpret Gould's statement to mean that humans are outliers with respect to a standard SLR model that assumes independent normal errors, there would be a major problem: the model does not provide an adequate description of the data. The assumption of independent errors is violated because species belonging to the same order tend to have similar residuals. For example, all observations for primates displayed in the figure on p. 182 of Gould (1977) lie above the regression line. Classifying an observation as an outlier with respect to an untenable model is questionable practice.

As evidenced by the case-deletion diagnostics, the Bayesian SLR model that we fit in Section 2 has similar deficiencies. Consideration of the residuals from that model should bear no relevance on the determination of whether or not humans have large brains. Instead, we argue that the two sets of average residuals computed from the VC model and displayed in Figure 9.5 are the ones that should be taken into consideration when answering the question at hand.

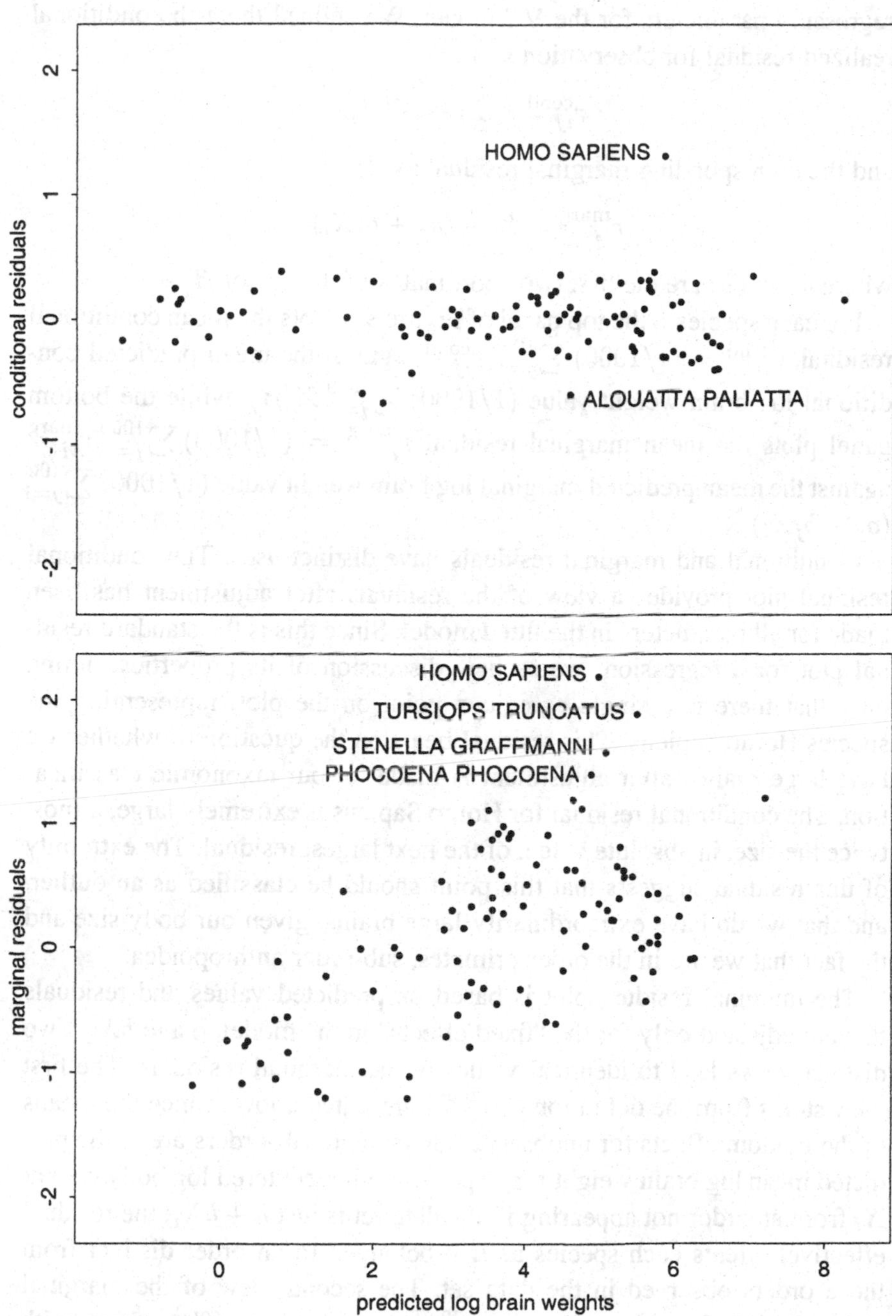

FIGURE 9.5. Plots of Conditional (top) and Marginal (bottom) Mean Residuals from the VC Model Fit vs. Predicted Log Brain Weights.

For $j = 1, \ldots, 1000$, let $\boldsymbol{\beta}_j$ denote the j-th BUGS draw of the vector of regression parameters for the VC model. We defined the j-th conditional realized residual for observation i as

$$r_{ij}^{\text{cond}} = Z_i - \boldsymbol{X}_i' \boldsymbol{\beta}_j,$$

and the corresponding marginal residual as

$$r_{ij}^{\text{marg}} = Z_i - (a_j + b_j X_i),$$

where a_j and b_j are the first two coordinates of the vector $\boldsymbol{\beta}_j$.

For each species i, the top panel of Figure 9.5 plots the mean conditional residual, $\bar{r}_i^{\text{cond}} = (1/1000) \sum_{j=1}^{1000} r_{ij}^{\text{cond}}$, against the mean predicted conditional log brain weight value $(1/1000) \sum_{j=1}^{1000} \boldsymbol{X}_i' \boldsymbol{\beta}_j$, while the bottom panel plots the mean marginal residual $\bar{r}_i^{\text{marg}} = (1/1000) \sum_{j=1}^{1000} r_{ij}^{\text{marg}}$ against the mean predicted marginal log brain weight value $(1/1000) \sum_{j=1}^{1000} (a_j + b_j X_i)$.

Conditional and marginal residuals have distinct uses. The conditional residual plot provides a view of the residuals after adjustment has been made for all parameters in the fitted model. Since this is the standard residual plot for a regression, we eschew discussion of its properties, noting only that there is a single egregious point on the plot, representing the species Homo Sapiens. This residual bears on the question of whether we have large brains, after adjustment is made for our taxonomic classification. The conditional residual for Homo Sapiens is extremely large, almost twice the size, in absolute value, of the next largest residual. The extremity of this residual suggests that this point should be classified as an outlier, and that we do have extraordinarily large brains, given our body size and the fact that we are in the order primates, sub-order anthropoidea.

The marginal residual plot is based on predicted values and residuals that are adjusted only for the "fixed effects" in the model, a and bX_i. Two distinct views lead to identical values of the marginal residuals. The first view stems from the definition of r_i^{marg} presented above. Since the means of the random effects for unobserved orders and sub-orders are 0, the predicted mean log brain weight for a species with recentered log body weight X_i from an order not appearing in the data set is just $a + bX_i$: the residual effectively treats each species as if it belonged to an order distinct from those orders observed in the data set. The second view of the marginal residual arises from the expression $r_i^{\text{marg}} = \gamma_\ell + \delta_m + r_i^{\text{cond}}$, where, with somewhat sloppy notation, γ_ℓ and δ_m represent the random effects associated with species i. Under this view, and ignoring the fact that a and b are fit to the data, each point is plotted as the sum of the random effects for the species, including the error term as a random effect, against the sum of the fixed effects.

The marginal residual plot shows an upward trend. While at first glance surprising, after some consideration, this trend is to be expected. It reflects the differences in the posterior distributions for b between the SLR model and the VC model. It also reflects the trend that species with larger body weights tend to be classified in orders (and sub-orders) with larger values of γ (and δ). This potential relationship between body weight and the random effects suggests a direction for further modeling.

The marginal residual plot enables us to look at whether an individual species has a large brain, unadjusted for taxonomic classification. The three species of dolphins/porpoise in the data set, labeled in the marginal residual plot, illustrate the difference between marginal and conditional questions. The dolphins stand out on the marginal residual plot. They have, as a whole, very large brains after adjustment is made for body weight. However, the dolphins do not stand out in the conditional residual plot. None of these species has a particularly large brain once we adjust for its classification (order cetacea, sub-order cetacea).

The species Alouatta Paliatta (order primates, sub-order anthropoidea) provides a counterpoint to the dolphins. The marginal residual for this species is slightly positive while the conditional residual is large and negative. This species has a typical to slightly large brain for a mammal, but it has a very small brain for a primate.

Homo Sapiens stands out on the marginal residual plot, with the largest positive residual. We have a large brain for a mammal, unadjusted for our taxonomic classification. While the absolute size of the marginal residual (2.191) is larger than the conditional residual (1.314), the visual impact of the conditional residual is much greater than that of the marginal residual. This is a consequence of the reduction in residual variability when the random effects are included in the fit rather than in the residual.

A comparison of the two plots elucidates further features of the data. First, we graphically see the impact of the variance components on the fit of the model. The predicted log brain weights on the conditional residual plot extend over a greater range than do those on the marginal residual plot, and the sizes of the conditional residuals are noticeably smaller than the sizes of the marginal residuals. The pair of plots together provides a visual illustration of the improvement in the model.

Second, we can interactively examine the two plots. Brushing techniques enable us to see how individual points or groups of points appear on the two plots and allow us to look for trends in collections of points. The illustrative cases of the dolphins and Alouatta Paliatta are best examined interactively, although the format of the printed page has limited us to a static presentation.

5 Discussion

The development of techniques and the verification that they can be used successfully is a much more difficult endeavor than is the development of new models or new techniques for fitting models. The very essence of EDA lies in discovering an unusual, perhaps unexpected feature of a data set that can in turn suggest a class of models which better explains the data. In a fortunate situation, the class of models would also suggest mechanisms which might give rise to the models, facilitating a better understanding of the phenomenon under study. The goal of discovery of unusual features of data renders simulation studies and the traditional analytic approaches to establishing the success of a technique artificial. Instead, success of the technique is best demonstrated empirically, by means of a case study.

The case study that we have designed examines the allometric relationship between brain weight and body weight across species of mammals. We set up an experiment where we, as naive users, approached the problem from the traditional presentation found in the statistical community—use of the SLR model to describe the relationship between (recentered) log body weight and log brain weight. On the basis of case deletion clustering, our Bayesian EDA tool, we uncovered a flaw in the standard analysis of allometric data, as found in our community. The assumption that the responses for species are independent, given a simple explanatory variable such as body weight, must be replaced by an assumption that is more plausible from both biological and data analytic perspectives.

The need to account for dependence among related species implies a need to model the structure of the dependence. We pursued this lead, refining the SLR model to account for the dependence in a rather coarse fashion, by creating the VC model based on taxonomic classification of species. The VC model not only provides a more satisfying fit to the data, but it is also similar to the models developed by the biological community over the past 15 years. Case deletion clustering has allowed us, as naive studiers of allometry, to identify deficiencies in the standard model and move toward a more sensible modeling scheme. The success of the technique is confirmed by the fact that our models closely resemble those proposed and fit by experts in the field of allometry (see Pagel, 1999): EDA has allowed us to replicate the behavior of those with far more expertise than we have.

The experiment that we have performed demonstrates the value of our EDA technique. The evidence for success lies in the adage that today's expert knowledge is tomorrow's naivete. Our success in examining these data indicates that, in the hands of a subject matter expert, case deletion clustering can suggest models that go beyond the current level of expert knowledge, leading to a genuine scientific contribution. For example, in the context of allometric studies, there is compelling evidence that environmental as well as evolutionary factors play a role in allometric relations (Tieleman and Williams, 2000). Case deletion clustering can be applied with environmental labels on the tree, perhaps suggesting the study of further environmental factors.

Following the thread of more realistically accounting for the dependence structure, improvements in the model would follow from either qualitatively or quantitatively improving the form of the dependence specified in the model. Replacement of the taxonomy with a more complete phylogenetic structure would provide a qualitative improvement in the model while specification of a more realistic prior distribution for the variance components would provide a quantitative improvement in the model.

Acknowledgments

The authors would like to thank Jerry Downhower, Irene Tieleman, and Joe Williams of the Department of Evolution, Ecology, and Organismal Biology at the Ohio State University for valuable suggestions and discussions.

References

Bennett, P.M. and Harvey, P.H. (1985.a). Relative brain size and ecology in birds, *Journal of the Zoological Society of London,* **207**, 151–169.

Bennett, P.M. and Harvey, P.H. (1985.b). Brain size, development and metabolism in birds and mammals, *Journal of the Zoological Society of London,* **207**, 5911–509.

Bradlow, E.T. and Zaslavsky, A.M. (1997). Case Influence Analysis in Bayesian Inference, *Journal of Computational and Graphical Statistics*, **6**, 314–331.

Calder III, W.A. (1984). *Size, Function, and Life History,* Cambridge, MA: Harvard University Press.

Felsenstein, J. (1985). Phylogenies and the Comparative Method, *The American Naturalist,* **125**, 1–15.

Garland Jr., T., Harvey, P.H. and Ives, A.R. (1992). Procedures for the Analysis of Comparative Data Using Phylogenetically Independent Contrasts, *Systematic Biology,* **41**, 18–32.

Gould S.J. (1977). *Ever Since Darwin,* New York: W. W. Norton & Co.

Hobart, J.P. and Casella, G. (1996). The Effect of Improper Priors on Gibbs Sampling in Hierarchical Linear Mixed Models, *Journal of the American Statistical Association*, **91**, 1461–1473.

Jerison, H.J. (1955). Brain to Body Ratios and the Evolution of Intelligence, *Science,* **121**, 447–449.

Pagel, M. (1999). Inferring the historical patterns of biological evolution. *Nature,* **401**, 877–884.

Peters, R.H. (1983). *The ecological implications of body size,* Cambridge, UK: Cambridge University Press.

Sacher, G.A. and Staffeldt, E.F. (1974). Relation of Gestation Time to Body Weight for Placental Mammals: Implications for the Theory of Vertebrate Growth, *The American Naturalist,* **108**, 593–615.

Spiegelhalter, D.J., Thomas, A., Best, N. and Gilks, W.R. (1995). BUGS, Bayesian inference Using Gibbs Sampling, Version 0.50, Medical Research Council Biostatistics Unit, Institute of Public Health, Cambridge University.

Tieleman, B.I. and Williams, J.B. (2000). The Adjustment of Avian Metabolic Rates and Water Fluxes to Desert Environments, *Physiological and Biochemical Zoology,* 73, 461–479.

von Bonin, G. (1937). Brain-Weight and Body-Weight of Mammals, *Journal of General Psychology,* **16**, 379–389.

Weisberg, S. (1985). *Applied Linear Regression* (2nd edition), New York: John Wiley & Sons.

Bayesian Protein Structure Prediction

Scott C. Schmidler
Jun S. Liu
Douglas L. Brutlag

ABSTRACT An important role for statisticians in the age of the Human Genome Project has developed in the emerging area of "structural bioinformatics". Sequence analysis and structure prediction for biopolymers is a crucial step on the path to turning newly sequenced genomic data into biologically and pharmaceutically relevant information in support of molecular medicine. We describe our work on Bayesian models for prediction of protein structure from sequence, based on analysis of a database of experimentally determined protein structures. We have previously developed segment-based models of protein secondary structure which capture fundamental aspects of the protein folding process. These models provide predictive performance at the level of the best available methods in the field (Schmidler et al., 2000). Here we show that this Bayesian framework is naturally generalized to incorporate information based on non-local sequence interactions. We demonstrate this idea by presenting a simple model for β-strand pairing and a Markov chain Monte Carlo (MCMC) algorithm for inference. We apply the approach to prediction of 3-dimensional contacts for two example proteins.

1 Introduction

The Human Genome Project estimates that sequencing of the entire complement of human DNA will be completed in the year 2003, if not sooner (Collins et al., 1998). At the same time a number of complete genomes for pathogenic organisms are already available, with many more under way. Widespread availability of this data promises to revolutionize areas of biology and medicine, providing fundamental insights into the molecular mechanisms of disease and pointing the way to the development of novel therapeutic agents. Before this promise can be fulfilled however, a number of significant hurdles remain. Each individual gene must be located within the 3 billion bases of the human genome, and the functional role of its associated protein product identified. This process of functional characterization, and subsequent development of pharmaceutical agents to affect that function, is greatly aided by knowledge of the 3-dimensional structure into which the protein folds. While the sequence of a protein can be determined directly from the DNA of the gene which encodes it, prediction of the 3-dimensional structure of the protein from that sequence remains one of the great open problems of science. Moreover, the scale of the problem (the human genome is projected to contain approximately 30,000-100,000 genes) necessitates the development of *computational* solutions which capitalize on the laboriously acquired experimental structure data. The field of research which has sprung up in support of these efforts is coming to be known as "structural bioinformatics", and poses a number of scientifically important and theoretically challenging problems involving data analysis and prediction. Emerging efforts to develop and make publicly available a large, structurally diverse set of experimental data as a structural analog of the Human Genome Project (Burley et al., 1999; Montelione and Anderson, 1999) promise to provide a multitude of statistical problems within this emerging research area.

2 Protein Structure Prediction

2.1 *Proteins and their structures*

A protein sequence is a linear heteropolymer, meaning simply that it is an unbranched chain of molecules with each "link" in the chain made up by one of the twenty *amino acids* (see Figure 10.1). Proteins perform the vast majority of the biochemistry required by living organisms, playing various catalytic, structural, regulatory, and signaling roles required for cellular development, differentiation, replication, and survival. The key to the wide variety of functions exhibited by individual proteins is not the linear sequence as shown in Figure 10.1 however, but the three dimensional configuration adopted by this sequence in its native environment. In order to understand protein function at the molecular level then, it is crucial to study the structure adopted by a particular sequence. Unfortunately the physical process by which a sequence achieves this structure, known as *protein folding*, remains poorly understood despite decades of study. In particular, serious difficulties present themselves when one attempts to *predict* the folded structure of a given protein sequence.

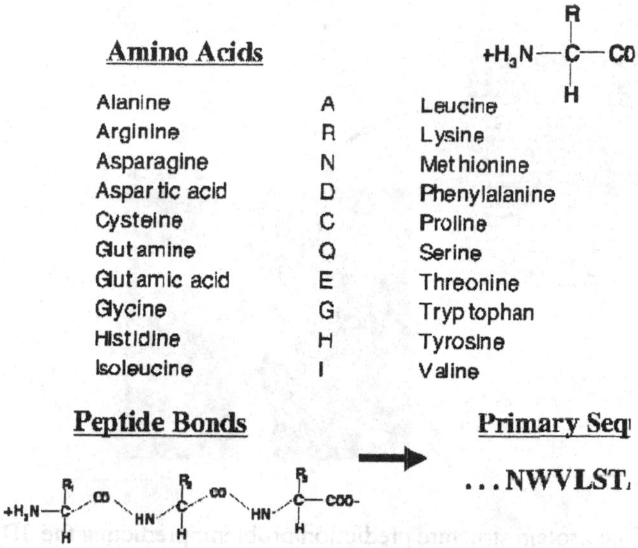

Amino Acids

+H₃N—C—CO (R above, H below)

Alanine	A	Leucine
Arginine	R	Lysine
Asparagine	N	Methionine
Aspartic acid	D	Phenylalanine
Cysteine	C	Proline
Glutamine	Q	Serine
Glutamic acid	E	Threonine
Glycine	G	Tryptophan
Histidine	H	Tyrosine
Isoleucine	I	Valine

Peptide Bonds → **Primary Seq**

$+H_3N$—C—CO—HN—C—CO—HN—C—COO-

...NWVLST.

FIGURE 10.1. The basic components of protein structure. Proteins are made up of twenty naturally occurring amino acids linked by peptide bonds to form linear polymers. Each amino acid is represented by a letter of the alphabet to produce a protein sequence.

2.2 *Protein structure prediction*

The basic problem of protein structure prediction is summarized in Figure 10.2. The goal is to take an amino acid sequence, represented as a sequence of letters as shown in Figure 10.1, and predict the three dimensional conformation adopted by the protein in its native (folded) state.

The difficulties in doing so are numerous, and significant effort has been directed towards developing approximate methods based on reduced representations of proteins (see Neumaier (1997) for a review from a mathematical perspective). Here we focus only on one such abstraction of the problem, which characterizes a protein structure by short segments of regular repeated conformation, known as *secondary structure*. Figure 10.3 shows the process of successive abstraction leading to a representation of protein structure in terms of secondary structure vectors in space. The secondary structure elements of greatest interest are helical regions known as α-*helices*, and extended regions known as β-*strands*, which join together to form β-*sheets*. Figure 10.4 shows both an α-helix and a β-sheet. The *secondary structure prediction problem* is the task of predicting the location of α-helices and β-strands in an amino acid sequence, in the absence

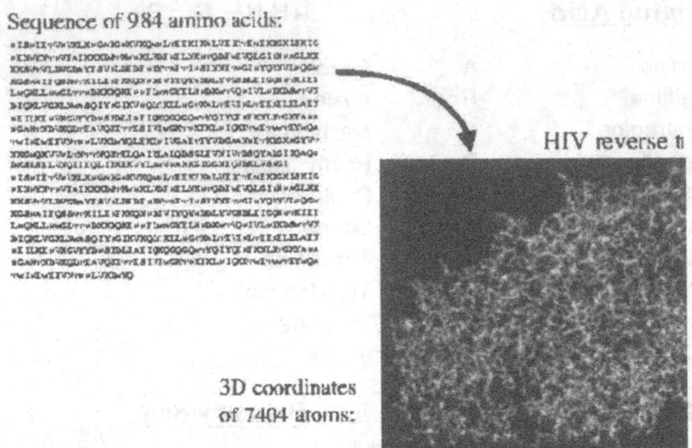

FIGURE 10.2. The protein structure prediction problem: predicting the 3D coordinates of a folded protein from the amino acid sequence. The example protein shown is HIV reverse transcriptase, a DNA polymerase required for HIV replication and therefore a target for pharmaceutical development.

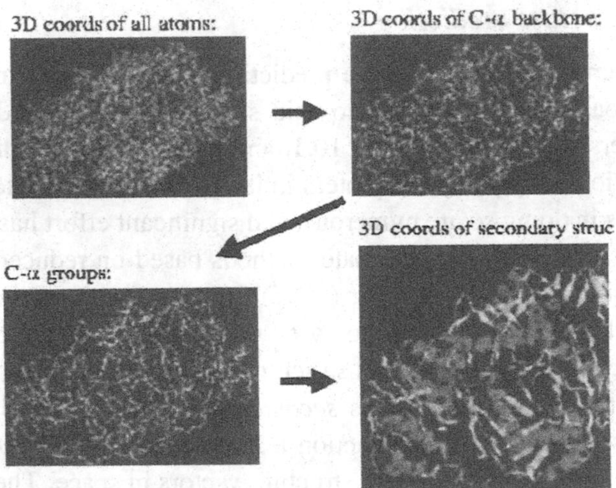

FIGURE 10.3. Successive abstraction of the problem: From atomic coordinates to α-carbon backbone to segments of secondary structure.

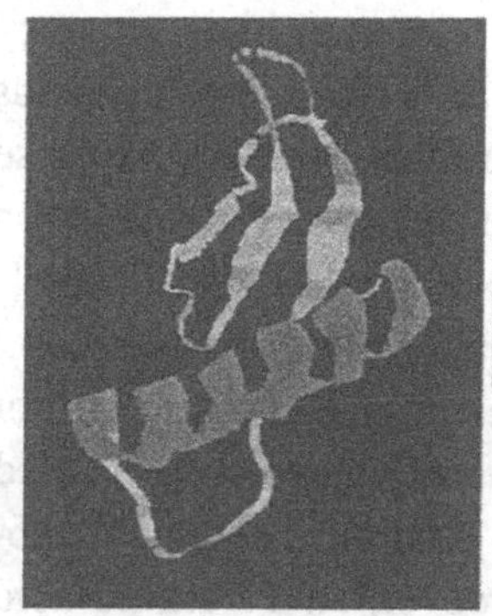

FIGURE 10.4. The secondary structure of a protein is defined by the local backbone conformation at each position. Secondary structure elements of greatest interest include α-helices (dark) and extended β-strands which come together to form β-sheets (light). These are represented as H and E respectively in the 1D summary. Remaining positions are represented by L for loop/coil.

of any knowledge of the tertiary structure of the protein. The task is thus to predict a 1-dimensional summary of the 3-dimensional folded structure, as shown in Figure 10.4. This 1D summary is typically formulated as a 3-state problem, with all positions classified as being in either α-helix (H), extended β-strand (E), or loop/coil (L) conformation. Accurate secondary structure predictions are of considerable interest, because knowledge of the location of secondary structure elements can be used for approximate folding algorithms (Monge et al., 1994; Eyrich et al., 1999) or to improve fold recognition algorithms (Fischer and Eisenberg, 1996; Russell et al., 1996), which can in many cases yield low-resolution 3D structures for the folded protein. Because of this, secondary structure prediction has received a great deal of attention over several decades, but remains a difficult problem (see Barton (1995) or references in King and Sternberg (1996); Schmidler et al. (2000) for a review). Standard approaches predict each sequence position independently based on a local surrounding subsequence. The most accurate such "window-based" methods currently use neural-networks or nearest-neighbor classifiers. A widely recognized drawback of these approaches is lack of interpretability of model parameters, yielding little insight into the important factors in protein folding.

2.3 Non-local effects

One of the difficulties in predicting secondary structure at high accuracy is the importance of *non-local* contacts in protein folding. Amino acids which are sequentially distant in the primary structure may be in close physical proximity in the tertiary structure, as the sequence folds back on itself in three dimensions. The relative importance of local vs. non-local effects in determining protein folds is still under debate (Baldwin and Rose, 1999; Dill, 1999), but it is clear that non-local effects can be important. For example, identical 5- and 6- amino acid subsequences have been located which take on different local conformations in different proteins (Kabsch and Sander, 1984; Cohen et al., 1993). Moreover, an 11 amino acid "chameleon" sequence has been designed which folds into an α-helical conformation when placed at one position in a particular protein, and a β-strand conformation when placed at a different position of the same protein (Minor and Kim, 1996). A possible explanation for such observations is the effect of non-local contacts in determining local structure.

Regardless of their importance for driving the physical folding process, non-local interactions induce correlations in the sequence which can provide useful information for protein structure prediction. For example, the side chains of adjacent β-strands in a β-sheet will experience a similar chemical environment, and therefore acceptable mutations in these strands will exhibit correlations (Lifson and Sander, 1980; Hutchinson et al., 1998). In general, positions which are in close physical proximity in the tertiary structure may be expected to exhibit correlated mutations, irrespective of their relative positions in sequence. In Section 4, we show how such information can be captured formally for use in prediction.

3 Bayesian Sequence Segmentation

We have developed a Bayesian framework for prediction of protein secondary structure from sequence. Our approach is based on the parameterization of protein sequence/structure relationships in terms of structural *segments*. An overview of the class of models developed is provided here; more details and relations to other statistical models can be found in

Schmidler (2000).

Let $R = (R_1, R_2, \ldots, R_n)$ be a sequence of n amino acid residues, $S = \{i \mid Struct(R_i) \neq Struct(R_{i+1})\}$ be the positions denoting the ends of m structural segments, and $T = (T_1, T_2, \ldots, T_m)$ be the secondary structure types for the segments. We refer to the set (m, S, T) as a *segmentation* of the sequence R. A segmentation defines an assignment of secondary structure to the sequence R, and we wish to infer the unobserved structure (m, S, T) for an observed sequence R.

We define a joint distribution over (R, m, S, T) of the form:

$$P(R, m, S, T) \propto P(m, S, T) \prod_{j=1}^{m} P(R_{[S_{j-1}+1:S_j]} \mid m, S, T) \qquad (10.1)$$

which factors the joint likelihood $P(R \mid m, S, T)$ by conditional independence of segments given their locations and structural types. Note that marginalization over latent variables (S, T) yields a complex dependency structure among the observed sequence. A special case of (10.1) is a hidden Markov model (HMM).

The segment likelihoods $P(R_{[S_{j-1}+1:S_j]} \mid m, S, T)$ may be of general form. Detailed segment models have been developed to account for experimentally and statistically observed properties of α-helices and β-strands (Schmidler et al., 2000). These models generalize existing stochastic models for secondary structure prediction based on HMMs (Asai et al., 1993; Stultz et al., 1993) in several important ways. The factorization in terms of segments allows modeling of non-independence and non-identity of amino acid distributions at varying positions in the segment. Both position-specific distributions and dependency among positions capture important structural signals such as helix-capping (Aurora and Rose, 1998) and side chain correlations (Klingler and Brutlag, 1994), and these advantages have been explored in detail in previous work.

Given a set of segment likelihoods, we wish to predict the secondary structure for a newly observed protein sequence R. Taking a Bayesian approach, we assign a prior $P(m, S, T)$ and base our predictions on $P(m, S, T \mid R)$, the posterior distribution over secondary structure assignments given the observed sequence. Choice of priors is discussed in Schmidler (2000); one possible approach is to factor $P(m, S, T)$ as a semi-Markov process:

$$P(m, S, T) = P(m) \prod_{j=1}^{m} P(T_j \mid T_{j-1}) P(S_j \mid S_{j-1}, T_j), \qquad (10.2)$$

which accounts for empirically observed differences in segment length distributions among structural types.

Under the model defined by (10.1) and (10.2), we consider two possible predictors of interest:

$$Struct_{MAP} = \arg\max_{(m,S,T)} P(m, S, T \mid R, \theta) \qquad (10.3)$$

$$Struct_{Mode} = \{\arg\max_{T} P(T_{R_{[i]}} \mid R, \theta)\}_{i=1}^{n} \qquad (10.4)$$

where $Struct_X$ is a segmentation of R, θ denotes the model parameters, and $P(T_{R_{[i]}} \mid R, \theta)$ is the marginal posterior distribution over structural types at a single position i in the sequence:

$$P(T_{R_{[i]}} \mid R, \theta) = \sum_{(m,S,T)} P(m, S, T \mid R, \theta) \mathbf{1}_{\{T_{R_i} = t\}}$$

(10.3) provides the *maximum a posteriori* segmentation of a sequence, while (10.4) provides the sequence of marginal posterior modes. Note that (10.4) involves marginalization over all possible segmentations. Efficient algorithms have been developed for computation of these estimators under the model defined by (10.1, 10.2) (Schmidler et al., 2000).

4 Incorporation of Inter-Segment Interactions

A fundamental assumption assumption of the class of models described by (10.1) is the conditional independence of amino acids which occur in distinct segments. This assumption enables the exact calculation posterior quantities as mentioned above. However, this assumption is clearly violated in the case of protein sequences, due to the non-local forces involved in protein folding described in Section 2.3. For example, β-sheets consist of β-strands linked by backbone hydrogen bonds (Figure 10.4). β-sheets are thus a major structural motif which involves interactions of sequentially distant segments to form a stable native fold. Other examples include disulfide bonds and helical bundles. The presence of correlated mutations in such motifs is well known (see Section 2.3). It is often suggested that the inability of window-based prediction algorithms to capture such non-local patterns is responsible for the low accuracy typically achieved in β-strand prediction.

In this section, we extend the framework of Section 3 by introducing joint segment models to account for such inter-segment residue correlations. We describe a MCMC algorithm for inference in this class of models, and demonstrate this approach with a simple model for β-strand pairing in β-sheets.

4.1 Joint segment likelihoods

Modeling of segment interactions may be achieved by definition of *joint* segment likelihoods. For two interacting segments j and k, we replace the terms

$$P(R_{[S_{j-1}+1:S_j]} \mid S_{j-1}, S_j, T_j) \text{ and } P(R_{[S_{k-1}+1:S_k]} \mid S_{k-1}, S_k, T_k)$$

in the product of (10.1) above with a joint term:

$$P(R_{[S_{j-1}+1:S_j]}, R_{[S_{k-1}+1:S_k]} \mid S_{j-1}, S_j, T_j, S_{k-1}, S_k, T_k) \qquad (10.5)$$

Hence we may include arbitrary joint segment distributions for segment pairs into the model. The extension to three or more segments (as may be required for 4-helix bundles or β-sheets, for example) is obvious. Such models contain pair potentials as a special case; see Schmidler (2000) for a more formal development of this class of models.

Inclusion of terms such as (10.5) leads to a joint distribution of the form:

$$P(R, m, S, T, \mathcal{P}) \propto P(m, S, T, \mathcal{P}) \prod_{j \notin \mathcal{P}} P(R_{[S_{j-1}+1:S_j]} \mid S, T, m, \mathcal{P}) \times \prod_{(j,k) \in \mathcal{P}} P(R_{[S_{j-1}+1:S_j]}, R_{[S_{k-1}+1:S_k]} \mid S, T, m, \mathcal{P}) \qquad (10.6)$$

where $\mathcal{P}$ is the set of pairs of interacting segments. For example, $\mathcal{P}$ might be the set of β-sheets, with each $p \in \mathcal{P}$ a set of β-strand segments participating in the sheet. Clearly elements $p \in \mathcal{P}$ may include > 2 segments, in which case (10.5) must be defined appropriately. It is also necessary to extend the prior $P(m, S, T)$ to include interactions $P(m, S, T, \mathcal{P})$. For the remainder of this paper we will take $P(m, S, T)$ as defined in (10.2) above, and take $P(\mathcal{P} \mid m, S, T) \propto 1$. This extends the previous semi-Markov prior by a conditionally uniform prior on segment interactions. More realistic priors for $(m, S, T, \mathcal{P})$ are developed in Schmidler (2000) .

This joint distribution (10.6) is easily evaluated for any fixed segmentation $(m, S, T, \mathcal{P})$ of a sequence R. However computation of posterior quantities such as (10.3) and (10.4) in the context of (10.6) involves maximization/marginalization over *all possible* segment interactions, an intractable computation.

4.2 Markov chain Monte Carlo segmentation

Despite the difficulty in exact calculation of posterior probabilities, approximate inference in models such as described by (10.6) is feasible using MCMC methods, now a standard tool in the Bayesian statistics community (Gilks et al., 1996). Because the problem has varying dimensionality (m and $\mathcal{P}$ are random variables), we use the reversible jump approach described by Green (1995).

To construct a Markov chain on the space of sequence segmentations, we define the following set of Metropolis proposals:

- *Type switching*:
  Given a segmentation (m, S, T), propose a move to segmentation (m, S, T^*) where $T_j^* = T_j, j \neq k$ for some k chosen uniformly at random or by systematic scan, and $T_k^* \sim U[\{H, E, L\}]$.
- *Position change*:
  Given (m, S, T), propose (m, S^*, T) with $S_j^* = S_j, j \neq k$ for some k and $S_k^* \sim U[S_{k-1} + 1, S_{k+1} - 1]$.
- *Segment split*:
  Given (m, S, T), propose (m^*, S^*, T^*) with $m^* = m + 1$ segments by splitting segment $1 \leq k \leq m$ into two new segments $(k^*, k^* + 1)$ where $k \sim U[1, m]$, $S_{k^*+1}^* = S_k$, and $S_{k^*}^* \sim U[S_{k-1} + 1, S_k - 1]$. With probability $\frac{1}{2}$, we set $T_{k^*} = T_k$ and $T_{k^*+1} = T_{new}$ with T_{new} chosen uniformly, and with probability $\frac{1}{2}$ do the reverse.
- *Segment merge*:
  Similar to *segment split*, but a randomly chosen segment is merged into a neighbor and $m^* = m - 1$.

All moves are accepted or rejected based on a reversible jump Metropolis criteria (Hastings, 1970; Green, 1995). Together, these steps are sufficient to guarantee ergodicity for models of the form (10.1). The factorization of (10.1) allows Metropolis ratios to be evaluated *locally* with respect to the affected segments. Often the above proposals can be replaced by Gibbs sampling steps which draw from the exact conditional distribution, although it may still be more efficient to *Metropolize* such moves (Liu, 1996).

For joint segment models such as (10.6), additional proposal moves must be added involving interacting segments:

- *Segment join*:
  Proposes a replacement of two non-interacting segments (S_j, T_j) and (S_k, T_k), $(j, k) \notin \mathcal{P}$ with an interaction (S_j, S_k, T_j, T_k), $(j, k) \in \mathcal{P}$. In Section 5 below, this corresponds to replacing two independent β-strands with a β-sheet consisting of the two strands joined.
- *Segment separate*:
  Reverse of *segment join*. For example, splits a 2-strand sheet into two independent strands.

Some care must be taken to realize these proposals for a particular set of joint models, such as those provided in Section 5, especially when interactions may involve more than 2 segments. This is discussed in greater detail by Schmidler (2000), who also provides additional Metropolis moves not required for ergodicity but helpful in improving mixing of the underlying Markov chain.

By defining (10.5) as a product of independent terms and choosing the prior appropriately, we can recover model (10.1) and hence compare this MCMC approach to exact calculations. Figure 10.5a shows that in this case convergence is quite rapid.

5 Application to Prediction of β-Sheets

As mentioned in Section 2.3, the existence of correlated mutations in β-sheets has been well studied in the protein structure literature. Some attempts have been made to incorporate such long-range sequence correlations into the prediction of protein structure (Hubbard and Park, 1995; Krogh and Riis, 1996; Frishman and Argos, 1996). Here, we show how these interactions are naturally modeled in the Bayesian framework provided by Sections 3 and 4, allowing the information to be formally included in the predictive model.

To demonstrate the application of (10.5) in this case, we define the following joint model for adjacent β-strands to incorporate pairwise side chain correlations:

$$\begin{aligned} &P(R_{[S_{j-1}+1:S_j]}, R_{[S_{k-1}+1:S_k]} \mid S, T, m, \mathcal{P}) = \\ &\prod_{(h_j,h_k)\in H} P(R_{[S_{j-1}+h_j]}, R_{[S_{k-1}+h_k]} \mid S, T, m, \mathcal{P}) \times \qquad (10.7) \\ &\prod_{h_j \notin H} P(R_{[S_{j-1}+h_j]} \mid S, T, m, \mathcal{P}) \prod_{h_k \notin H} P(R_{[S_{k-1}+h_k]} \mid S, T, m, \mathcal{P}) \end{aligned}$$

where H is the set of (ordered) cross-strand neighboring pairs. This model is simply a product distribution over pairs of neighboring amino acids, the simplest possible model which captures some notion of inter-strand correlation. More detailed models are currently being developed.

This approach has been applied to the prediction of contacts for two test proteins, bovine pancreatic trypsin inhibitor (BPTI) shown in Figure 10.5b and flavodoxin (not shown). Results for BPTI are shown in Figure 10.5c,d, where strand pairing is well predicted. Results for flavodoxin are shown in Figure 10.5e,f, where it is seen that strands are well identified but their interaction pattern has high uncertainty. More accurate interaction models and priors may help resolve this uncertainty. In each case the simulations shown restrict the orientation (parallel vs. anti-parallel) of the interactions to be correct, eliminating a further source of variability. A more extensive evaluation of this approach on a large database is underway, and will be reported elsewhere.

6 Discussion

We have discussed the problem of protein structure prediction, and presented a Bayesian formulation. Models based on factorization of the joint distribution in terms of structural segments naturally capture important properties of proteins, permit efficient algorithms, and produce accurate predictions. Moreover, we have shown here that the Bayesian framework is naturally generalized to model non-local interactions in protein folding. As an example, we have presented a simple model for β-strand pairing, and a Markov chain Monte Carlo algorithm for inference, and have demonstrated this approach on example sequences. Further work on modeling and evaluation for this problem is underway (Schmidler, 2000). The ability to predict tertiary contacts between β-sheets represents a potentially important step in going beyond traditional secondary structure prediction towards the goal of full 3D structure prediction.

Acknowledgments

SCS was partially supported by NLM training grant LM-07033 and NCHGR training grant HG-00044-04 during portions of this work. JSL is partially supported by NSF grants DMS-9803649 and DMS-0094613. DLB is supported by NHGRI grant HGF02235-07.

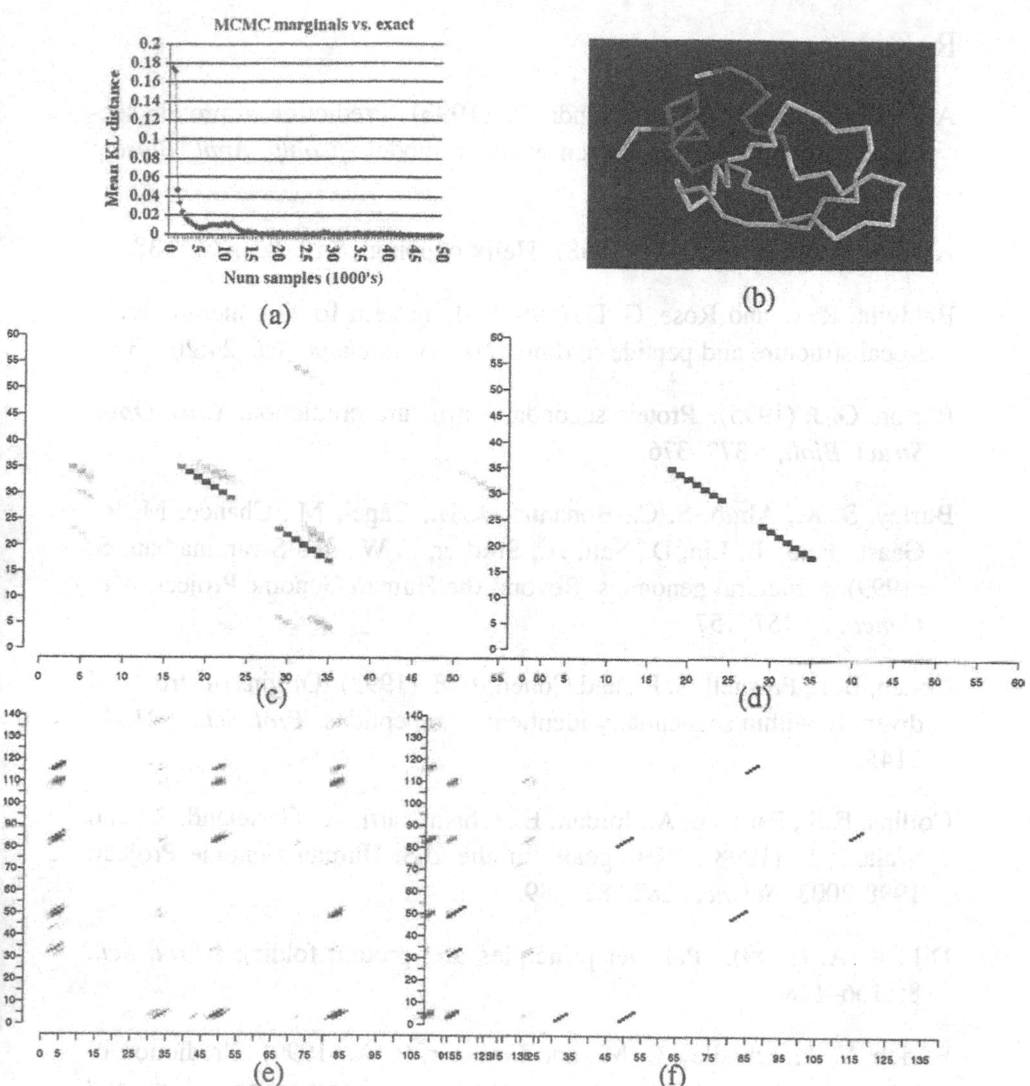

FIGURE 10.5. (a) Convergence of MCMC simulation to exact calculations. Plot is mean Kullback-Leibler (KL) divergence between marginal distributions $P(T_{R_{[i]}} \mid R, \theta)$ obtained from exact and MCMC calculations for a protein sequence, against number iterations (each iteration 1 full scan). KL divergence between two probability distributions $\mathbf{p}$ and $\mathbf{q}$ is defined as $KL(\mathbf{p}, \mathbf{q}) = \sum_i p_i \log(\frac{p_i}{q_i})$. (b) True structure of bovine pancreatic trypsin inhibitor (BPTI). (c) Predicted and (d) true β-strand contacts for BPTI. Axes are sequence position, and shading of (x, y) is proportional to predicted probability of contact for positions x, y. The β-hairpin contacts are predicted with high probability. The *maximum a posteriori* sheet topology correctly identifies β-strand locations and register (not shown). Pairings representing register shifts are also observed with lower probability. (e) Predicted and (f) true contacts for flavodoxin, showing significant uncertainty in correct pairing of strand segments.

References

Asai, K., Hayamizu, S., and Handa, K. (1993). Prediction of protein secondary structure by the hidden Markov model. *Comp. Appl. Biosci.*, 9(2):141–146.

Aurora, R. and Rose, G. D. (1998). Helix capping. *Prot. Sci.*, 7:21–38.

Baldwin, R. L. and Rose, G. D. (1999). Is protein folding hierarchic? I. Local structure and peptide folding. *Trends Biochem. Sci.*, 24:26–33.

Barton, G. J. (1995). Protein secondary structure prediction. *Curr. Opin. Struct. Biol.*, 5:372–376.

Burley, S. K., Almo, S. C., Bonanno, J. B., Capel, M., Chance, M. R., Gaasterland, T., Lin, D., Sali, A., Studier, F. W., and Swaminathan, S. (1999). Structural genomics: Beyond the Human Genome Project. *Nat. Genet.*, 23:151–157.

Cohen, B. I., Presnell, S. R., and Cohen, F. E. (1993). Origins of structural diversity within sequentially identical hexapeptides. *Prot. Sci.*, 2:2134–2145.

Collins, F. S., Patrinos, A., Jordan, E., Chakravarti, A., Gesteland, R., and Walters, L. (1998). New goals for the U.S. Human Genome Project: 1998-2003. *Science*, 282:682–689.

Dill, K. A. (1999). Polymer principles and protein folding. *Prot. Sci.*, 8:1166–1180.

Eyrich, V. A., Standley, D. M., and Friesner, R. A. (1999). Prediction of protein tertiary structure to low resolution: Performance for a large and structurally diverse test set. *J Mol. Biol.*, 288:725–742.

Fischer, D. and Eisenberg, D. (1996). Protein fold recognition using sequence-derived predictions. *Prot. Sci.*, 5:947–955.

Frishman, D. and Argos, P. (1996). Incorporation of non-local interactions in protein secondary structure prediction from the amino acid sequence. *Prot. Eng.*, 9(2):133–142.

Gilks, W. R., Richardson, S., and Spiegelhalter, D. J., editors (1996). *Markov Chain Monte Carlo in Practice*. Chapman & Hall.

Green, P. J. (1995). Reversible jump Markov chain Monte Carlo computation and Bayesian model determination. *Biometrika*, 82(4):711–32.

Hastings, W. K. (1970). Monte Carlo sampling methods using Markov chains and their applications. *Biometrika*, 57:97–109.

Hubbard, T. J. and Park, J. (1995). Fold recognition and ab initio structure predictions using hidden Markov models and β-strand pair potentials. *Proteins: Struct. Funct. Genet.*, 23:398–402.

Hutchinson, E. G., Sessions, R. B., Thornton, J. M., and Woolfson, D. N. (1998). Determinants of strand register in antiparallel β-sheets of proteins. *Prot. Sci.*, 7:2287–2300.

Kabsch, W. and Sander, C. (1984). On the use of sequence homologies to predict protein structure: Identical pentapeptides can have completely different conformations. *Proc. Natl. Acad. Sci. USA*, 81(4):1075–1078.

King, R. D. and Sternberg, M. J. E. (1996). Identification and application of the concepts important for accurate and reliable protein secondary structure prediction. *Prot. Sci.*, 5:2298–2310.

Klingler, T. M. and Brutlag, D. L. (1994). Discovering structural correlations in α-helices. *Prot. Sci.*, 3:1847–1857.

Krogh, A. and Riis, S. K. (1996). Prediction of beta sheets in proteins. In Touretzky DS, Mozer MC, H. M., editor, *Advances in Neural Information Processing Systems 8*. MIT Press.

Lifson, S. and Sander, C. (1980). Specific recognition in the tertiary structure of β-sheets of proteins. *J Mol. Biol.*, 139:627–639.

Liu, J. S. (1996). Peskun's theorem and a modified discrete-state Gibbs sampler. *Biometrika*, 83:681–682.

Minor, D. L. J. and Kim, P. S. (1996). Context-dependent secondary structure formation of a designed protein sequence. *Nature*, 380:730–734.

Monge, A., Friesner, R. A., and Honig, B. (1994). An algorithm to generate low-resolution protein tertiary structures from knowledge of secondary structure. *Proc. Natl. Acad. Sci. USA*, 91:5027–5029.

Montelione, G. T. and Anderson, S. (1999). Structural genomics: Keystone for a Human Proteome Project. *Nat. Struct. Biol.*, 6:11–12.

Neumaier, A. (1997). Molecular modeling of proteins and mathematical prediction of protein structure. *SIAM Rev.*, 39(3):407–460.

Russell, R. B., Copley, R. R., and Barton, G. J. (1996). Protein fold recognition by mapping predicted secondary structures. *J Mol. Biol.*, 259:349–365.

Schmidler, S. C. (2000). *Statistical Models and Monte Carlo Methods for Protein Structure Prediction*. PhD thesis, Stanford University.

Schmidler, S. C., Liu, J. S., and Brutlag, D. L. (2000). Bayesian segmentation of protein secondary structure. *J. Comp. Biol.*, 7(1):233–248.

Stultz, C. M., White, J. V., and Smith, T. F. (1993). Structural analysis based on state-space modeling. *Prot. Sci.*, 2:305–314.

Bayesian Analysis of Sensory Inputs of a Mobile Robot

Paola Sebastiani
Marco Ramoni
Paul Cohen

ABSTRACT This paper applies a novel Bayesian clustering method to identify characteristic dynamics of sensory inputs of a mobile robot. The method starts by transforming the sensory inputs into Markov chains and then applies our new agglomerative clustering procedure to discover the most probable set of clusters describing the robot's experiences. To increase efficiency, the method uses an entropy-based heuristic search strategy.
KEYWORDS: Model-based clustering; Markov Chains; Robotics.

1 Introduction

The goal of this work is to enable mobile robots to learn the characteristic dynamics of their activities. Our robot interacts with the world via its sensors and, during its activities, it records their values every 1/10 of a second. In an extended period of wandering in the laboratory, the robot will engage in several different activities — moving toward an object, loosing sight of an object, bumping into something — and all this activities will have different sensory signatures. If we regard the sequence of sensory inputs of the robot as a time series, different sensory signatures can be identified with different dynamic processes generating the series. It is important, to the goals of our project, that the robot's learning should be *unsupervised*, which means we do not tell the robot when it has switched from one activity to another. Instead, we define a simple *event marker* — a simultaneous change of at least three sensors — and we define an *episode* as the time series between two consecutive event markers. The available data are then a set of episodes for each sensor and the statistical problem is to cluster episodes having the same dynamics and to learn the dynamic process generating all the episodes in a cluster.

The solution we have developed is an algorithm for Bayesian clustering by dynamics (BCD). We model the dynamic of each episode as a Markov

Chain (MC) and our algorithm learns MC representations of the episode dynamics and cluster those episodes that have a high posterior probability of being generated from the same MC. The character of our method is to regard the clustering task as a Bayesian model selection problem and the model we look for is the best way of partitioning episodes, conditional on the data at hand and prior information about the problem. Hence, in principle, we just need to evaluate all possible partitions of the episodes and select that with largest posterior probability. However, the number of possible partitions grows exponentially with the number of episodes and a heuristic search is needed to make the method realistic. The solution we have developed is to use a measure of similarity between MCs to drive the search process in a subspace of all possible partitions.

Bayesian model based clustering was originally considered to cluster static data (Banfield and Raftery, 1993; Cheeseman and Stutz, 1996). Recent work (Smyth, 1999) attempted to extend the idea to dynamic processes without, however, succeeding in finding a closed form solution as the one we have identified. Furthermore, an important novelty of our method is its heuristic search that makes the algorithm very efficient.

FIGURE 11.1. The Pioneer 1 robot.

The rest of the paper is organized as follows. After a brief description of the robot we use in our work, we describe the BCD algorithm and then show its application to learn characteristic dynamics of the robot episodes.

2 The Robot and its Sensors

Our robot is the Pioneer 1 and it is depicted in Figure 11.1. It is a small platform with two drive wheels and a trailing caster, and a two degree of freedom paddle gripper (the two metal arms coming out of the platform). For sensors, the Pioneer 1 has shaft encoders, stall sensors, which signal when the right and left wheel velocities are discordant, five forward pointing and two side pointing sonars (recognizable as the circles in the top half of Figure 11.1), bump sensors and a pair of infra-red sensors at the front and back of its gripper. The bump sensors signal when the robot has touched an object so, for example, they go on when the robot pushes or bumps into something. The infra-red sensors at the front and back of the gripper signal when the robot is in contact with an object and is, for example, grasping it. The robot has also a simple vision system that reports the location and size of color-coded objects via three sensors. The robot can be trained to recognize objects of a particular color and has a black and white vision that is created using the area of the object seen (expressed as number of pixels) and the horizontal and vertical location in its two-way vision field. Our configuration of the Pioneer 1 has roughly forty sensors, though the values returned by some are derived from others.

We will focus attention to only 8 sensors, that are described in Table 11.1. Figure 11.2 shows an example of sensor values recorded during 30 seconds of activity of the robot. The velocity related sensors r.vel, l.vel, as well as the sensors of the vision system vis.a, vis.x and vis.y take continuous values and were discretized into 5 bins of equal length, labeled between .2 and 1. The vis.a sensor has a highly skewed distribution so that the square root of the original values were discretized. Hence, the category .2 for both sensors r.vel l.vel represents values between -600 and -340, while .4 represents values between -360 and -120, .6 represents values between -120 and 120 and so on. Negative values of the velocities of both wheels result in the robot moving backward, while positive velocities of both wheels represent forward movements. Both negative and positive velocities of the wheels result in the robot turning. The first two plots show sensory values of the left and right wheel velocity from which we can deduce that the robot is probably not moving during the first 5 seconds (steps 1 to 50) or moving slowly (the bin labeled 0.6 represents range of velocity between -120 and 120). After the 5th second, the robot turns (the values of the velocity of the two wheels are discordant) stops and then moves forward, first at low velocity then at increasing velocity until is stops, begins

Sensor name	Interpretation	Range of values
r.vel	velocity of right wheel	-600–600
l.vel	velocity of left wheel	-600–600
grip.f	infra-red sensor at the front of the gripper	0=off; 1=on
grip.r	infra-red sensor at the rear of the gripper	0=off; 1=on
grip.b	bumper sensor	0=off; 1=on
vis.a	number of pixels of object in the visual field	0–40,000
vis.x	horizontal location of object in the visual field	-140 = nearest 140= furthest
vis.y	position of object in the visual field	0 = most left, 256= most right

TABLE 11.1. Some of the robot's sensors and their range of values.

moving backward (as the velocity of both wheels is negative) and then forward again. Note that the sensors grip.f and grip.b go on and stay on in the same time interval. Furthermore, the dynamic of the sensor vis.a shows the presence of an object of increasing and decreasing size in the visual field, with maximum size corresponding to the time in which the sensors of the wheel velocity record a change of trend. The trend of the other two sensors of the vision system, vis.x and vis.y, both support the idea that the robot is moving toward an object, bumps into it, and then moves away.

During an extensive period of wandering, the robot will engage in several similar activities. The goal is to enable it to recognize similar activities on the basis of the similarity of the dynamics of the sensor inputs. The next section will describe the BCD algorithm to achieve this goal.

3 Bayesian Clustering by Dynamics

Suppose we have a batch of m time series that record the values $1, 2, ..., s$ of a variable X. The goal is to identify time series that exhibit similar dynamics. To cluster time series by their dynamics, we model the dynamics of time series as Markov chains (MCs). For each time series, we estimate a transition matrix from data and then we cluster similar transition matrices.

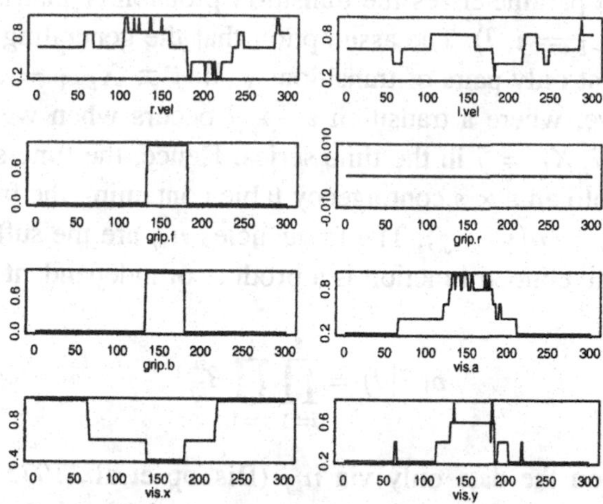

FIGURE 11.2. Sensory inputs recorded by the robot during 30 seconds. The x-axes report the time, measured every 1/10 of a second.

3.1 Learning Markov Chains

Suppose we observe a time series $S = (x_0, x_1, x_2, ..., x_{i-1}, x_i, ..)$, where each x_i is one of the states $1, ..., s$ of a variable X. The process generating the sequence S is a MC if the conditional probability that the variable X visits state j at time t, given the sequence $(x_0, x_1, x_2, ..., x_{t-1})$, is only a function of the state visited at time $t-1$. Hence, we write $p(X_t = j|(x_0, x_1, x_2, ..., x_{t-1})) = p(X_t = j|x_{t-1})$, with X_t denoting the variable X at time t. In other words, the probability distribution of the variable X at time t is *conditionally independent* of the values $(x_0, x_1, x_2, ..., x_{t-2})$, once we know x_{t-1}. This conditional independence assumption allows us to represent a MC as a vector of probabilities $p_0 = (p_{01}, p_{02}, ..., p_{0s})$, denoting the distribution of X_0 (the initial state of the chain) and a matrix P of transition probabilities, where $p_{ij} = p(X_t = j|X_{t-1} = i)$, so that

$$P = (p_{ij}) = \begin{array}{c|cccc} & & X_t & & \\ X_{t-1} & 1 & 2 & \cdots & s \\ \hline 1 & p_{11} & p_{12} & \cdots & p_{1s} \\ 2 & p_{21} & p_{22} & \cdots & p_{2s} \\ \vdots & & & \cdots & \\ s & p_{s1} & p_{s2} & \cdots & p_{ss} \end{array}$$

Given a time series generated from a MC, we can estimate the probabilities p_{ij} from the data by Bayesian conjugate analysis. We define $\theta = (\theta_{ij})$ as the vector that parameterizes the transition probability matrix, and $\theta_{ij} = p(X_t = j|X_{t-1} = i, \theta)$. The assumption that the generating process is a MC implies that only pairs of transitions $i \to j \equiv X_{t-1} = i \to X_t = j$ are informative, where a transition $i \to j$ occurs when we observe the pair $X_{t-1} = i, X_t = j$ in the time series. Hence, the time series can be summarized into an $s \times s$ contingency table containing the frequencies of transitions $n_{ij} = n(i \to j)$. The frequencies n_{ij} are the sufficient statistics since the likelihood function is a product of independent multinomial distributions

$$p(S|\theta) = \prod_{i=1}^{s} \prod_{j=1}^{s} \theta_{ij}^{n_{ij}}$$

and depends on the data only via n_{ij} (Bishop et al., 1975). To choose the prior, we suppose to have some background knowledge that can be represented in terms of a hypothetical time series of length $\alpha - s^2 + 1$ in which the $\alpha - s^2$ transitions are divided into $\alpha_{ij} - 1$ transitions of type $i \to j$. This background knowledge gives rise to a $s \times s$ contingency table, homologous to the frequency table, containing these hypothetical transitions $\alpha_{ij} - 1$ that are used to formulate a conjugate prior with density function $p(\theta) \propto \prod_{i=1}^{s} \prod_{j=1}^{s} \theta_{ij}^{\alpha_{ij}-1}$ which corresponds to assigning independent *Dirichlet* distributions, with hyper-parameters α_{ij}, to the parameters θ_i associated with each row conditional distribution of the matrix P. We will adopt the standard notation of denoting one Dirichlet distribution associated with the conditional probabilities $(\theta_{i1}, ..., \theta_{is})$ by $D(\alpha_{i1}, ..., \alpha_{is})$. A distribution given by independent Dirichlet is also known as a Hyper-Dirichlet distribution (Dawid and Lauritzen, 1993). We will denote such a Hyper-Dirichlet distribution by $HD(\alpha_{ij})_s$ where the index s denotes the number of independent Dirichlet distributions defining the Hyper-Dirichlet.

A Bayesian estimation of the probabilities p_{ij} is the posterior expectation of θ_{ij}. By conjugate analysis, the posterior distribution of θ is still Hyper-Dirichlet with updated hyper-parameters $\alpha_{ij} + n_{ij}$ and the posterior expectation of θ_{ij} is

$$\hat{p}_{ij} = \frac{\alpha_{ij} + n_{ij}}{\alpha_i + n_i} \qquad (11.1)$$

where $\alpha_i = \sum_j \alpha_{ij}$ and $n_i = \sum_j n_{ij}$. Thus, α_i and n_i are the numbers of times the variable X visits state i in a process consisting of α and n transitions, respectively.

N=

	0.2	0.4	0.6	0.8	1
0.2	0	0	0	0	0
0.4	0	10	1	0	0
0.6	0	1	89	1	0
0.8	0	0	32	0	0
1	0	0	0	0	0

$\hat{P}$=

	0.2	0.4	0.6	0.8	1
0.2	0.20	0.20	0.20	0.20	0.20
0.4	0.02	0.84	0.10	0.02	0.02
0.6	0.00	0.01	0.98	0.01	0.00
0.8	0.00	0.00	1.00	0.00	0.00
1.0	0.20	0.20	0.20	0.20	0.20

TABLE 11.2. Observed and learned transition matrices for the left.vel first 135 values displayed in Figure 11.2.

Table 11.2 reports the frequencies of transition n_{ij} $i, j = 1, ..., 5$ observed in the first 135 values recorded by the sensor left.vel in Figure 11.2, and the learned transition matrix when the prior global precision is $\alpha = 5$ and $\alpha_{ij} = 1/5$. The matrix $\hat{P}$ describes a dynamic process with transitions among states 0.4, 0.6 and 0.8, while states 0.2 and 1.0 are never visited. The category 0.6 corresponds to values of the velocity between -120 and 120, hence the large probability of visiting state 0.6, given that the velocity is in state 0.6, represents either a situation in which the robot is not moving or it is moving slowly and this activity can be followed by a short moving backward, or a very short moving forward.

3.2 *Clustering*

The second step of the learning process is an unsupervised agglomerative clustering of MCs on the basis of their dynamics. The available data are a set $S = \{S_i\}$ of m time series and the task of the clustering algorithm is two-fold: finding the set of clusters that gives the best partition of the data and assigning each time series S_i to one cluster.

Formally, the clustering is done by regarding a partition as a discrete variable C with states $C_1, \ldots, C_c$ that are not observed. Each state C_k of the variable C labels, in the same way, the time series generated from the same MC with transition probability matrix P_k and, hence, it represents a cluster of time series. The number c of states of the variable C is unknown but it is bounded above by the total number of time series in the data set S because, initially, each time series in the data set S has its own label. The clustering algorithm then tries to relabel those time series that are likely to have been generated from the same MC and thus merges the initial states $C_1, \ldots, C_m$ into a subset $C_1, \ldots, C_c$, with $c \leq m$. Figure 11.3 provides an example of three different relabeling of a data set S of four time series.

Original Data

Cluster	Time Series
C_1	S_1
C_2	S_2
C_3	S_3
C_4	S_4

Partitions

$M_1 =$

Cluster	Time Series
C_1	S_1
C_1	S_2
C_2	S_3
C_3	S_4

$M_2 =$

Cluster	Time Series
C_1	S_1
C_2	S_2
C_2	S_3
C_1	S_4

$M_3 =$

Cluster	Time Series
C_1	S_1
C_1	S_2
C_1	S_3
C_2	S_4

FIGURE 11.3. Three models corresponding to different re-labeling of a data set of four time series.

Each relabeling determines a model so that, for example, model M_1 is characterized by the variable C having three states C_1, C_2 and C_3 with C_1 labeling the time series S_1 and S_2, C_2 labeling S_3 and C_3 labeling S_4. In models M_2 and M_3, the variable C has only two states but they correspond to different labeling of the time series and hence different clusters.

The specification of the number c of states of the variable C and the assignment of one of its states to each time series S_i define a statistical model M_c. Thus, we can regard the clustering task as a Bayesian model selection problem, in which the model we are looking for is the most probable way of re-labeling time series according to their similarity, given the data. We denote by $p(M_c)$ the prior probability of each model M_c and then we use Bayes' theorem to compute its posterior probability and we select the model with maximum posterior probability. The posterior probability of M_c, given the sample S, is $p(M_c|S) \propto p(M_c)p(S|M_c)$ where the marginal likelihood $p(S|M_c)$ is computed by averaging out the parameters from the likelihood function. We show next that, under reasonable assumptions on the sample space, the adoption of a particular parameterization for the model M_c and the specification of a conjugate prior lead to a simple, closed-form expression for the marginal likelihood $p(S|M_c)$.

Conditional on the model M_c and hence on a specification of the number of states of the variable C and of the labeling of the original time series (or, equivalently, conditional on the specification of c clusters of time series), we suppose that the marginal distribution of the variable C is multinomial, with cell probabilities parameterized as $\theta_k = p(C = C_k|\theta)$. Furthermore, we suppose that, conditional on C, the MCs generating the time series assigned to different clusters C_k are independent and, also, that time series generated from the same MC are independent. We denote by $P_k = (p_{kij})$ the transition probability matrix of the MC generating the time series in cluster C_k and parameterize the cell probabilities as $\theta_{kij} = p(X_t|X_{t-1} = i, C_k, \theta)$. Therefore, the overall likelihood function is $p(S|\theta) = \prod_{k=1}^{c} \theta_k^{m_k} \prod_{ij=1}^{s} \theta_{kij}^{n_{kij}}$, where n_{kij} denotes the observed frequency of transitions $i \to j$ observed in all time series assigned to cluster C_k, and m_k is the number of time series that are assigned to cluster C_k. We now define a prior density for θ as a product of $c \times s+1$ Dirichlet densities. One Dirichlet is the prior distribution assigned to (θ_k), say $D(\beta_1, ..., \beta_c)$. The other $c \times s$ densities correspond to c independent Hyper-Dirichlet distribution $HD(\alpha_{kij})_s$, each distribution $HD(\alpha_{kij})_s$ being assigned to the parameters θ_{kij} of the MC generating thc time series in cluster C_k. The marginal likelihood is then given by $p(S|M_c) = \int p(S|\theta)p(\theta)d\theta$, and it is easy to show (by using the same integration techniques in Cooper and Herskovitz (1992)) that

$$p(S|M_c) = \frac{\Gamma(\beta)}{\Gamma(\beta+m)} \prod_{k=1}^{c} \frac{\Gamma(\beta_k + m_k)}{\Gamma(\beta_k)} \prod_{i=1}^{s} \frac{\Gamma(\alpha_{ki})}{\Gamma(\alpha_{ki} + n_{ki})} \prod_{j=1}^{s} \frac{\Gamma(\alpha_{kij} + n_{kij})}{\Gamma(\alpha_{kij})}$$

where $\Gamma(\cdot)$ denotes the Gamma function, $n_{ki} = \sum_j n_{kij}$ is the number of transitions from state i observed in cluster C_k, $\sum_k m_k = m$ and $\beta = \sum_k \beta_k$. Once the, *a posteriori*, most likely partition has been selected, the transition probability matrix P_k associated with the cluster C_k can be estimated as $(\hat{p}_{kij}) = ((\alpha_{kij} + n_{kij})/(\alpha_{ki} + n_{ki}))$ and the probability of $C = C_k$ can be estimated as $(\hat{p}_k) = ((\beta_k + n_k)/(\beta + m))$.

In practice, we use symmetric prior distributions for all the transition probabilities considered at the beginning of the search process. The initial $m \times s \times s$ hyper-parameters α_{kij} are set equal to $\alpha/(ms^2)$ and, when two time series are assigned to the same cluster and the corresponding observed frequencies of transitions are summed up, their hyper-parameters are summed up. Thus, the hyper-parameters of a cluster corresponding to the merging of m_k initial MCs will be $m_k\alpha/(ms^2)$. In this way, the specification of the prior hyper-parameters requires only the prior global precision α, which measures the confidence in the prior model. An analogous procedure can be applied to the hyper-parameters β_k associated with the prior estimates of p_k. Empirical evaluations have shown that the magnitude of the α value has the effect of zooming out differences between dynamics of different time series, so that, increasing the value of α yields an increasing number of clusters.

3.3 A Heuristic Search

To implement the clustering method described in the previous section, we might search all possible partitions and return that one with the highest posterior probability. Unfortunately, the number of possible partitions grows exponentially with the number of MCs and a heuristic method is required to make the search feasible.

The BCD algorithm uses a measure of similarity between estimated transition probability matrices to reduce the search process to a subset of all possible partitions. The similarity measure that guides the search process can be any distance between probability distributions. The current implementation of the BCD algorithm is based on the Kullback-Liebler distance and it is computed as follows. Let P_1 and P_2 be matrices of transition probabilities of two MCs. Because each matrix is a collection of s row conditional probability distributions, and rows with the same index are probability distributions conditional on the same event, the measure of similarity that BCD uses is an average of the Kullback-Liebler distances between row conditional distributions. Let p_{1ij} and p_{2ij} be the probabilities of the transition $i \to j$ in P_1 and P_2. The Kullback-Liebler distance of the two probability distributions in row i is $D(p_{1i}, p_{2i}) = \sum_{j=1}^{s} p_{1ij} \log(p_{1ij}/p_{2ij})$. The average distance between P_1 and P_2 is then $D(P_1, P_2) = \sum_i D(p_{1i}, p_{2i})/s$.

Initially, BCD transforms the time series in S in a set of m MCs, using the procedure described in the previous sections, and computes the set of pairwise distances between the transition probability matrices. Then BCD sorts the generated distances, merges the two closest MCs and evaluates whether the resulting model M_c, in which the two MCs are replaced by the MC resulting from their merging, is more probable than the model M_s in which these MCs are different. If the probability $p(M_c|S)$ is larger than $p(M_s|S)$, BCD updates the set of MCs by replacing the two MCs with the cluster resulting from their merging. Consequently, BCD updates the set of

Sensor	α 1	5	10	20	40
vis.a	34	34	{ 18 16	18 { 13 3	{ 11 7 { 10 3 3
	2	2	2	2	2

TABLE 11.3. Size of clusters found by the BCD algorithm for the sensor vis.a. Brackets before a pair of numbers indicate that the clusters were produced by splitting the episodes belonging to one cluster only for a smaller value of α.

ordered distances by removing all the ordered pairs involving the merged MCs, and by adding the distances between the new MC and the remaining MCs in the set and the procedure is iterated on the new set of MCs. If the probability $p(M_c|S)$ is not larger than $p(M_s|S)$, BCD tries to merge the second best, the third best, and so on, until the set of pairs is empty and, in this case, returns the most probable partition found so far. The rationale of this search is that merging closest MCs first should result in better models and increase the posterior probability sooner. Empirical evaluations of the methods in simulated data appear to support this intuition (Ramoni et al., 2000b).

4 Clustering the Robot Sensory Inputs

In this section we describe the results obtained with the BCD algorithm on a data set of 11,118 values recorded, for each of the 8 sensors in Table 11.1, during an experimental trial that lasted about 30 minutes. The event marker led to split the original time series into 36 episodes, of average length 316 time steps. The shortest episode was 6 time steps long and the longest episode was 2917 time steps.

We run our implementation of the BCD algorithm on the set of 36 episodes for each sensor, using different values for the precision α while β was set equal to 1. Table 11.3 shows the number of clusters created by the BCD algorithm for the sensor vis.a, for some of the values of α used in this experiment. A small value of the precision α leads to identify two clusters, one merging 34 episodes, the other one merging two episodes. Increasing values of α make the BCD algorithm create an increasing number of clusters by monotonically splitting the cluster merging 34 episodes.

For example, this cluster is split into two clusters of 18 and 16 episodes, when $\alpha = 10$. The cluster collecting 16 episodes is then split into two of 13 and 3 episodes, for $\alpha = 20$, and then, when $\alpha = 40$, the cluster of 18 episodes is split into two clusters of 11 and 7 episodes, while the cluster of 13 episodes is split into two clusters of 10 and 3 episodes. Larger values of α make BCD create an even larger number of clusters, some of which represent MC learned from very sparse frequency matrices, so that we decide to stop the algorithm for $\alpha = 40$, to avoid overfitting.

The pictures in Table 11.4 represent the *essential* dynamics of the MCs induced from the BCD algorithm with the 36 episodes of the sensor vis.a. By essential dynamics we mean that neither transitions with probability inferior to 0.01 are represented in the chains nor are the uniform transition probabilities of visiting all states that are computed only from the prior hyper-parameters, since no transition were observed. We stress here that the interpretation of the dynamics represented by the MCs is our own one — we looked at the transition probability matrices and labeled the dynamics according to our knowledge of the robot's perception system — and it is by no means knowledge acquired by the robot. So, for example, the first chain represents a dynamic process concentrated on the first three states of the sensor vis.a and state 0.2 represents the presence, in the robot's visual field, of an object of size varying between 0 and 1600 pixels, state 0.4 represents the presence of an object of size between 1600 and 6400 pixels, while state 0.6 represents the presence of an object of size between 6400 and 14,400 pixels. The maximum size is given by 40,000 pixels so that values between 0 and 14,400 represent an object that, at most, takes 1/4 of the visual field. Now the dynamics between these three states can be that either the sensor value is constant or decreases because it visits a state preceding itself, so that the overall dynamics is that of an object of decreasing size in the visual field that eventually disappears. The interpretation of the other dynamics was deduced in a similar way. Interestingly, the last chain represents essentially a deterministic process, in which the sensor value is constant in state 0.2 showing that there is no object in the robot's visual field.

Similar results were found for the other sensors related to the robot's vision system. For example, the values of the sensors vis.x and vis.y produce two clusters for $\alpha = 1$ and eight clusters for $\alpha = 40$. The values of the other sensors tend to produce a smaller number of clusters and to need a much higher prior precision to induce clusters: the minimum value of the precision to obtain at least 2 clusters was 20 for the sensor l.vel; 40 for the sensors r.vel, grip.f and grip.r; and it was 45 for the sensor grip.b.

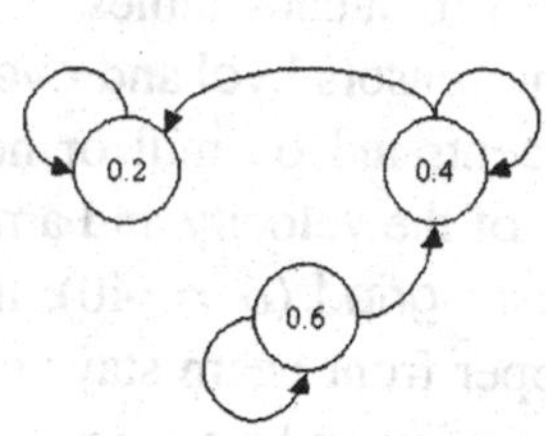

Cluster 1: Object disappears from the visual field

Cluster 2: Object of increasing size in the visual field

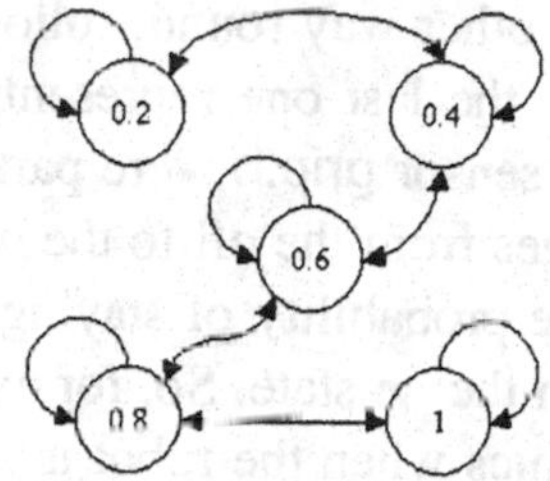

Cluster 3: Object of varying size in the visual field

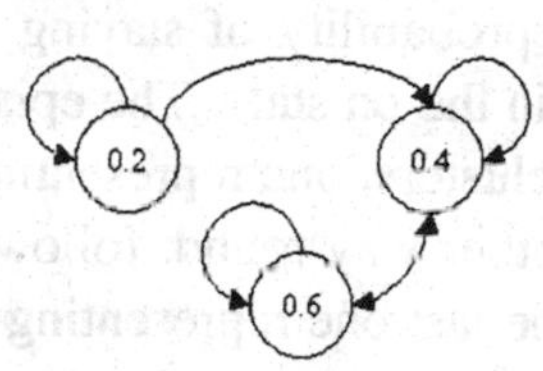

Cluster 4: Object appearing in the visual field

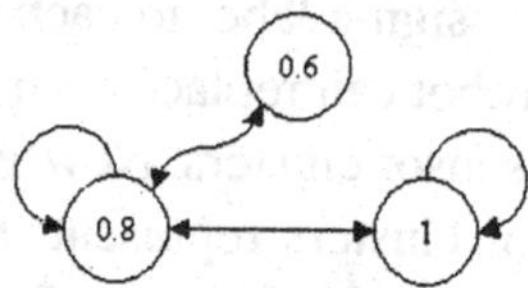

Cluster 5: Object of increasing size and encounter

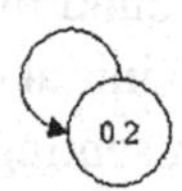

Cluster 6: no object in the visual field.

TABLE 11.4. MCs extracted with the BCD algorithm from the 36 episodes of the sensor vis.a.

We used the same approach described above to choose a value of α: we run the BCD algorithm for increasing values of α and stopped the algorithm when it began producing MCs with very sparse frequency tables.

We found three clusters of MCs for both the sensors l.vel and r.vel ($\alpha =$ 45 and 50), which represent dynamics concentrated on null or negative values of the velocity, null or positive values of the velocity and a mixture of those. We found two clusters for the sensor grip.f ($\alpha = 40$), the first one representing a process in which the gripper front beam stays off with high probability and with small probability goes on and stays on, while, in the second one, there is a larger probability of changing from the off to the on state. Hence, the second cluster represents more frequent encounters with an object. The episodes for the sensor grip.r were partitioned in three clusters ($\alpha = 40$), one representing rapid changes from the on to the off state, followed by a large probability of staying off; one representing rare changes from the off to the on state, or the other way round, followed by a large probability of staying in that state; the last one representing the sensor in the on state. The episodes for the sensor grip.b were partitioned in two clusters, one representing rare changes from the off to the on state, or the other way round, followed by a large probability of staying in that state; the last one representing the sensor in the on state. So, for example, the first cluster represents the sensor dynamics when the robot is not near an object but, when it does, it pushes it for some time. The second cluster is the sensor dynamics when the robot is pushing an object. Finally, the three clusters found for the sensor vis.x ($\alpha = 5$) distinguish the sensor dynamics when an object is far from the robot, or near the robot or a mixture of the two, while the four clusters of dynamics for the sensor vis.y ($\alpha = 1$) distinguish among an object moving from the left of the robot's visual field to the center, from the left to the right, from the right to the left, or in front of the robot.

The clusters found by the BCD algorithm assign a label to each episode so that, after this initial cluster analysis, the robot can replace each episode with a label representing a combination of 8 sensor clusters. Now, episodes labeled with the same combination of sensor clusters represent the same "activity" characterized by the same sensor dynamic signature. For example, one such activity is characterized by the combination cluster 1 for r.vel, cluster 3 for l.vel, cluster 1 for grip.f, grip.r and grip.b, cluster 2 for vis.a and vis.x and cluster 1 for vis.y. This activity is repeated in 7 of the 36 episodes. By using our human interpretation of the dynamics represented by the clusters, we can deduce that this activity represents the robot that rotates and moves far from an object (the velocity of the wheels are discordant, and the size of the object in the visual field decreases and becomes null) and hence we have a confirmation that this activity is meaningful. However, as far as the robot's world in concerned, this activity is simply a combination of sensory dynamics. This process of labeling the episodes in activities by replacing each sensor episode by the cluster membership reduces the initial 36 episodes into 22 different activities, some of which are experience more than once. Thus, the robot learned 22 different activities characterized by different dynamic signatures.

5 Conclusions

This paper presented a Bayesian analysis of sensory inputs of a mobile robot. The sensory inputs were modeled as MCs and then we used a new algorithm for Bayesian clustering by dynamic to cluster episodes characterized by a similar dynamic. This process allowed the robot to learn different activities represented by different sensor dynamic signatures. There are several open questions that are currently under investigation.

One such question is whether the use of Bayesian methods can coexist with the robot's learning being unsupervised. For example, the analysis presented in this paper was based on our choice of the prior hyperparameters and, in order to make the method completely unsupervised, there is need to provide the robot with decision rules that allow it to reason without human intervention. Another problem that we are currently investigating is the effect of the definition of event-marker on the whole analysis. To avoid the need to split the sequence of sensory inputs into episodes, we are developing a Bayesian method for incremental clustering.

The model selection strategy of BCD seeks the clustering model with maximum posterior probability. Other choices here would be possible as, for example, selecting the median posterior probability model Barbieri and Berger (2000). An open question is to compare these different model choices and to see whether a similar heuristic search can be developed when the algorithm seeks for the median posterior probability model. In order to simplify the analysis, we discretized the continuous values of the robot's sensors and we clustered the univariate time series independenlty. In Ramoni et al. (2000a) we presented a multivariate generalization of BCD which is still limited to clustering multivariate discrete MCs. We are currently developing a generalization of BCD which clusters time series of continuous variables using auto-regressive models (Sebastiani and

Ramoni, 2000), and future work will focus on the development of a multivariate clustering algorithm by dynamics which could treat continuous and discrete values time series.

Acknowledgements

This research is supported by DARPA/AFOSR under contract(s) No(s) F49620-97-1-0485. The U.S. Government is authorized to reproduce and distribute reprints for governmental purposes notwithstanding any copyright notation hereon. The views and conclusions contained herein are those of the authors and should not be interpreted as necessarily representing the official policies or endorsements either expressed or implied, of DARPA/AFOSR or the U.S. Government.

References

Banfield, J. D., and Raftery, A. E. (1993). "Model-based gaussian and non-gaussian clustering," *Biometrics*, *49*, 803–821.

Barbieri, M., and Berger, J. O. (2000). "Optimal predictive variable selection," Isds discussion paper, Duke University.

Bishop, Y. M. M., Fienberg, S. E., and Holland, P. W. (1975). *Discrete Multivariate Analysis: Theory and Practice*. MIT Press, Cambridge, MA.

Cheeseman, P. and Stutz, J. (1996). "Bayesian classification (autoclass): Theory and results," in *Advances in Knowledge Discovery and Data Mining*, 153–180. MIT Press, Cambridge, MA.

Cooper, G. F. and Herskovitz, E. (1992). "A Bayesian method for the induction of probabilistic networks from data," *Machine Learning*, 9, 309–347.

Dawid, A. P. and Lauritzen, S. L. (1993). "Hyper Markov laws in the statistical analysis of decomposable graphical models," *Annals of Statistics*, 21, 1272–1317. Correction ibidem, (1995), 23, 1864.

Ramoni, M., Sebastiani, P., and Cohen, P. R. (2000a). "Multivariate clustering by dynamics," in *Proceedings of the Seventeeth National Conference on Artificial Intelligence.* San Francisco, CA. Morgan Kaufmann.

Ramoni, M., Sebastiani, P., Cohen, P. R., Warwick, J., and Davis, J. (2000b). "Bayesian clustering by dynamics," *Machine Learning*. To appear.

Sebastiani, P. and Ramoni, M. (2000). "Bayesian clustering by dynamics of European school population," in *Proceedings of the ISBA2000 Conference*. To appear.

Smyth, P. (1999). "Probabilistic model-based clustering of multivariate and sequential data," in *Uncertainty 99: Proceedings of the Seventh International Workshop on Artificial Intelligence and Statistics*, 299–304. Morgan Kaufmann, San Francisco, CA.

Hidden Markov Models for Analysis of Biological Rhythm Data

Howard J. Seltman

ABSTRACT Analysis of biological rhythm data presents special challenges when the observed quantity is a concentration measurement. Neither deconvolution nor cosinor approaches sufficiently account for the net result of periodic secretion and an appropriate elimination mechanism. This paper discusses methods for data analysis that achieve improved modeling. The underlying biological rhythm for a group of subjects is modeled as periodically varying transition probabilities that control a two-state (quiescent vs. secreting) hidden Markov chain in each subject. The measured concentration is then modeled as the net result of secretion and elimination plus Gaussian error. Results from a Bayesian implementation of the model as applied to a serum cortisol study are presented.

1 Introduction

Cyclic biological phenomena are ubiquitous in both the plant and animal kingdoms. Those rhythms that exhibit approximately daily period are called circadian. Biological rhythms are controlled by endogenous cellular time-keepers in various locations in the body. The major pacemakers in the circadian timing system in mammals appear to be located in the brain (Moore-Ede, et al., 1982). Interest in biological rhythm is motivated by both the desire to understand basic physiological mechanisms and by evidence that this knowledge may have health and safety benefits.

The cyclic phenomenon which stimulated this work is the human cortisol biological rhythm. Cortisol is a key regulatory hormone with myriad effects throughout the body (Felig et al., 1981). It participates in the regulation of carbohydrate, fat, and protein metabolism, and is a key hormone involved in maintenance of blood sugar levels. It is also involved in stress response, immunity, and mineral balance. Cortisol is secreted by the adrenal glands in response to adrenocorticotropic hormone (ACTH) from the pituitary. The level of cortisol in the blood stream exhibits complex patterns. Over a 24 hour period there is a large early morning peak and a smaller afternoon peak. At a finer scale, an average of 15 daily pulses of about 15 minutes duration are superimposed upon the circadian pattern (Iranmanesh, et al. 1990).

Based on the practical considerations of subject health and comfort, as well as expense, the available human cortisol data are usually characterized by 24 to 48 hours of sampling with a 15 to 30 minute sampling interval. Few studies have more than 30 patients, and most include a complex mixture of age, sex, menstrual stage, and diagnosis categories (VanCauter, 1996). Thus there are major analytic hurdles related to the fact that there are only one or two periods measured for each subject, and to the necessity to account for the covariates.

Most previous attempts to model the biological rhythm of blood cortisol levels have used either "cosinor" or "deconvolution" analyses. Cosinor analysis is based on harmonic regression techniques which assume that the residual errors are independent and identically distributed (iid) (Radomski, et al., 1995; Iranmanesh, et al. 1990). Typically results are reported as mean of the fitted curve (mesor), amplitude of the peak relative to the mesor, and time of the peak. Greenhouse, Kass, and Tsay (1987) go beyond the iid approach, and fit a non-linear harmonic regression model with general ARMA residual error structure including fitting of the fundamental frequency.

Cosinor analysis suffers from the fact that it ignores the physiological basis (secretion and elimination) of the measured concentration, does not take into account serial correlation, and has no mechanism for combining data from multiple subjects. A major step forward is the work of Greenhouse, Kass, Lam, and Tsay (1993), which extends the hierarchical mixed-models approach of Laird and Ware (1982) to harmonic regression with ARMA error structure and random intercepts. Although extremely general, the fitting of these models requires complex, specialized computer programs. A more important criticism is that residual plots indicate that, due to the inherent pulsatile nature of the secretion curve, conditional normality assumptions are untenable, casting doubt on the validity of any hypothesis testing (unpublished results).

Deconvolution analysis attempts to provide reasonable estimates of the input function (secretion) under the assumption that measured concentrations are the convolution of input and a known excretion function. Methods are well described in DeNicolao and Liberati (1993). This technique has seen extensive use in recent years. Despite its usefulness and appeal, the disadvantages include the fact that it focuses on the secretion pulses rather that the underlying biological rhythm, the need to fix a specific elimination profile, and the lack of a direct method for application to multiple subjects.

2 Model and Methods

The model developed here combines the use of the physiological knowledge of secretion and elimination with focus on the underlying biological rhythm. The concentration of any substance in the bloodstream (or and other compartment) is the result of a dynamic process of input of the substance into the compartment through secretion or synthesis, and removal of the substance from the compartment through degradation or transport. For cortisol, the input is secretion from the adrenal glands, and the output is removal by the liver and kidneys. In general, for any functional form of secretion, $S(t)$, and elimination, $E(t)$, the net concentration, $\mu(t)$, is based on the convolution formula:

$$\mu(t) = \int_{-\infty}^{t} S(u)E(t-u)du$$

Elimination is assumed to follow a first order mechanism, $E(t) = \exp(-\lambda t)$, with unknown decay constant, λ. Note that this corresponds to half life, $HL = \log(2)/\lambda$.

The model follows the discretized form of the convolution formula where the n observations are $Y_0, \ldots, Y_{n-1}$. We also use the simplifying assumption that secretion can be represented as a step function with time interval equal to the measurement interval and with constant height, H. Then the hidden secretion state is called S_t where $S_t = 1$ indicates secretion at rate H during the time interval $[t, t+1)$. This means that the increase in concentration during one time interval due to constant secretion is

$$H^* = \int_0^1 He^{-\lambda(1-u)}du = \frac{H(1-e^{-\lambda})}{\lambda}.$$

Given the secretion state, it is assumed that the measured concentrations are subject to iid measurement error, $\epsilon_t \sim \mathrm{N}(0, \sigma^2)$, based on the conjecture that the serial correlation seen in hormone measurements relates to ignoring the presence of the secretion pulses. An initial non-zero true concentration is called I. This leads to the following model:

$$Y_t = Ie^{-\lambda t} + \sum_{i=0}^{t-1} H^* S_i e^{-\lambda(t-i-1)} + \epsilon_t. \qquad (12.1)$$

To construct a model of biological rhythm for M subjects in a group expected to have a similar underlying rhythm, it is assumed that the secretion states for each subject, $\mathbf{S}_i \equiv \{S_{i,0}, \ldots, S_{i,n-2}\}$, are individual hidden Markov chains (HMCs) subject to common transition probabilities. A flexible model that results in simulated data similar to actual cortisol concentrations can be constructed with a fixed parameter for the probability of ending a secretion pulse, $\beta \equiv \Pr(\mathrm{S_{t+1}} = 0|\mathrm{S_t} = 1)$, and a periodically varying probability of beginning a secretion, $\alpha_t \equiv \Pr(\mathrm{S_{t+1}} = 1|\mathrm{S_t} = 0)$. Here we use

$$\log\left(\frac{\alpha_t}{1-\alpha_t}\right) = \gamma_0 + \sum_{f=1}^{F}\left[\gamma_{(2f-1)}\cos(2\pi\omega t) + \gamma_{(2f)}\sin(2\pi\omega t)\right] \tag{12.2}$$

where ω is the fundamental frequency and F is a small positive integer specifying the number of harmonic frequencies. This allows for an arbitrarily complex periodic function with $2F+1$ gamma parameters and appropriate restrictions for the values of the transition probabilities. Recent evidence (Szeisler, et al., 1999) confirms that it is reasonable to take ω to be fixed at 1/24 hours, because the cortisol pattern is truly diurnal in humans. The work presented here takes $F = 1$, because there was no improvement in the model fit when $F = 2$ was tried.

The complete hierarchical model combines the log-cosinor model, (12.2) with the convolution model, (12.1). Each subject has an individual HMC and initial concentration, but the values of $\gamma, \beta, \lambda, H$ and σ^2 are common to all subjects. Because the initial concentration is a random effect, the model specifies that $I_i \sim \text{Log-normal}(\mu_I, \sigma_I^2)$.

A frequentist approach to this model, e.g. using expectation-maximization, is intractable. Results based on producing a posterior sample with Markov Chain Monte Carlo (MCMC) methods are presented here.

Because the physiology of human cortisol has been extensively studied, proper, but weak, prior distributions are placed on all of the parameters. MCMC implementation is based on the Metropolis-Hastings algorithm embedded within the Gibbs sampler. Each parameter is visited in turn alternately with an update of the HMC.

The most difficult part of the MCMC implementation is the proposal distribution for the HMC update step. One successful approach is to repeat the following step several dozen times per HMC update. A random time, $t \in \{0, 1, \ldots, n-1\}$ is chosen, along with a random step size $s \in \{0, 1, \ldots, s_{max}\}$ where s_{max} is e.g. 15. Then the Metropolis-Hastings algorithm is used to decide whether to accept a proposal to swap the states at time t and at time $t + s$ if $s > 0$. This type of proposal mechanism is needed to avoid getting stuck in many local minima.

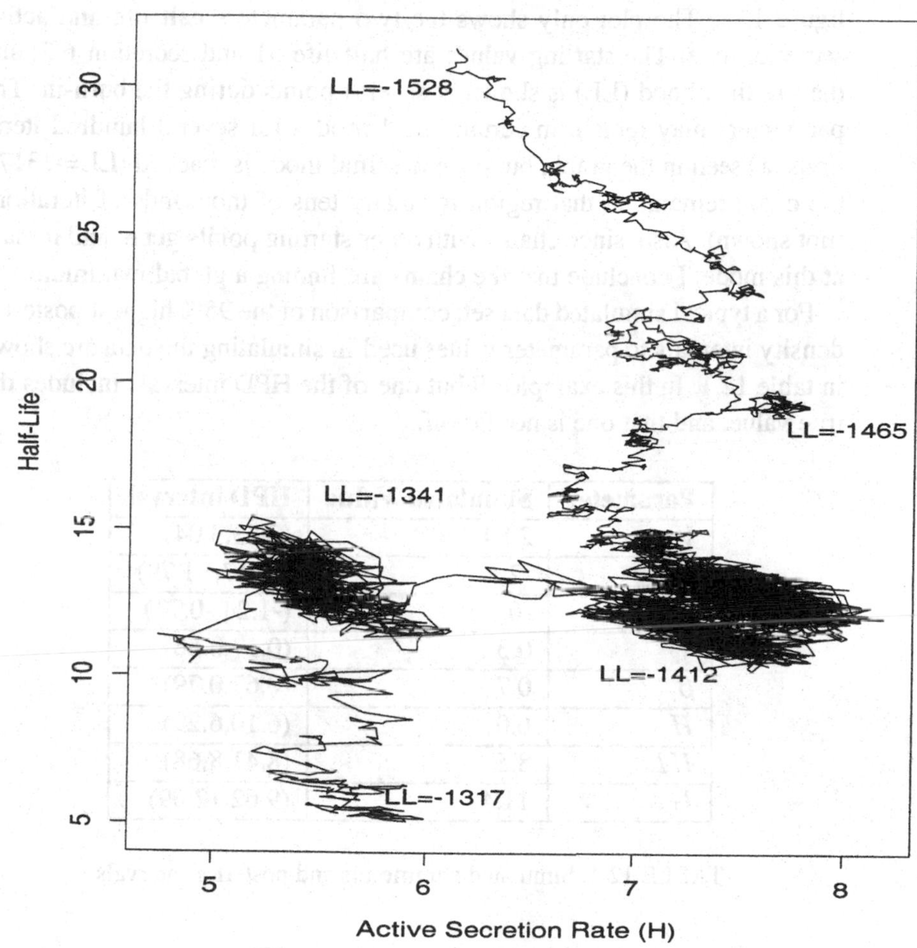

FIGURE 12.1. Posterior sampling with a long burn-in

3 Results and Discussion

The log-cosinor/convolution model described here is used to fit both simulated data and data from a patient study of cortisol biological rhythm. In either case, the Markov chain representing the (correlated) sample from the posterior converge to the same region from varying starting points, although sometimes after a long and circuitous path through the parameter space. An example of such a path with a long burn-in time is shown in figure 12.1. The plot only shows the two parameters half-life and active secretion rate. The starting values are half-life 31 and secretion 6.2, and the log-likelihood (LL) is shown at several points during the burn-in. The parameters may remain in certain local modes for several hundred iterations, as seen in the graph, but once the final mode is reached (LL=-1317), the chain remains in that region for many tens of thousands of iterations (not shown). Also, since chains with other starting points get to and remain at this mode, I conclude that the chains are finding a global maximum.

For a typical simulated data set, comparison of the 95% highest posterior density intervals to parameter values used in simulating the data are shown in table 12.1. In this example all but one of the HPD intervals includes the true value, and that one is not far off.

Parameter	**Simulated Value**	**HPD interval**
V	2.84	(2.63,3.04)
γ_0	-2	(-2.17, -1.79)
γ_1	-0.8	(-1.24,-0.72)
γ_2	0.3	(0.17,0.66)
β	0.7	(0.67,0.79)
H	6.0	(6.10,6.29)
HL	8.5	(8.41,8.68)
I_1	11.4	(9.62,12.09)

TABLE 12.1. Simulated parameters and posterior intervals

In addition, the posterior distribution of the HMCs are consistent with the known states. The MCMC values for the HMC state matches the known state in at least 80% the MCMC iterations, with the most frequent discrepancies due to an occasional MCMC state equal to one rather than zero when adjacent to a state that is truly one.

The discrepancy between fitted and observed values is of course greater for the human data than the simulated data, because the model is only an approximation in the former case. Nevertheless the model does fit the observed data quite well as seen in figure 12.2 for the final MCMC iteration for one of twelve subjects.

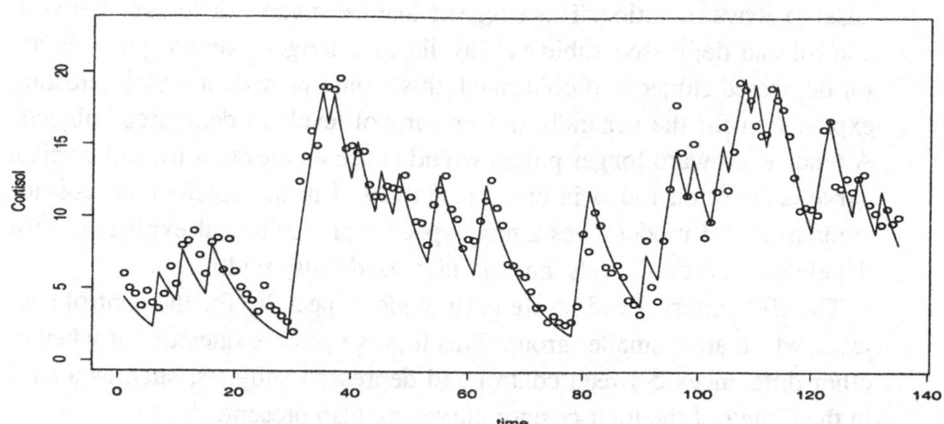

FIGURE 12.2. Model fit for one subject

MCMC results for the 6 control subjects and the 14 depressed subjects are given in table 12.2. The highest posterior density intervals are shown for each of the parameters for both control and depressed subjects.

Subjects	γ_0	γ_1	γ_2	β
Control	(-2.30,-1.77)	(-1.14,-0.45)	(-1.03,-0.40)	(0.64,0.83)
Control*	(-2.30,-1.79)	(-1.12,-0.44)	(-1.05,-0.38)	(0.65,0.82)
Depressed	(-2.12,-1.79)	(-0.88,-0.41)	(-1.03,-0.63)	(0.51,0.63)
Depressed*	(-2.08,-1.75)	(-0.84,-0.38)	(-1.02,0.58)	(0.50,0.62)

Subjects	B	H	$t_{\frac{1}{2}}$	σ^2
Control	(0.22,0.34)	(5.48,6.39)	(3.82,4.34)	(2.58,3.20)
Control*	(0.19,0.33)	(5.42,6.40)	(3.69,4.50)	(2.63,3.24)
Depressed	(0.25,0.33)	(4.42,4.75)	(4.24,4.59)	(2.10,2.42)
Depressed*	(0.22,0.324)	(4.26,4.74)	(4.28,4.85)	(2.08,2.42)

TABLE 12.2. 95% HPD intervals for clinical subjects. (*) indicates use of the weak instead of the very weak prior distribution.

All of the intervals overlap except β, the transition probability from basal to active secretion. This suggests that the biggest difference between control and depressed subjects may lie in a longer average pulse width for depressed subjects. If confirmed, this would provide a very interesting explanation for the generally higher cortisol levels in depressed subjects. A tendency toward longer pulses would cause an elevated overall cortisol curve, as is often found in cosinor analysis. But the HMM logit-cosinor compartmental model gives a more specific physiological explanation for the elevated cortisol levels, namely increased pulse width.

The HPD interval widths are fairly wide, especially for the control subjects, which are a smaller group. This leaves open the question of whether other differences between control and depressed subjects, such as a shift in the timing of the logit-cosinor curve, are also present.

The sensitivity of the posterior distribution to the prior distribution is also shown in table 12.2. The lines in the table that are marked with "*" indicates results using a weak prior that is specifically based on knowledge of cortisol physiology, while the other lines are for a very weak prior with about ten times as much spread. The small size of the differences in HPD intervals between these two prior distributions, even with only six subjects, suggests that there is little problem with sensitivity to choice of prior distributions.

The model described here provides a practical new approach to analysis of biological rhythm data measured as concentrations. Several aspects of the model indicate that it is an improvement over existing models. First, data simulated from this model are quite similar to real data, while data simulated under the deconvolution model have no periodicity, and data generated from the cosinor model have no pulsatility. Second, this model is able to model realistic subject to subject variability based on a common underlying periodic secretion method. And third, the results from the patient analysis provide a more clinically useful interpretation than do results from the other methods.

Future work will be aimed at incorporating appropriate covariate effects, and at finding the posterior distribution of differences in the γ, β, and H parameters between physiologically distinct categories of subjects, e.g. control and depressed patients. The latter will achieve one of the chief goals of researchers interested in biological rhythms, namely characterization of rhythm changes due to disease states.

References

DeNicolao, G. and Liberati, D., (1993). Linear and nonlinear techniques for the deconvolution of hormone time-series. *IEEE Transactions in Biomedical Engineering* **40**, 440-455.

Felig, P., Baxter, J.D., Broadus, A.E., and Frohman, L.A., (1981). *Endocrinology and Metabolism.* McGraw-Hill Book Company, New York.

Greenhouse, J.B., Kass, R.E., and Tsay, R.S., (1987). Fitting nonlinear models with ARMA errors to biological rhythm data. *Statistics in Medicine* **6**, 167-183.

Greenhouse, J.B., Kass, R.E., Lam, T. and Tsay, R.S., (1993). A hierarchical model for serially correlated data: Analysis of biological rhythm data. Technical Report, Department of Statistics, Carnegie Mellon University, Pittsburgh, PA. (Available at http://lib.stat.cmu.edu.)

Iranmanesh, A., Lizarralde, G., Johnson, M.L., and Veldhuis, J.D, (1990). Dynamics of 24-hour endogenous cortisol secretion and clearance in primary hypothyroidism assessed before and after partial thyroid hormone replacement. *Journal of Clinical Endocrinology and Metabolism* **70**, 155-61.

Laird, N.M. and Ware, H.H., (1982). Random-Effects Models for Longitudinal Data. *Biometrics* **38**,.963-974.

Radomski, M.W., Buguet, A., Montmayeur, A., Bogui, P., Bourdon, L., Doua, F., Lonsdorfer, A., Tapie, P. and Dumas, M., (1995). Twenty-four-hour plasma cortisol and prolactin in human African trypanosomiasis patients and healthy African controls. *American Journal of Tropical Medicine and Hygiene* **52**, 281-6.

Moore-Ede, M., Sulzman, F., and Fuller, C., (1982). *The Clocks That Time Us*. Harvard University Press, Cambridge, MA.

Szeisler, C., Duffy, J., Shanahan, T., Brown, E., Mitchell, J., Rimmer, D., Ronda, J., Silva, E., Allan, J., Emens, J., Dilk, D. & Kronauer, R. (1999). Stability, precision, and near-24-hour period of the human circadian pacemaker'. *Science* **284**, 2177-2181.

Van Cauter, E., Leproult, R. and Kupfer, D.J., (1996). Effects of gender and age on the levels and circadian rhythmicity of plasma cortisol. *Journal of Clinical Endocrinology and Metabolism* **81**, 2468-73.

Using Prior Opinions to Examine Sample Size in Two Clinical Trials

Chin-Pei Tsai
Kathryn Chaloner

ABSTRACT Two examples of large clinical trials for the treatment of advanced HIV disease are described. Chaloner and Rhame (2001) elicited prior opinions about the outcomes of the two trials from over 50 HIV clinicians. Their prior opinions are used here for design: the sample size for reaching consensus with high probability is calculated. Consensus is said to occur when all clinicians have posterior opinions which would lead to prescribing the same treatment. Posterior beliefs are calculated using a simple linear Bayes approximation. In addition plots are given for determining parameter values for which a particular sample size is sufficient for consensus to be reached with high probability. These calculations are useful tools at the design stage and are simple to implement.

1 Introduction

The design of two large AIDS trials is examined. The trials were designed by the Community Program for Clinical Research in AIDS (CPCRA), a collaborative group, sponsored by the U.S. National Institutes of Health. Both trials examined long term treatment for the prevention of a common opportunistic infection, *Pneumocystis carinii* pneumonia (PCP), in patients with advanced HIV disease. For the two trials, Chaloner and Rhame (2001) elicited prior opinions from over 50 clinicians by surveying both the CPCRA clinicians enrolling patients and also a group of clinicians in Minnesota treating patients with HIV. Details of the elicitation, with other uses of the opinions, are in Chaloner and Rhame (2001).

Both these trials were large pragmatic trials aimed to influence clinical practice. The approach taken here, therefore, is that, when the trial is done, the sample size should be sufficient so that clinicians will probably agree on which of the two treatments to prescribe.

2 Assumptions and Notation

Suppose the two trials are both balanced designs with n independent Bernoulli observations on each treatment denoted $\{X_n\}$ and $\{Y_n\}$. The probabilities of PCP in two years are denoted θ_X and θ_Y, respectively. $\bar{X}$ and $\bar{Y}$ are the sample proportions for the two treatments. A normal approximation to the marginal posterior distribution of $\theta_X - \theta_Y$ given the data, $\bar{X}$ and $\bar{Y}$, will be used. A different approximation, due to Hartigan (1969), was used in Tsai (1999).

Specifically, denote a prior mean of $\theta_X - \theta_Y$ as $m_X - m_Y$ and the corresponding prior variance as τ^2. The sampling distribution of $\bar{X} - \bar{Y}$, given θ_X and θ_Y, is approximately normal with mean $\theta_X - \theta_Y$ and variance $n^{-1}\sigma^2$ where $\sigma^2 = \theta_X(1-\theta_X)+\theta_Y(1-\theta_Y)$. Approximating both the prior distribution and the likelihood by a normal distribution gives a posterior mean for $\theta_X - \theta_Y$, denoted $\mu(\bar{X}, \bar{Y})$, of

$$\begin{aligned} &\mu(\bar{X}, \bar{Y}) \\ \approx\; & \frac{n\tau^2}{n\tau^2+\sigma^2}(\bar{X}-\bar{Y}) + \frac{\sigma^2}{n\tau^2+\sigma^2}(m_X - m_Y). \end{aligned} \tag{13.1}$$

For any k, and $n \geq 1$, define

$$K = \frac{k(n\tau^2+\sigma^2) - \sigma^2(m_X - m_Y)}{n\tau^2}, \tag{13.2}$$

then given θ_X and θ_Y and a prior distribution,

$$\begin{aligned} & Pr\left[\mu(\bar{X},\bar{Y}) \leq k|\theta_X, \theta_Y, m_X, m_Y, \tau^2\right] \\ = \;& Pr\left[\bar{X}-\bar{Y} \leq \frac{k(n\tau^2+\sigma^2)-\sigma^2(m_X-m_Y)}{n\tau^2}|\theta_X, \theta_Y, m_X, m_Y, \tau^2\right] \\ \approx\; & \Phi\left[\frac{K-(\theta_X-\theta_Y)}{\sqrt{n^{-1}\sigma^2}}\right]. \end{aligned} \tag{13.3}$$

This is an approximation to the sampling probability, given θ_X, θ_Y and a prior distribution, that the posterior mean will be less than or equal to k. Expressions (13.1), (13.2) and (13.3) depend on m_X, m_Y and τ^2 and will therefore be different for each clinician.

For the two trials only the means for θ_X and θ_Y and a 95% prior probability interval for $\theta_X - \theta_Y$ were elicited. It is therefore assumed that, if L is the length of the 95% interval, $L = 4\tau$, or $\tau^2 = L^2/16$.

2.1 Consensus

Suppose opinions are documented from t clinicians. Given data $\bar{X}$ and $\bar{Y}$, denote the posterior mean for the ith clinician as $\mu_i(\bar{X}, \bar{Y})$. Then $\mu_i(\bar{X}, \bar{Y})$ will be as in (13.1) with $m_X = m_{iX}$, $m_Y = m_{iY}$ and $\tau^2 = \tau_i^2$. Consensus for prescribing X is defined, in the two examples, as all t clinicians having posterior means, $\mu_i(\bar{X}, \bar{Y}), i = 1 \ldots t$, less than or equal to a value k. Similarly they will agree in prescribing Y if $\mu_i(\bar{X}, \bar{Y}) > k, i = 1 \ldots t$. The value of k must be determined from the context: $k = 0$ if treatments X and Y are similar in cost, toxicity and ease of adherence. In some situations there may be a range of equivalence (Spiegelhalter, Freedman and Parmar, 1994) where either treatment may be prescribed, but in the two trials considered here this does not apply (see Sections 4 and 5).

For fixed θ_X and θ_Y, where $\theta_X - \theta_Y \leq k$, let n_X be thc sample size needed to convince all clinicians to prescribe X, with probability at least $1 - \beta$. Similarly, for $\theta_X - \theta_Y > k$, let n_Y be the sample size needed to convince all clinicians to prescribe Y with probability $1 - \beta$. Define n_X to be the smallest sample size such that the probability, under the joint sampling distribution of $\bar{X}$ and $\bar{Y}$, is at least $1 - \beta$ that $\mu_i(\bar{X}, \bar{Y}) \leq k$ for all $i = 1, \ldots, t$. Similarly n_Y is the smallest sample size such that the probability is at least $1 - \beta$ that $\mu_i(\bar{X}, \bar{Y}) > k$ for all $i = 1, \ldots, t$.

Specifically, for the ith clinician, define $K_i(n)$ to be as in (13.2):

$$K_i(n) = \frac{k(n\tau_i^2 + \sigma^2) - \sigma^2(m_{iX} - m_{iY})}{n\tau_i^2}.$$

Then for $\theta_X - \theta_Y \leq k$, n_X is the smallest n such that

$$Pr\left[\mu_i(\bar{X}, \bar{Y}) \leq k, \text{ for all } i = 1, 2, \cdots, t | \theta_X, \theta_Y\right] \geq 1 - \beta, \quad (13.4)$$

under the sampling distribution of $\bar{X}, \bar{Y}$. For $n \geq 1$ this is equivalent to

$$Pr\left[\bar{X} - \bar{Y} \leq K_i(n), \text{ for all } i = 1, 2, \cdots, t | \theta_X, \theta_Y\right] \geq 1 - \beta,$$

which, in turn is equivalent to

$$Pr\left[\bar{X} - \bar{Y} \leq \min_i\{K_i(n)\} | \theta_X, \theta_Y\right] \geq 1 - \beta.$$

Note that if $m_{iX} - m_{iY} > k$ then $K_i(n)$ is increasing with n and that if there is at least one clinician with $m_{iX} - m_{iY} > k$ then $\min\{K_i(n), i = 1, 2, \cdots, t\}$ will be a $K_s(n)$ where s corresponds to one of these clinicians.

If all $m_{iX} - m_{iY} \leq k$ then the sample size is zero because all clinicians already agree and (13.4) is satisfied for $n = 0$.

Note that the value σ^2 depends on the values of θ_X and θ_Y. Both n_X and n_Y depend on θ_X and θ_Y. Also the sampling distribution of $\bar{X}, \bar{Y}$ for fixed θ_X, θ_Y is used, not the predictive distribution of $\bar{X}, \bar{Y}$. The predictive distribution would require further averaging over a prior distribution which raises the question of which prior distribution to use (see Section 8).

3 The PCP Prophylaxis trials

The CPCRA PCP-TMS trial compares two dosing regimens of trimethoprim sulfamethoxazole (TMP-SMX) in HIV infected patients who are not known to be intolerant of the drug (see El-Sadr et al, 1999). TMP-SMX is believed to be the most effective drug for the prevention of PCP, the most common infection in patients with advanced HIV disease. The standard dose was one double strength tablet daily (D) but a lower dose of a double strength tablet three times a week (T) was considered and used by some. The three times a week dosing could be associated with fewer side effects and toxicities, making it tolerable for a longer period and therefore more effective. Alternatively, it could be less effective as first, bioavailability may be less, and second, patients might find it harder to remember to take a dose just three times a week.

A second trial was designed for patients who develop intolerance to TMP-SMX. This PCP-INT2 trial compares two drugs, atovaquone and dapsone, in patients who are intolerant of TMP-SMX. Dapsone was the standard treatment for patients intolerant of TMP-SMX and atovaquone was a newly licensed drug, approved, and sometimes used, for treatment of PCP, but rarely used for preventive treatment. Insurance carriers typically would not cover atovaquone for the prevention of PCP and so clinicians would not prescribe it because of its cost.

For each trial, 58 clinicians provided opinions about the proportion of PCP after two years for each treatment, (see Chaloner and Rhame, 2001, for more details of the elicitation).

Figure 13.1 is a plot of the elicited quantities: a prior mean (denoted $\times$) for the difference in the two proportions and also a 95% prior probability interval for the difference. The mean was not necessarily in the middle of the interval. Some clinicians only provided a mean and, in this case, only the mean is shown. Some clinicians only answered the questions about the TMS trial and so their belief for the INT2 trial is left blank. Note that a complete joint probability distribution for the two probabilities was not specified.

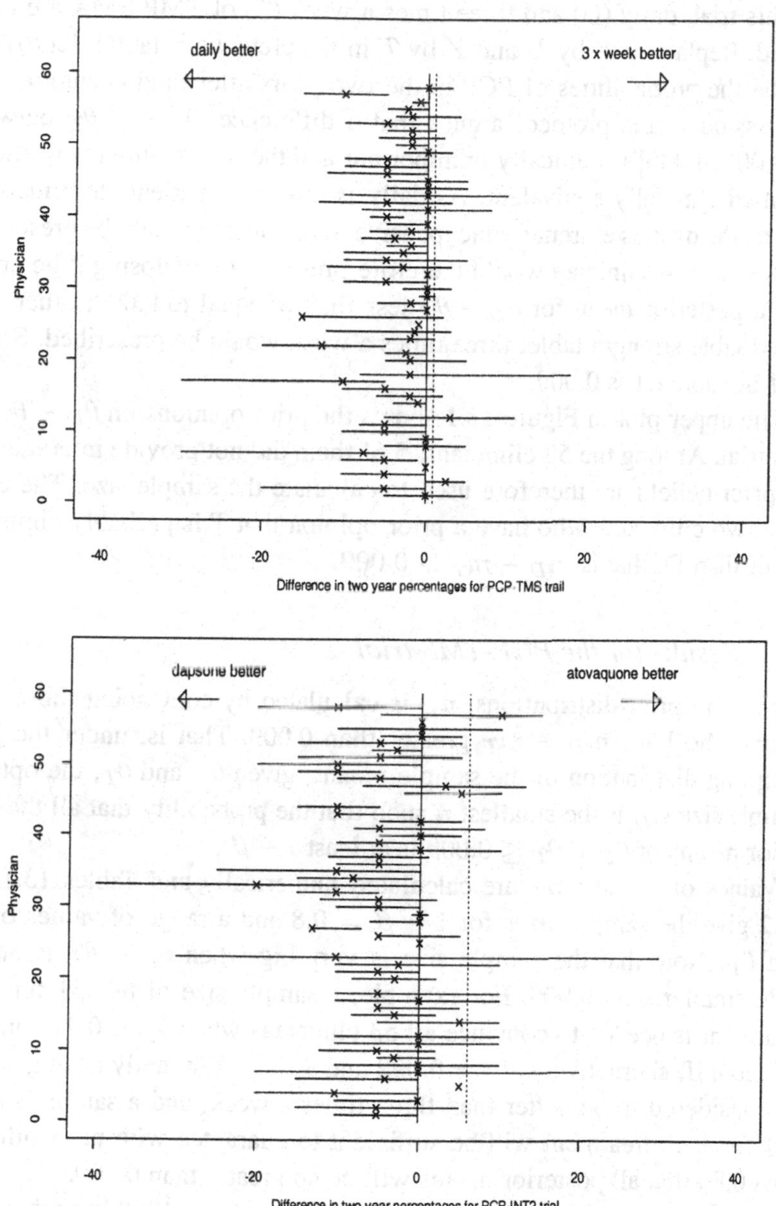

FIGURE 13.1. 95% prior belief intervals for both PCP-TMS trial and PCP-INT2 trial for all clinicians. The solid vertical line is $\theta_X - \theta_Y = 0$ and the dotted vertical line $\theta_X - \theta_Y = k$. The symbol $\times$ denotes the prior mean and the horizontal line the 95% prior probability interval for the difference of the two proportions.

4 The PCP-TMS Trial

In this trial, daily (D) and three times a week (T) of TMP-SMX are compared. Replacing X by D and Y by T in the previous notation, let θ_D and θ_T be the probabilities of PCP in the two years after randomization. The discussion in the protocol argues that a difference of $\theta_D - \theta_T$ between $[-0.009, 0.009]$ is clinically unimportant and the two treatments would be deemed clinically equivalent. As daily is easier for patients to remember and as the drug is extremely inexpensive, daily dosing would be prescribed in this case. A clinician would therefore prescribe daily dosing if he or she had a posterior mean for $\theta_D - \theta_T$ less than or equal to 0.009; otherwise, one double strength tablet three times a week would be prescribed. So the k of Section 2.1 is 0.009.

The upper plot in Figure 13.1 shows the prior opinions on $\theta_D - \theta_T$ for this trial. Among the 58 clinicians, 5 of them did not provide intervals and 53 prior beliefs are therefore used to calculate the sample size. There are only two clinicians who have a prior opinion that T is probably clinically better than D: that is $m_D - m_T > 0.009$.

4.1 *Results for the PCP-TMS trial*

For the 53 prior distributions, n_D is calculated by convincing those clinicians who have $m_D - m_T$ greater than 0.009. That is, under the joint sampling distribution of the sample means, given θ_D and θ_T, the optimal sample size n_D is the smallest n such that the probability that all the posterior means of $\theta_D - \theta_T \leq 0.009$ is at least $1 - \beta$.

Values of n_D and n_T are calculated numerically and Tables 13.1 and 13.2 give the sample sizes for $1 - \beta = 0.8$ and a range of values of θ_D and θ_T. Note that the sample size is very big when $\theta_D - \theta_T$ is only a little smaller than 0.009. For example, a sample size of 89,031 for each treatment is needed to convince all 53 clinicians when θ_D is 0.212 and θ_T is 0.205. If, alternatively, $\theta_D = 0.092$ and $\theta_T = 0.123$, daily dosing would be considered to be better than three times a week, and a sample size of 741 for each treatment will be sufficient to guarantee with probability at least 0.80 that all posterior means will be no greater than 0.009.

The actual sample size in this trial was chosen to be 1250 in each group, using frequentist hypothesis testing calculations. For this sample size, Figure 13.2 gives the values of θ_D and θ_T for which consensus will be reached on prescribing with probability at least 0.80. The region shaded with horizontal lines is the region in which the actual sample size of 1250 is too

		θ_T							
		0.041	0.082	0.123	0.164	0.205	0.305	0.405	0.505
	0.012	211	152	130	117	107	89	74	61
	0.032	737	260	185	153	133	104	84	68
	0.052	-	497	270	201	165	121	95	75
	0.072	-	1388	418	269	207	140	107	84
θ_D	0.092	-	-	741	374	263	163	120	92
	0.112	-	-	1896	556	342	190	135	102
	0.212	-	-	-	-	89031	494	249	165
	0.312	-	-	-	-	-	115090	594	284
	0.412	-	-	-	-	-	-	130359	641
	0.512	-	-	-	-	-	-	-	134837

TABLE 13.1. The Optimal Sample Sizes for Prescribing Daily Dosing in the PCP-TMS Trial for Possible Values of θ_D and θ_T

		θ_T							
		0.041	0.082	0.123	0.164	0.205	0.305	0.405	0.505
	0.012	-	-	-	-	-	-	-	-
	0.032	-	-	-	-	-	-	-	-
	0.052	65810	-	-	-	-	-	-	-
	0.072	4503	-	-	-	-	-	-	-
θ_D	0.092	2580	306329	-	-	-	-	-	-
	0.112	1923	7804	-	-	-	-	-	-
	0.312	782	1064	1464	2088	3229	-	-	-
	0.412	622	794	1007	1285	1666	3884	-	-
	0.512	498	616	752	915	1114	1899	4197	-

TABLE 13.2. The Optimal Sample Sizes for Prescribing One Double Strength Tablet Three Times a Week in the PCP-TMS Trial for Possible Values of θ_D and θ_T

small to reach consensus with probability at least 0.80.

5 The PCP-INT2 trial

The PCP-INT2 trial was designed to compare two alternative drugs, dapsone (100mg PO daily) and atovaquone (1500mg PO daily). Details are in El-Sadr et al (1998). The index X is therefore replaced by D (dapsone) and Y is replaced by A (atovaquone) in this example. In this trial, the protocol specifies that dapsone is clinically better than atovaquone if $\theta_D - \theta_A$ is less than -0.06 and the range of clinical equivalence is [-0.06,0.06]. Since dapsone is much less expensive than atovaquone, dapsone would be prescribed if the effects of these two drugs were clinically equivalent. The optimal sample size for prescribing dapsone, therefore, denoted by n_D, would be the smallest n such that with probability at least $1-\beta$, all posterior means of $\theta_D - \theta_A$ will be less than or equal to 0.06. Similarly, n_A is the / smallest n such / that all the posterior means of / $\theta_D - \theta_A$ are / greater

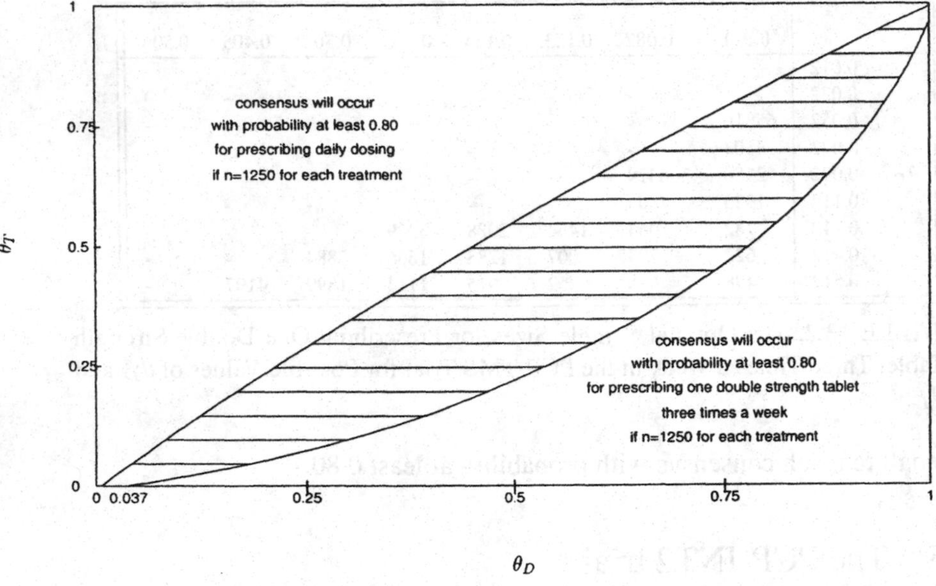

FIGURE 13.2. Values of θ_D and θ_T for which the probability of consensus on prescribing is at least 0.80 when n is 1250 for each group in the PCP-TMS trial. For n=1250, the probability of consensus is less than 0.80 if θ_D and θ_T lie in the area shaded with horizontal lines.

than 0.06 with probability at least $1 - \beta$. In this trial, 7 clinicians did not provide information about the results of this trial. Only 51 prior beliefs are used to calculate the sample size.

5.1 *Results for the PCP-INT2 trial*

In the PCP-INT2 trial, there was only one clinician who believed prior to the trial that atovaquone was probably clinically better than dapsone: that is $m_D - m_A$ was bigger than 0.06.

For illustration of the sample size calculation, suppose $\theta_D = 0.24$, the probability of PCP on dapsone is 0.24. Further suppose $\theta_A = 0.36$, then dapsone is better than atovaquone, and (13.4) gives that a sample size of 60 for each treatment is sufficient for the 51 clinicians to reach consensus with probability 0.8. In contrast, suppose $\theta_D = 0.24$ and now $\theta_A = 0.12$, then atovaquone is now better than dapsone and the sample size is 12,357 for each treatment for a 0.8 probability of consensus.

Tables 13.3 and 13.4 show additional sample sizes for other values of θ_D and θ_A for $1 - \beta = 0.8$. Note that if $\theta_D - \theta_A$ is less than but close to 0.06, the sample size is very big. Note also that the sample size for reaching a consensus for prescribing dapsone is smaller than the sample size needed for reaching consensus for prescribing atovaquone if the values of θ_D and θ_A are reversed.

In this trial, the planned sample size was 700 (350 in each group). For this sample size, $n = 350$, Figure 13.3 shows values of θ_D and θ_A for which consensus will be reached on prescribing with probability at least 0.80. In this plot the probability of consensus is less than 0.80 when θ_D and θ_A are in the large shaded area. Note that with a sample size of 350 for each group, there is only a very small region (the region shaded with "+"), where θ_A is close to 0 and θ_D is close to 1, where consensus will be reached, with probability at least 0.80, to prescribe atovaquone. This is useful to know, and calls into question the usefulness of this trial as dapsone was already close to being the standard treatment.

6 Outlying Beliefs

The sample sizes needed for prescribing the "non-standard" treatments are big for both trials. For prescribing the treatment atovaquone, in the PCP-INT2 trial, the very big sample sizes are caused by a few prior opinions which could be argued to be unreasonable. For example, one clinician has

	θ_A											
θ_D	0.04	0.09	0.14	0.19	0.24	0.29	0.34	0.39	0.44	0.49	0.54	0.59
0.04	48	33	27	24	21	19	17	15	14	13	11	10
0.09	1207	101	55	40	32	27	24	21	18	16	14	13
0.14	-	2029	148	75	51	40	32	27	24	21	18	16
0.19	-	-	2751	190	92	61	46	37	31	26	22	19
0.24	-	-	-	3373	225	106	69	51	40	33	28	23
0.29	-	-	-	-	3894	253	117	75	55	43	35	29
0.34	-	-	-	-	-	4316	276	126	80	57	44	36
0.39	-	-	-	-	-	-	4636	293	132	83	59	45
0.44	-	-	-	-	-	-	-	4857	303	135	84	59
0.49	-	-	-	-	-	-	-	-	4977	308	136	83
0.54	-	-	-	-	-	-	-	-	-	4997	306	134
0.59	-	-	-	-	-	-	-	-	-	-	4917	298

TABLE 13.3. The Optimal Sample Sizes for Prescribing Dapsone in the PCP-INT2 Trial for Possible Values of θ_D and θ_A

	θ_A									
θ_D	0.04	0.09	0.14	0.19	0.24	0.29	0.34	0.39	0.44	0.49
0.04	-	-	-	-	-	-	-	-	-	-
0.09	-	-	-	-	-	-	-	-	-	-
0.14	10383	-	-	-	-	-	-	-	-	-
0.19	5431	15417	-	-	-	-	-	-	-	-
0.24	3963	7464	19798	-	-	-	-	-	-	-
0.29	3210	5166	9215	23524	-	-	-	-	-	-
0.34	2722	4025	6189	10683	26597	-	-	-	-	-
0.39	2361	3313	4708	7033	11869	29016	-	-	-	-
0.44	2071	2806	3799	5260	7697	12773	30781	-	-	-
0.49	1824	2413	3164	4183	5681	8181	13394	31893	-	-
0.54	1606	2090	2682	3438	4462	5970	8486	13733	32351	-
0.59	1408	1813	2292	2878	3626	4638	6127	8612	13789	32154

TABLE 13.4. The Optimal Sample Sizes for Prescribing Atovaquone in the PCP-INT2 Trial for Possible Values of θ_D and θ_A

a 95% interval of [-0.01,0.01] and this length seems too small to be reasonable given that little was known about the effectiveness of the two drugs at the time. Similarly two clinicians have means for the difference of -0.13 and -0.1 with corresponding intervals [-0.13,-0.07] and [-0.1,-0.01] respectively. These two prior opinions do not look reasonable as, again, very little data was available on the two drugs at the time the beliefs were specified but these opinions correspond to very strong opinions. It may therefore be appropriate not to consider these prior beliefs. If these three outlying beliefs are omitted, the sample sizes become much smaller.

Outlying beliefs should be omitted with caution and justifying such omission is difficult. Pragmatically, a follow up discussion with each clinician with apparently outlying beliefs would be helpful. If such discussion indicated that the clinician did not understand the question being asked or did not understand the nature of uncertainty, then omitting that clinician's beliefs could be justified. If, alternatively, the clinician provided a carefully thought out scientific rationale for their beliefs, and defended them, then it would be harder to justify omitting them. This is clearly a difficult issue.

7 Conclusions

Spiegelhalter, Freedman and Parmar (1994) and others argue that a clinical trial should stop when consensus would be reached in the scientific community about the primary result of a trial. The two trials considered here were trials with an objective of answering two questions: first what dose of TMP-SMX should be prescribed for PCP prophylaxis for patients who can tolerate TMP-SMX and, second, should dapsone or atovaquone be prescribed for patients who are intolerant of TMP-SMX. The methods described here are simple tools which add to considerations of sample size. They indicate that for the first question, unless the two dosages are very similar in effect, the sample size of 1250 on each treatment is large enough to probably reach consensus in the PCP-TMS trial. For the second question, if the answer is that dapsone should be used then consensus will probably be reached, but the sample size is too small to give a high probability of reaching consensus if the answer is atovaquone. This is an important consideration for design.

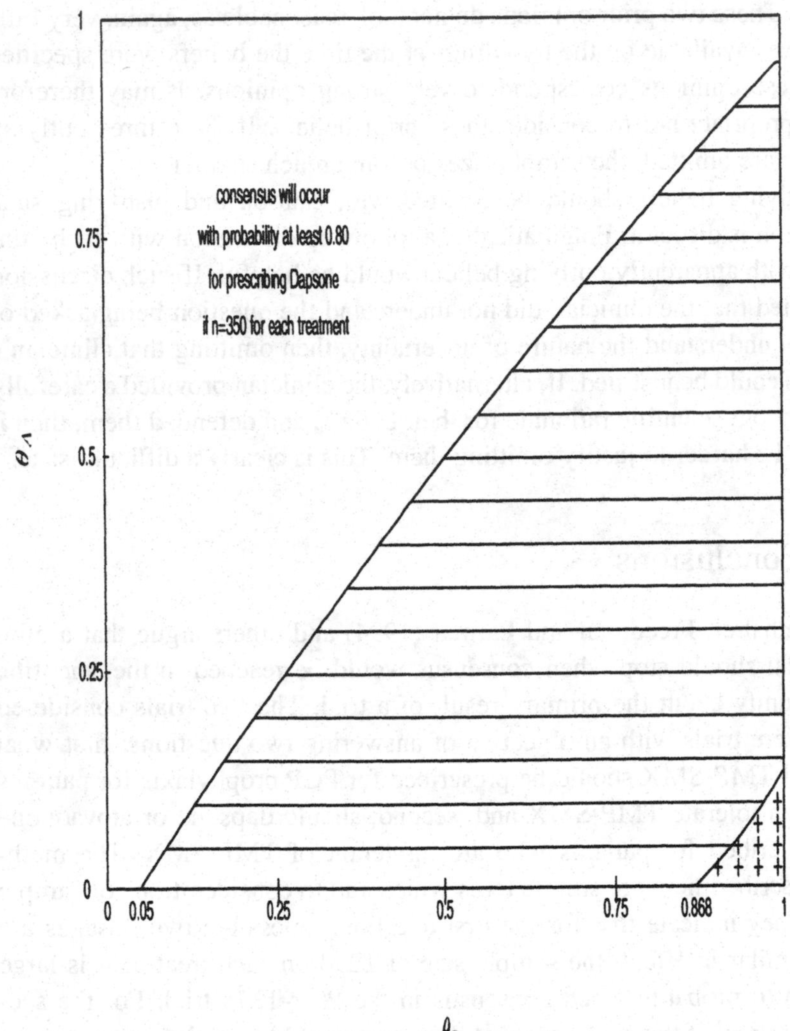

FIGURE 13.3. Values of θ_D and θ_A for which the probability of consensus on prescribing is at least 0.80 when n is 350 for each group in the PCP-INT2 trial. For n=350, the probability of consensus on prescribing Atovaquone is at least 0.80 if θ_D and θ_T lie in the area shaded with +. If θ_D and θ_A lie in the area shaded with horizontal lines, the actual sample size of 350 is too small to reach consensus with probability at least 0.80.

8 Discussion

The proposed method for choosing and examining sample size involves the sampling distribution of each clinician's posterior mean, conditional on the parameters θ_X and θ_Y. For a proposed sample size n and for each value of θ_X and θ_Y, Figures 13.2 and 13.3 summarize calculations of the probability that all clinicians agree on whether the posterior mean of the difference, $\theta_X - \theta_Y$, is bigger than, or smaller than, some critical value k. The use of the sampling distribution, conditional on θ_X and θ_Y may be controversial.

Each clinician has their own predictive distribution for the data, $\bar{X}$ and $\bar{Y}$, based on their own prior distribution for the parameters. Each clinician would therefore have a predictive probability of reaching each of the two conclusions: use X or use Y. If $n \to \infty$ these predictive probabilities will approach the prior probabilities. It is not clear that these are relevant in choosing the sample size when a large number of prior distributions are available and the goal is to reach consensus.

In this paper the probabilities conditional on each combination of the underlying parameters are used. These can be thought of as corresponding to a prediction about what the experts will conclude, from a prior distribution which is point mass on a single parameter value. An objective observer can interpret plots such as Figures 13.2 and 13.3, based on these sampling distribution probabilities, as "if the value of θ_X, θ_Y is this value, say, then for a sample size of n the probability that these clinicians will agree to use X is at least 80%." Or, an objective observer could perhaps construct their own prior distribution, possibly by combining all the individual opinions, and average over that.

Many additional difficult issues are raised in this case study such as how to choose the clinicians and how many to include. It would seem important to include the clinicians enrolling patients in the trial, and possibly others who will act on the results.

As discussed at more length in Chaloner and Rhame (2001), documenting a collection of prior beliefs is important to document the uncertainty required to ethically justify a trial. Beliefs can also be used for many other purposes: for other aspects of design, for monitoring and possible early stopping, and for analysis. Other uses of subjective information in clinical trials can be found elsewhere: for example in Spiegelhalter, Freedman and Parmar (1994), Berry and Stangl (1995), Carlin, Chaloner, Louis and Rhame (1995), Kadane (1996), and Fayers, Ashby and Parmar (1997).

Extensions to this work are in Tsai (1999). For example, an alternative approach is given there where just three prior distributions are specified, representing optimistic, pessimistic and skeptical opinions. The sample size is calculated for reaching consensus for these three prior opinions. For the two trials discussed here, however, given the wide diversity of opinion, using all the prior opinions elicited seems more appropriate.

Acknowledgments

A referee provided helpful comments which greatly improved this paper. This research was supported in part by grants from the National Security Agency and National Institutes of Health.

References

Berry, D. A. and Stangl, D. K. (eds) (1995), *Bayesian Biostatistics,* Marcel-Dekker, New York.

Carlin, B. P., Chaloner, K., Louis, T. A. and Rhame, F. S. (1995), "Elicitation, Monitoring and Analysis of an AIDS Clinical Trial," In *Case Studies of Bayesian Statistics in Science and Industry, Volume II,* Gatsonis C, Hodges J, Kass R, Singpurwalla N (eds), Springer-Verlag, New York, 48-89.

Chaloner, K. and Rhame, F. S. (2001), "Quantifying and Documenting Prior Beliefs in Clinical Trials," *Statistics in Medicine,* **20**, 581–600.

El-Sadr W. M., Luskin-Hawk R., Yurik, T. M., et al. (1999), "A Randomized Trial of Daily and Thrice-weekly Trimethoprim-sulfamethoxazole for the Prevention of *Pneumocystis Carinii* Pneumonia in Human Immunodeficiency Virus-Infected Persons," *Clin Infect Dis,* **29**, 775-783.

El-Sadr W. M., Murphy, R. L., Yurik, T. M., et al. (1999), "Atovaquone Compared with Dapsone for the Prevention of *Pneumocystis Carinii* Pneumonia in Patients with HIV Infection Who Cannot Tolerate Trimethoprim, Sulfonamides, or Both," *N Eng J Med,* **339**, 1889-1895.

Fayers, P. M., Ashby, D. and Parmar, M. K. B. (1997), "Bayesian Data Monitoring in Clinical Trials," *Statistics in Medicine,* **16**, 1413-1430.

Hartigan, J. A. (1969), "Linear Bayesian Methods," *Journal of the Royal Statistical Soc, Ser B* **31**, 446-454.

Kadane, J. B. (1996), *Bayesian Methods and Ethics in a Clinical Trial Design,* Wiley, New York.

Spiegelhalter, D. J., Freedman, L. S. and Parmar, M. K. B. (1994), "Bayesian Approaches to Randomized Trials (with discussion)," *Journal of the Royal Statistical Soc, Ser A* **157**, 357-416.

Tsai, C. (1999), *Bayesian Experimental Design with Multiple Prior Distributions*, PhD Thesis, School of Statistics, University of Minnesota.

Author Index

Subject Index

Lecture Notes in Statistics

For information about Volumes 1 to 108, please contact Springer-Verlag

109: Helmut Rieder (Editor), Robust Statistics, Data Analysis, and Computer Intensive Methods. xiv, 427 pp., 1996.

110: D. Bosq, Nonparametric Statistics for Stochastic Processes. xii, 169 pp., 1996.

111: Leon Willenborg and Ton de Waal, Statistical Disclosure Control in Practice. xiv, 152 pp., 1996.

112: Doug Fischer, Hans-J. Lenz (Editors), Learning from Data. xii, 450 pp., 1996.

113: Rainer Schwabe, Optimum Designs for Multi-Factor Models. viii, 124 pp., 1996.

114: C.C. Heyde, Yu. V. Prohorov, R. Pyke, and S. T. Rachev (Editors), Athens Conference on Applied Probability and Time Series Analysis, Volume I: Applied Probability In Honor of J.M. Gani. viii, 424 pp., 1996.

115: P.M. Robinson and M. Rosenblatt (Editors), Athens Conference on Applied Probability and Time Series Analysis, Volume II: Time Series Analysis In Memory of E.J. Hannan. viii, 448 pp., 1996.

116: Genshiro Kitagawa and Will Gersch, Smoothness Priors Analysis of Time Series. x, 261 pp., 1996.

117: Paul Glasserman, Karl Sigman, and David D. Yao (Editors), Stochastic Networks. xii, 298 pp., 1996.

118: Radford M. Neal, Bayesian Learning for Neural Networks. xv, 183 pp., 1996.

119: Masanao Aoki and Arthur M. Havenner, Applications of Computer Aided Time Series Modeling. ix, 329 pp., 1997.

120: Maia Berkane, Latent Variable Modeling and Applications to Causality. vi, 288 pp., 1997.

121: Constantine Gatsonis, James S. Hodges, Robert E. Kass, Robert McCulloch, Peter Rossi, and Nozer D. Singpurwalla (Editors), Case Studies in Bayesian Statistics, Volume III. xvi, 487 pp., 1997.

122: Timothy G. Gregoire, David R. Brillinger, Peter J. Diggle, Estelle Russek-Cohen, William G. Warren, and Russell D. Wolfinger (Editors), Modeling Longitudinal and Spatially Correlated Data. x, 402 pp., 1997.

123: D. Y. Lin and T. R. Fleming (Editors), Proceedings of the First Seattle Symposium in Biostatistics: Survival Analysis. xiii, 308 pp., 1997.

124: Christine H. Müller, Robust Planning and Analysis of Experiments. x, 234 pp., 1997.

125: Valerii V. Fedorov and Peter Hackl, Model-Oriented Design of Experiments. viii, 117 pp., 1997.

126: Geert Verbeke and Geert Molenberghs, Linear Mixed Models in Practice: A SAS-Oriented Approach. xiii, 306 pp., 1997.

127: Harald Niederreiter, Peter Hellekalek, Gerhard Larcher, and Peter Zinterhof (Editors), Monte Carlo and Quasi-Monte Carlo Methods 1996. xii, 448 pp., 1997.

128: L. Accardi and C.C. Heyde (Editors), Probability Towards 2000. x, 356 pp., 1998.

129: Wolfgang Härdle, Gerard Kerkyacharian, Dominique Picard, and Alexander Tsybakov, Wavelets, Approximation, and Statistical Applications. xvi, 265 pp., 1998.

130: Bo-Cheng Wei, Exponential Family Nonlinear Models. ix, 240 pp., 1998.

131: Joel L. Horowitz, Semiparametric Methods in Econometrics. ix, 204 pp., 1998.

132: Douglas Nychka, Walter W. Piegorsch, and Lawrence H. Cox (Editors), Case Studies in Environmental Statistics. viii, 200 pp., 1998.

133: Dipak Dey, Peter Müller, and Debajyoti Sinha (Editors), Practical Nonparametric and Semiparametric Bayesian Statistics. xv, 408 pp., 1998.

134: Yu. A. Kutoyants, Statistical Inference For Spatial Poisson Processes. vii, 284 pp., 1998.

135: Christian P. Robert, Discretization and MCMC Convergence Assessment. x, 192 pp., 1998.

136: Gregory C. Reinsel, Raja P. Velu, Multivariate Reduced-Rank Regression. xiii, 272 pp., 1998.

137: V. Seshadri, The Inverse Gaussian Distribution: Statistical Theory and Applications. xii, 360 pp., 1998.

138: Peter Hellekalek and Gerhard Larcher (Editors), Random and Quasi-Random Point Sets. xi, 352 pp., 1998.

139: Roger B. Nelsen, An Introduction to Copulas. xi, 232 pp., 1999.

140: Constantine Gatsonis, Robert E. Kass, Bradley Carlin, Alicia Carriquiry, Andrew Gelman, Isabella Verdinelli, and Mike West (Editors), Case Studies in Bayesian Statistics, Volume IV. xvi, 456 pp., 1999.

141: Peter Müller and Brani Vidakovic (Editors), Bayesian Inference in Wavelet Based Models. xiii, 394 pp., 1999.

142: György Terdik, Bilinear Stochastic Models and Related Problems of Nonlinear Time Series Analysis: A Frequency Domain Approach. xi, 258 pp., 1999.

143: Russell Barton, Graphical Methods for the Design of Experiments. x, 208 pp., 1999.

144: L. Mark Berliner, Douglas Nychka, and Timothy Hoar (Editors), Case Studies in Statistics and the Atmospheric Sciences. x, 208 pp., 2000.

145: James H. Matis and Thomas R. Kiffe, Stochastic Population Models. viii, 220 pp., 2000.

146: Wim Schoutens, Stochastic Processes and Orthogonal Polynomials. xiv, 163 pp., 2000.

147: Jürgen Franke, Wolfgang Härdle, and Gerhard Stahl, Measuring Risk in Complex Stochastic Systems. xvi, 272 pp., 2000.

148: S.E. Ahmed and Nancy Reid, Empirical Bayes and Likelihood Inference. x, 200 pp., 2000.

149: D. Bosq, Linear Processes in Function Spaces: Theory and Applications. xv, 296 pp., 2000.

150: Tadeusz Caliński and Sanpei Kageyama, Block Designs: A Randomization Approach, Volume I: Analysis. ix, 313 pp., 2000.

151: Håkan Andersson and Tom Britton, Stochastic Epidemic Models and Their Statistical Analysis. ix, 152 pp., 2000.

152: David Ríos Insua and Fabrizio Ruggeri, Robust Bayesian Analysis. xiii, 435 pp., 2000.

153: Parimal Mukhopadhyay, Topics in Survey Sampling. x, 303 pp., 2000.

154: Regina Kaiser and Agustín Maravall, Measuring Business Cycles in Economic Time Series. vi, 190 pp., 2000.

155: Leon Willenborg and Ton de Waal, Elements of Statistical Disclosure Control. xvii, 289 pp., 2000.

156: Gordon Willmot and X. Sheldon Lin, Lundberg Approximations for Compound Distributions with Insurance Applications. xi, 272 pp., 2000.

157: Anne Boomsma, Marijtje A.J. van Duijn, and Tom A.B. Snijders (Editors), Essays on Item Response Theory. xv, 448 pp., 2000.

158: Dominique Ladiray and Benoît Quenneville, Seasonal Adjustment with the X-11 Method. xxii, 220 pp., 2001.

159: Marc Moore (Editor), Spatial Statistics: Methodological Aspects and Some Applications. xvi, 282 pp., 2001.

160: Tomasz Rychlik, Projecting Statistical Functionals. viii, 184 pp., 2001.

161: Maarten Jansen, Noise Reduction by Wavelet Thresholding. xxii, 224 pp., 2001.

162: Constantine Gatsonis, Bradley Carlin, Alicia Carriquiry, Andrew Gelman, Robert E. Kass Isabella Verdinelli, and Mike West (Editors), Case Studies in Bayesian Statistics, Volume V. xiv, 448 pp., 2001.